Process Management in Spinning

Process Management in Spinning

R. Senthil Kumar

CRC Press
Taylor & Francis Group
Boca Raton London New York

CRC Press is an imprint of the
Taylor & Francis Group, an **informa** business

CRC Press
Taylor & Francis Group
6000 Broken Sound Parkway NW, Suite 300
Boca Raton, FL 33487-2742

First issued in paperback 2019

ISBN-13: 978-1-4822-0836-8 (hbk)
ISBN-13: 978-0-367-37833-2 (pbk)

Library of Congress Cataloging-in-Publication Data

Kumar, R. Senthil.
   Process management in spinning / R. Senthil Kumar.
      pages cm
   Includes bibliographical references and index.
   ISBN 978-1-4822-0836-8 (hardback)
   1. Spinning. 2. Process control. I. Title.

TS1480.K855 2014
677'.02822--dc23
                                             2014024306

I dedicate this work

to

all my teachers and my lovely daughter, Janani

Contents

Foreword xxi
Preface.. xxiii
Author.. xxv

1. Process Control in Mixing...1
   1.1   Significance of Process Control in Mixing 1
   1.2   Fiber Quality Index...3
   1.3   Essential Properties of Fiber...3
          1.3.1   Fiber Length...5
          1.3.2   Fiber Strength and Elongation..6
          1.3.3   Fiber Fineness and Maturity 7
          1.3.4   Trash..10
          1.3.5   Color...10
          1.3.6   Neps..11
   1.4   Properties of Cotton: Global Scenario...11
   1.5   Process Sequence: Short Staple Spinning 12
   1.6   Bale Packing and Dimensions...13
   1.7   Bale Management...15
          1.7.1   Selection of Cottons for Mixing..16
          1.7.2   Points to Follow for Effective Bale Management 17
   1.8   Bale Lay Down Planning 17
          1.8.1   Procedure for Bale Lay Down Planning...18
   1.9   Linear Programming Technique for Cotton Mixing 19
          1.9.1   Formulation of LPT Model 19
          1.9.2   LPT in the Optimization of Cotton Mixing 20
   1.10 Cotton Property Evaluation Using HVI and AFIS......................................21
          1.10.1 High Volume Instruments..21
                 1.10.1.1 Modules of HVI...22
          1.10.2 Advanced Fiber Information System 23
   1.11 Cotton Stickiness..23
          1.11.1 Effect of Stickiness on Various Processes......................................24
                 1.11.1.1 Ginning..24
                 1.11.1.2 Spinning 24
                 1.11.1.3 Effect of Stickiness on Weaving.....................................25
          1.11.2 Stickiness Detection and Measurement 25
   1.12 Contamination and Its Impact 26
          1.12.1 Effects of Contamination 26
          1.12.2 Measures to Reduce Contamination..26
          1.12.3 Contamination Cleaning Methods..27
                 1.12.3.1 Hand Picking Method 27
                 1.12.3.2 Blow Room Equipped with Contamination Detection
                             and Ejecting Units 27
   1.13 Soft Waste Addition in Mixing 29
   References 30

2. Process Control in Blow Room 31
  2.1   Significance of Blow Room Process 31
  2.2   Intensity of Fiber Opening 32
  2.3   Cleaning Efficiency 34
         2.3.1   Classification of Cleaning Efficiency 35
         2.3.2   Determination of Cleaning Efficiency 36
         2.3.3   Points Considered for Attaining Better Cleaning Efficiency 36
  2.4   Neps in Blow Room 37
  2.5   Lint Loss 39
  2.6   Blending Homogeneity 41
  2.7   Fiber Rupture and Its Measurement 43
  2.8   Microdust 44
         2.8.1   Classification of Dust 45
         2.8.2   Problems Associated with Microdust 45
         2.8.3   Microdust Extraction 45
  2.9   Lap Uniformity 47
  2.10 Technological Developments in Blow Room Machinery 48
         2.10.1 Automatic Bale Openers 48
         2.10.2 Openers and Cleaners 49
         2.10.3 Blenders or Mixers 51
         2.10.4 Contamination Sorting 53
  2.11 Chute Feed System and Its Control Mechanism 54
  2.12 Process Parameters in Blow Room 54
  2.13 Defects Associated with the Blow Room Process 56
         2.13.1 Lap Licking 56
         2.13.2 Conical Lap 56
         2.13.3 Soft Lap 57
         2.13.4 Curly Cotton 57
         2.13.5 Nep Formation in Blow Room 57
         2.13.6 High Lap C.V% or Tuft Size Variation 57
         2.13.7 Other Defects 58
  2.14 Work Practices in Blow Room 58
  2.15 General Considerations in Blow Room Process 58
  References 59

3. Process Control in Carding 61
  3.1   Significance of the Carding Process 61
  3.2   Neps in Carding 62
         3.2.1   Nep Monitoring and Control 63
  3.3   Influence of Licker-In Zone on the Carding Process 64
         3.3.1   Feeding 64
         3.3.2   Licker-In: Opening and Cleaning 66
         3.3.3   Quality of Blow Room Lap (in Lap Feeding System) 68
  3.4   Influence of the Carding Zone on the Carding Process 68
         3.4.1   Cylinder 68
         3.4.2   Flats 69
         3.4.3   Precarding and Postcarding Segments 70
  3.5   Doffer Zone 71
         3.5.1   Fiber Transfer Efficiency 72

3.6	Sliver Formation	72
3.7	Wire Geometry in Licker-In, Cylinder, Flats, and Doffer	72
	3.7.1 Point Density	73
	3.7.2 Wire Point Profile	74
	3.7.3 Wire Angle	74
	3.7.4 Tooth Depth	74
	3.7.5 Basic Maintenance of Card Wire	75
	3.7.5.1 Grinding	75
	3.7.5.2 Stripping	75
	3.7.5.3 Stationary Flats	75
3.8	Autoleveler in Carding	76
	3.8.1 Benefits Associated with Autoleveler	76
3.9	Process Parameters in Carding	77
3.10	Defects Associated with the Carding Process	78
	3.10.1 Patchy Web	78
	3.10.2 Singles	79
	3.10.3 Sagging Web	79
	3.10.4 Higher Card Waste	79
	3.10.5 Low Nep Removal Efficiency	80
	3.10.6 Higher Unevenness of Sliver	80
	3.10.7 Higher Sliver Breaks	80
3.11	Ambient Conditions	81
3.12	Cleaning Efficiency and Lint Loss%	81
3.13	Control of Waste	83
	3.13.1 Control of Nonusable Waste	83
	3.13.2 Control of Soft Waste	83
	3.13.3 Automatic Waste Evacuation System	84
3.14	Productivity and Quality for Different End Uses	84
	3.14.1 Points for Effective Control of Quality in the Carding Process	85
3.15	Technological Developments in Carding	85
	3.15.1 Developments in Rieter Card: C70®	85
	3.15.2 Developments in Trutzschler TC11®	86
	3.15.3 Developments in Marzoli C701®	86
References		86
4. Process Control in Drawing		**89**
4.1	Significance of the Drawing Process	89
4.2	Fundamentals of Drafting System	90
4.3	Fiber Control in Roller Drafting	91
4.4	Doubling	92
4.5	Influence of Draw Frame Machine Elements on Process	94
	4.5.1 Creel	94
	4.5.2 Drafting Zone	95
	4.5.2.1 Bottom Drafting Rollers	96
	4.5.2.2 Top Rollers	96
	4.5.2.3 Top Roller Weighing System	97
	4.5.3 Web Condenser	98
	4.5.4 Coiler Trumpet	98
4.6	Roller Setting	98

4.7   Top Roller Maintenance 99
        4.7.1   Measurement of Shore Hardness 99
        4.7.2   Berkolization of Top Roller.. 100
        4.7.3   Testing of Top Roller Concentricity and Surface Roughness 101
4.8   Draw Frame: Speeds and Draft Distribution.. 102
4.9   Count C.V% and Irregularity *U%*... 103
        4.9.1   Causes and Control of *U%* in Draw Frame.................................. 104
        4.9.2   Autoleveler... 104
4.10   Defects Associated with Draw Frame Process 105
        4.10.1   Roller Lapping... 105
        4.10.2   Sliver Chocking in Trumpet.. 106
        4.10.3   Creel Breakage.. 106
        4.10.4   More Sliver Breakages... 106
        4.10.5   Improper Sliver Hank 106
        4.10.6   Singles.. 106
4.11   Process Parameters in Draw Frame.. 107
4.12   Work Practices 109
4.13   Technological Developments in Draw Frame... 110
        4.13.1   Rieter Draw Frame.. 110
        4.13.2   Trützschler Draw Frame... 110
References 111

5. Process Control in Comber and Its Preparatory... 113
    5.1   Significance of Combing Process.. 113
    5.2   Lap Preparation... 114
            5.2.1   Lap Preparation Methods... 114
            5.2.2   Precomber Draft.. 115
            5.2.3   Degree of Doubling 116
    5.3   Factors Influencing the Combing Process 116
            5.3.1   Fiber Properties.. 116
            5.3.2   Lap Preparation.. 116
            5.3.3   Machine Factors 117
    5.4   Setting Points in Comber Machine... 118
            5.4.1   Feed Setting 118
                     5.4.1.1   Type of Feed 118
                     5.4.1.2   Amount of Feed per Nip 119
            5.4.2   Detachment Setting... 119
            5.4.3   Point Density and Wire Angle of Comb.................................... 120
            5.4.4   Top Comb Parameters (Depth of Penetration
                     and Needle Density) 120
            5.4.5   Timing 121
            5.4.6   Nips per Minute (Comber Speed).. 121
            5.4.7   Piecing 122
    5.5   Draft... 122
    5.6   Noil Removal 122
            5.6.1   Combing Efficiency 123
            5.6.2   Degrees of Combing.. 123

5.7	Nep Removal in Combing	124
5.8	Hook Straightening in Comber	124
5.9	Sliver Uniformity	125
5.10	Control of Feed Lap Variation	126
5.11	Defects and Remedies	126
	5.11.1 Inadequate Removal of Short Fibers and Neps	126
	5.11.2 Short-Term Unevenness	127
	5.11.3 Hank Variations	127
	5.11.4 Higher Sliver Breaks at Coiler	127
	5.11.5 Coiler Choking	127
	5.11.6 Web Breakages at Drafting Zone	128
	5.11.7 Breakages in Comber Heads	128
	5.11.8 Excessive Lap Licking and Splitting	129
5.12	Technological Developments in Comber and Its Preparatory	129
	5.12.1 Rieter® Comber and Lap Former	129
	5.12.2 Trutzschler® Comber	130
	5.12.3 Marzoli® Comber	131
	5.12.4 Toyota® Comber	131
5.13	Work Practices	131
	References	132
6.	**Process Control in Speed Frame**	**133**
6.1	Significance of Speed Frame	133
6.2	Tasks of Speed Frame	134
6.3	Importance of Machine Components in Speed Frame	135
	6.3.1 Creel Zone	135
	6.3.2 Drafting System	136
	6.3.2.1 Bottom Rollers	136
	6.3.2.2 Top Rollers	138
	6.3.2.3 Aprons, Cradle, Condensers, and Spacer	138
	6.3.2.4 Top Arm Loading	141
	6.3.3 Flyer and Spindle	142
6.4	Draft Distribution	143
6.5	Twist	145
6.6	Bobbin Formation	147
	6.6.1 Taper Formation	148
6.7	Quality Control of Roving	148
	6.7.1 Ratching	148
	6.7.1.1 Procedure to Determine Ratching% in Roving	149
	6.7.2 Roving Strength	149
	6.7.3 Count C.V%	149
	6.7.4 Unevenness	149
6.8	Defects in Roving	150
	6.8.1 Higher U% of Rove	150
	6.8.2 Higher Roving Breakages	150
	6.8.3 Soft Bobbins	150
	6.8.4 Lashing-In	151

6.8.5   Hard Bobbins.. 151
6.8.6   Oozed-Out Bobbins 151
6.8.7   High Roving Count C.V%... 152
6.8.8   Roller Lapping.. 152
6.8.9   Slubs.. 153
6.9   Technological Developments in Speed Frame....................................... 153
References 155

7. Process Control in Ring Spinning.. 157
7.1   Significance of Ring Spinning Process 157
    7.1.1   Ring Spinning Machine... 157
7.2   Influence of Ring Spinning Machine Components on Spinning Process 158
    7.2.1   Creel.. 158
    7.2.2   Roving Guide.. 160
    7.2.3   Drafting Elements.. 160
        7.2.3.1   Bottom Rollers 161
        7.2.3.2   Top Roller Cots 162
        7.2.3.3   Top Arm Loading.. 164
        7.2.3.4   Spacer–Apron Spacing.. 165
    7.2.4   Lappet.. 166
    7.2.5   Balloon Control Ring... 166
    7.2.6   Ring and Traveler.. 167
        7.2.6.1   Load on Ring and Traveler 168
        7.2.6.2   Shape of the Traveler.. 169
        7.2.6.3   Traveler Friction... 169
        7.2.6.4   Traveler Mass 170
        7.2.6.5   Traveler Speed and Yarn Count 171
        7.2.6.6   Traveler and Spinning Tension 172
        7.2.6.7   Traveler Clearer 172
        7.2.6.8   Prerequisites for Smooth and Stable Running
                of Traveler on Ring.. 173
        7.2.6.9   Traveler and Spinning Geometry 173
        7.2.6.10 Traveler Fly... 176
        7.2.6.11 Impact of Ring and Traveler on Yarn Quality............. 176
    7.2.7   Roller Setting 177
        7.2.7.1   Top Roller Overhang... 178
        7.2.7.2   Spinning Geometry 179
    7.2.8   Spindle and Its Drive... 180
        7.2.8.1   Spindle Speed 182
        7.2.8.2   Ring Spinning Empties or Tubes 182
7.3   End Breakage Rate 183
    7.3.1   Occurrence of End Breakage... 183
    7.3.2   Conditions in the Spinning Triangle 184
        7.3.2.1   Forces in the Yarn during the Spinning Process 184
    7.3.3   Causes of Yarn Breaks... 185
        7.3.3.1   Breakage during Doffing.. 186
        7.3.3.2   Breaks during Spinning Process.................................... 186

		7.3.4	Effects of End Breakage	186
		7.3.5	Influence of Various Parameters on End Breakage Rate in Spinning	187
			7.3.5.1 Yarn Count	187
			7.3.5.2 Traveler Wear	188
			7.3.5.3 Defective Feed Bobbin	188
			7.3.5.4 Operator Assignment	188
		7.3.6	Control of End Breakage in Ring Frame	189
		7.3.7	End Breakage and Economics	190
	7.4	Draft Distribution		191
	7.5	Twist		192
	7.6	Nonconformities in the Ring Spinning Process		194
		7.6.1	Hard Twisted Yarn	194
		7.6.2	Unevenness	194
		7.6.3	Soft Twisted Yarn	194
		7.6.4	Hairiness	194
		7.6.5	Undrafted Ends	194
		7.6.6	Higher Thick and Thin Places	194
		7.6.7	Idle Spindles	196
		7.6.8	Slub	196
		7.6.9	Neps	196
		7.6.10	Snarl	196
		7.6.11	Crackers	198
		7.6.12	Bad Piecing	198
		7.6.13	Kitty Yarn	198
		7.6.14	Foreign Matters	199
		7.6.15	Spun-In Fly	199
		7.6.16	Corkscrew Yarn	199
		7.6.17	Oil-Stained Yarn	200
		7.6.18	Slough-Off	200
		7.6.19	Low Cop Content	200
		7.6.20	Improper Cop Build	201
		7.6.21	Ring Cut Cops	201
		7.6.22	Lean Cops	203
	7.7	Package Size or Cop Content		203
		7.7.1	Coil Spacing	204
		7.7.2	Cop Bottom or Base Building Attachment	205
	7.8	Count C.V% and Evenness		205
		7.8.1	Drafting Waves	207
	7.9	Tenacity and Tenacity C.V%		208
	7.10	Roller Lapping		209
		7.10.1	Factors Influencing Roller Lapping	209
		7.10.2	Measures to Prevent Lapping Tendency	209
	7.11	Yarn Quality Requirements for Different Applications		210
		7.11.1	Weaving	210
		7.11.2	Knitting	210
	7.12	Technological Developments in Ring Spinning		212

     7.12.1 Developments in Rieter® Spinning Machines 212
     7.12.2 Developments in Toyota® Spinning Machines......................................214
     7.12.3 Developments in Zinser® Spinning Machines 215
     7.12.4 Compact Spinning System...215
  References 216

8. Process Control in Winding 219
  8.1   Significance of the Winding Process...219
     8.1.1 Objectives of Winding Process 219
     8.1.2 Types of Wound Packages 220
     8.1.3 Cross-Winding Technology: Terminologies 220
  8.2   Demands of Cone Winding Process...220
     8.2.1 Quality Requirements of Ring Bobbins for Winding Operation..........221
     8.2.2 Bobbin Rejection in Automatic Winding Machine.....................222
     8.2.3 Acceptable Deterioration in Quality from Ring Bobbin to Cone..........223
  8.3   Factors Influencing Process Efficiency of Automatic Winding Machine 223
  8.4   Winding Speed..224
     8.4.1 Slough-Off in High-Speed Unwinding 224
  8.5   Yarn Tension 225
     8.5.1 Tension Control in Unwinding Zone.......................................226
     8.5.2 Tension Control in Winding Zone..226
  8.6   Package Density of Cone..227
  8.7   Yarn Clearing...229
     8.7.1 Yarn Faults 229
     8.7.2 Yarn Clearers 230
     8.7.3 Yarn Clearer Setting in Automatic Winding Machine...........231
  8.8   Waxing...231
     8.8.1 Factors Influencing Correct Waxing 232
  8.9   Knotting and Splicing 232
     8.9.1 Factors Influencing Quality of Knot 234
     8.9.2 Quality Assessment of Yarn Splicing 234
     8.9.3 Factors Influencing Properties of Spliced Yarn 235
  8.10 Package Defects in Winding..235
     8.10.1 Missing Tail End 235
     8.10.2 Cut Cone..236
     8.10.3 Yarn Entanglement 236
     8.10.4 Hard Waste/Bunch...236
     8.10.5 Stitch/Jali Formation 236
     8.10.6 Patterning/Ribbon Formation 237
     8.10.7 Sloughing Off 237
     8.10.8 Wrinkle/Cauliflower-Shaped Cone 237
     8.10.9 Hard/Soft Cones 238
     8.10.10 Double End 238
     8.10.11 Missing End/Cob Web..238
     8.10.12 Drum Lap..238
     8.10.13 Ring in Cone 239
     8.10.14 Oily/Greasy Stains on Cone 239
  8.11 Determination of Shade Variation in Yarn Package 239

8.12 Control of Hard Waste...240
    8.12.1 Practices to Be Adopted to Control Hard Waste 240
8.13 Wrong Work Practices in the Winding Department............................240
    8.13.1 Poor Work Practices in Manual Winding Process 240
    8.13.2 Poor Work Practices in Automatic Winding Process...............244
8.14 Yarn Conditioning 246
    8.14.1 Benefits of Yarn Conditioning..247
8.15 Technological Developments in Winding Machine.............................247
    8.15.1 Antiribboning or Ribbon Breaker Mechanism.........................248
    8.15.2 Hairiness Reduction 248
    8.15.3 Tension Control 248
References 249

9. Process Control in Rotor Spinning...251
9.1 Significance of Rotor Spinning...251
    9.1.1 Tasks of Rotor Spinning Machine 253
9.2 Raw Material Selection...253
    9.2.1 Fiber Strength 254
    9.2.2 Fiber Fineness 254
    9.2.3 Fiber Length...254
    9.2.4 Cotton Cleanliness..254
    9.2.5 Fiber Friction...255
9.3 Sliver Preparation...255
9.4 Influence of Machine Components on Rotor Spinning Process 256
    9.4.1 Opening Roller...256
        9.4.1.1 Opening Roller: Wire Profile 257
        9.4.1.2 Opening Roller: Speed..258
    9.4.2 Rotor..258
        9.4.2.1 Rotor Diameter 259
        9.4.2.2 Rotor Groove..260
        9.4.2.3 Rotor Speed..261
        9.4.2.4 Slide Wall Angle..263
    9.4.3 Navel or Withdrawal Tube or Draw-Off Nozzle 264
    9.4.4 Winding Zone 265
        9.4.4.1 Tension Control 266
        9.4.4.2 Cradle Pressure..266
        9.4.4.3 Stop Motion...266
        9.4.4.4 Ribbon Breaker 266
9.5 Draft..267
9.6 Twist..267
9.7 Doubling Effect..269
9.8 Winding Angle 270
9.9 Waxing...271
9.10 End Breakage in Rotor Spinning 271
9.11 Relative Humidity in Rotor Spinning Process.......................................273
9.12 Defects Associated with Rotor Spinning Process 274
    9.12.1 Neppy and Uneven Yarn 274
    9.12.2 Stitches..274
References 274

10. Energy Management in the Spinning Mill..277
   10.1 Significance of Energy Management in the Spinning Mill....................277
   10.2 Manufacturing Cost of Yarn in Spinning Mill 277
   10.3 Energy Distribution in Ring Spinning Process 278
   10.4 Calculation of Energy Consumption of Ring Frame Machines........................279
   10.5 Energy Management Programs 279
   10.6 Energy Conservation in the Spinning Mill..281
         10.6.1 Spinning Preparatory Process 281
               10.6.1.1 High-Speed Carding Machine 281
               10.6.1.2 Installation of Electronic Roving End-Break Stop-Motion
                         Detectors Instead of Pneumatic Systems.................................281
               10.6.1.3 Ring Frame 281
         10.6.2 Energy Conservation in Postspinning Process 285
               10.6.2.1 Intermittent Modes of the Movement of Empty Bobbin
                         Conveyors in Autoconer/Cone Winding Machines.............285
               10.6.2.2 Two for One (TFO) Twister 285
               10.6.2.3 Yarn Conditioning Process 285
         10.6.3 Energy Conservation in Humidification System 285
               10.6.3.1 Replacement of Aluminum Fan Impellers with
                         High-Efficiency FRP Fan Impellers in Humidification
                         Plants and Cooling Tower Fans......................................285
               10.6.3.2 Installation of Variable Frequency Drive on
                         Humidification System Fan Motors for Flow Control 286
               10.6.3.3 Other Areas in Humidification System...........................287
         10.6.4 Overhead Traveling Cleaners..287
               10.6.4.1 Attachment of Control Systems in OHTC 287
               10.6.4.2 Provision of Energy-Efficient Fan Instead of Blower Fan
                         in OHTC 288
         10.6.5 Electrical Distribution Network 288
               10.6.5.1 Cable Losses 288
               10.6.5.2 Power Factor..288
   10.7 Lighting 288
         10.7.1 Replacement of T-12 Tubes with T-8 Tubes.................................289
         10.7.2 Replace Magnetic Ballasts with Electronic Ballasts...................290
         10.7.3 Optimization of Lighting (Lux) in Production
               and Nonproduction Areas..290
         10.7.4 Optimum Use of Natural Sunlight..290
   10.8 Compressed Air System...290
   10.9 Energy Demand Control..291
         10.9.1 Calculating the Load Factor 291
   10.10 Motor Management Plan 292
         10.10.1 Motor Maintenance 292
         10.10.2 Energy-Efficient Motors 293
         10.10.3 Rewinding of Motors 293
         10.10.4 Motor Burnouts 294
         10.10.5 Power Factor Correction 294
         10.10.6 Minimizing Voltage Unbalances 294
   References 294

11. Humidification and Ventilation Management 297
   11.1   Importance of Maintaining Humidity in Spinning Process 297
         11.1.1 Humidification: Terms 298
   11.2   Humidity and Working Conditions 298
   11.3   Humidity and Yarn Properties 299
   11.4   Humidity and Static Electricity 300
   11.5   Humidity and Hygiene 301
   11.6   Humidity and Human Comfort 302
   11.7   Humidity and Electronic Components 302
   11.8   Humidity and Dust Control 303
   11.9   Moisture Management in Ginning 303
   11.10 Humidification Management in Spinning Mill 304
         11.10.1 Air Washer 304
         11.10.2 Determination of Department Heat Load 304
                 11.10.2.1 Solar Heat Gain through Insulated Roof 305
                 11.10.2.2 Heat Dissipation from the Machines 306
                 11.10.2.3 Heat of Air 306
                 11.10.2.4 Heat Load from Lighting 307
                 11.10.2.5 Occupancy Heat Load 307
          11.10.3 Determination of Supply Air Quantity 307
         11.10.4 Water Quality 308
         11.10.5 Types of Humidifiers 309
                 11.10.5.1 Steam Humidification 309
                 11.10.5.2 Atomizing Humidifiers 309
                 11.10.5.3 Air Washer Humidifiers 309
   11.11 Conventional Humidification System 309
         11.11.1 Merits 309
         11.11.2 Demerits 309
   11.12 Modern Humidification System 310
         11.12.1 Merits 310
         11.12.2 Demerits 310
   References 310

12. Pollution Management in Spinning Mill 313
   12.1   Significance of Pollution in Spinning Mill 313
         12.1.1 Types of Pollutant in the Spinning Process 313
   12.2   Cotton Dust 313
         12.2.1 Classification of Cotton Dust 314
         12.2.2 Types of Dust 314
         12.2.3 Generation of the Cotton Dust during Manufacturing 315
         12.2.4 Health Hazards Associated with Cotton Dust Exposure 315
                 12.2.4.1 Byssinosis 315
         12.2.5 Permissible Exposure Limits for Cotton Dust
               for Different Work Areas 316
         12.2.6 Medical Monitoring 316
         12.2.7 Environmental Exposure Monitoring 317
         12.2.8 Vertical Elutriator 317
         12.2.9 Dust Control Measures 318

          12.2.10 Preventive Measures to Be Followed during Manufacturing Process 318
                  12.2.10.1 General Practices.. 318
                  12.2.10.2 Work Practices during Material Handling and Cleaning 318
    12.3 Pollution in Cotton Cultivation and Processing 319
          12.3.1 Impact of Chemical Used in Cotton Cultivation 319
          12.3.2 Cotton Usage 320
          12.3.3 Organic Cotton 320
          12.3.4 Necessity to Shift to Organic Production 321
          12.3.5 Comparison of Conventional Cotton and Organic Cotton
                  Production 321
          12.3.6 Limitations of Organic Cotton Production 322
    12.4 Significance of Noise Pollution 322
          12.4.1 Noise: Terminologies 323
          12.4.2 Ambient Air Quality Standards in Terms of Noise 323
          12.4.3 Perceived Change in Decibel Level 323
          12.4.4 Noise in the Textile Industry 323
          12.4.5 Method of Noise Evaluation 324
          12.4.6 Effect of Noise Pollution 325
          12.4.7 Suggestion to Eradicate Noise Pollution in Textile Industry 325
          12.4.8 Preventive Measures to Control Noise Pollution 326
    References 326

13. **Process Management Tools** 329
    13.1 Significance of Process Management 329
          13.1.1 Process Management in Spinning Mill 329
    13.2 Process Management Tools 330
          13.2.1 5S 330
          13.2.2 Application of 5S in Spinning Mill 331
                  13.2.2.1 Bale Godown 331
                  13.2.2.2 Preparatory Department 331
                  13.2.2.3 Spinning Department 331
                  13.2.2.4 Maintenance Department 332
          13.2.3 Advantages of 5S 332
          13.2.4 Implementation Program of 5S 333
          13.2.5 5S Radar Chart 333
    13.3 Total Productive Maintenance 334
          13.3.1 Conventional Maintenance System 334
          13.3.2 TPM: Definition 335
          13.3.3 Objectives of TPM 335
          13.3.4 TQM versus TPM 335
          13.3.5 Different Modules in the Implementation of TPM
                  in a Spinning Mill 336
                  13.3.5.1 Preparatory Module 336
                  13.3.5.2 Introduction Module 336
                  13.3.5.3 Implementation Module 336
          13.3.6 Eight Pillars of TPM 337
    13.4 Lean Manufacturing Concepts 338
          13.4.1 Lean Manufacturing: Definition 339
          13.4.2 Principle of Lean Manufacturing 339

13.4.3 Goals of Lean Manufacturing...339
13.4.4 Lean Concepts in the Spinning Mills 339
13.4.5 Implementation of Lean Concept 341
13.4.6 Lean Manufacturing Tools 341
13.4.7 Advantages of Lean Manufacturing......................................342
References 342

14. Productivity, Waste Management, and Material Handling..................343
14.1 Productivity in the Spinning Process 343
14.1.1 Factors Influencing Productivity of a Spinning Mill............344
14.1.2 Productivity Measurement..345
14.1.2.1 Production per Spindle...345
14.1.2.2 Labor Productivity 346
14.1.2.3 Operatives per 1000 Spindles (OHSAM)...............346
14.1.3 Measures to Improve Productivity in the Spinning Mill 346
14.2 Yarn Realization..346
14.2.1 Measures to Improve Yarn Realization 347
14.2.2 Waste Management in Spinning Mill.....................................348
14.2.2.1 Waste Investigation...349
14.2.2.2 Recording of Waste 349
14.2.2.3 Waste Reduction and Control................................349
14.2.3 Invisible Loss 350
14.3 Material Handling 351
14.3.1 Principles of Material Handling 352
14.3.2 Factors Governing Selection of Material Handling Equipment 352
14.3.3 Material Handling Equipment in Spinning Mills 353
14.3.3.1 Bale Godown...353
14.3.3.2 Mixing and Blow Room Department....................353
14.3.3.3 Trolleys Used in Carding, Drawing, and Comber 354
14.3.3.4 Roving Bobbin Trolley..355
14.3.3.5 Ring Bobbin Trolley 356
14.3.3.6 Winding and Packing..356
14.3.4 Automation in Roving Bobbin Transportation.....................356
14.3.5 Automation in Ring Bobbin Transportation.........................357
References 357

15. Case Studies..359
15.1 Mixing-Related Problems 359
15.1.1 Higher Needle Breakage in Knitting 359
15.1.2 Barre Effect in the Woven Fabric 359
15.1.3 Poor Fabric Appearance due to Black Spots in the Knitted Fabric 360
15.1.4 Higher Sliver Breakage in Carding 360
15.1.5 Higher Roller Lapping in Spinning Preparatory Process.....360
15.1.6 Higher Polypropylene Contamination in Yarn 361
15.2 Blow Room–Related Problems 361
15.2.1 Higher Sliver Breakages in Card Sliver 361
15.2.2 Higher End Breakage in Rotor Groove...................................362
15.2.3 Holes in the Blow Room Lap...362
15.2.4 High Short Thick Places in Yarn...362

15.3   Carding-Related Problems..363
       15.3.1   Higher Yarn Imperfections..363
       15.3.2   Higher Creel Breakages in Drawing....................................363
15.4   Draw Frame–Related Problems 363
       15.4.1   Poor Fabric Appearance..363
       15.4.2   High Yarn Count C.V% 364
15.5   Comber 364
       15.5.1   High Yarn Unevenness 364
15.6   Speed Frame..365
       15.6.1   High Level of Thin Places in the Yarn................................365
15.7   Ring Frame..365
       15.7.1   Higher Hard Waste in Winding 365
       15.7.2   Shade Variation in Cone 366
       15.7.3   Higher Yarn Breakages in Weaving...................................366
       15.7.4   Barre in Fabric 366

Bibliography...367
Index 379

Foreword

As one definition goes, "process management" is the application of knowledge, skills, tools, techniques, and systems to define, visualize, measure, control, report, and improve processes with the objective to meet customer requirements profitably. Process management comprises, among others, effective utilization of resources, selection of appropriate processes and process variables involved therein, and involves fine-tuning of processes/ machinery involved in the manufacturing of a product on a continual basis.

The yarn manufacturing sector is a critical segment in the textile manufacturing chain as it is a resource-intensive process. Therefore, around the world, this segment of manufacturing faces numerous challenges in maintaining competitive manufacturing costs while meeting quality requirements. It is prudent to mention here that consumers' expectations on the quality of yarn has also risen considerably over a period of time—principally on account of increasing speeds of the manufacturing processes brought about by constant improvements in technology, precision, and sophistication of machinery, and perhaps also on account of increased awareness on the part of the consumers with regard to quality.

It is essential for both the technicians and management personnel in a spinning mill to have a thorough understanding of the various factors contributing to effective process management. This book, *Process Management in Spinning*, is a timely addition to the vast repository of published textile knowledge available today. The book consists of 15 chapters widely encompassing process control from mixing to winding, energy management, pollution control, humidification management, process management tools, waste management, material handling and case studies. It discusses elaborately the various aspects of process management in spinning.

I appreciate the author, R. Senthil Kumar, for putting sincere efforts in bringing out this book successfully. The author's industrial exposure has helped him to articulate all his experiences in a lucid manner in the book. I am confident that this book will be an interesting introduction to the student community and a useful manual for textile technologists and researchers.

Dr. Prakash Vasudevan
Director, South India Textile Research Association (SITRA)
Coimbatore, Tamil Nadu, India

Preface

Textile occupies a significant position among the basic necessities of human beings. There is always a considerable demand for high-quality textiles at optimum cost around the world. According to recent statistics, the global textile market is worth more than $400 billion. It is predicted that global textile production will increase by up to 50% by 2014. The textile industry is providing employment to numerous people around the world. The applications of textiles are versatile and now extends to technical textiles, where function rather than aesthetics is of primary concern. Among the different sectors of textile manufacturing processes, the yarn manufacturing process is an important one, as it influences the final quality of the fabric significantly, as well as the cost of the fabric.

The yarn quality expectations, irrespective of end-use applications, improved to a great extent over a period of time due to the technological developments witnessed in machineries and testing equipment. The world's total installed spindleage for yarn manufacturing was around 200 million. The control of quality and productivity at minimum cost is a highly challenging task for any manufacturing industry.

This book provides a broad perspective and insight to the reader on process control procedures and methods. Substandard quality and/or defects have a major impact on the process, and thus it is necessary to control the quality and to check for defects during the spinning process. This book covers defects associated with each and every process with causes, effects, and control measures. It acquaints the reader with the defects and its remedial measures, which in turn minimizes the occurrence of substandard quality product. The effect of crucial machinery elements in the machines of each intermediate and final process of yarn manufacturing on process and quality were not discussed in elaborative manner previously. The impact of each machinery elements and the effects of its interaction with several machineries as well as process parameters on process and quality are discussed in this book in an interesting manner. This book will serve as a wonderful guide for technicians in the fine-tuning of machines for obtaining expected quality and effective process control. The experimental work carried out by researchers worldwide with relevance to quality and process control of yarn manufacturing processes is also discussed in every chapter.

Chapters 1 through 9 of this book cover the process control of various stages, including the intermediate and the final stages, of yarn manufacturing processes, ranging from mixing to winding. Chapter 10 broadly discusses the energy management methods adopted in the spinning mill. Chapter 11 covers the basics of humidification and ventilation systems. Chapter 12 discusses pollution management in spinning mills. The various process management tools adopted in the spinning mill are covered in Chapter 13. Chapter 14 elaborates on the various important aspects of spinning mills such as productivity, waste control, and material handling. A separate chapter (Chapter 15) is dedicated to real-time case studies, where typical problems that arise in spinning processes are discussed and various control strategies adopted are devised. Overall, this book will be very useful for industry technicians and students of textile technology at various levels.

The industrial experience I gathered over six years in various departments of the yarn manufacturing process provided the impetus for this book. I thank the management and my colleagues at Kumaraguru College of Technology, Coimbatore, for their support throughout this work. I am extremely grateful to companies like Rieter, Trutzschler, LMW,

Toyota, Zinser, Schlafhorst, Murata, Uster, LCC, LRT, and Rieners for granting permission to use images and technologies from their brochures and technical literature for this book. I would like to express my gratitude to Dr. Prakash Vasudevan, Director, SITRA, Coimbatore, India, for writing the foreword for this book. I acknowledge P. Kaleeswaran, final year student, for his time and effort in preparing the majority of the graphs in this book. Finally, I am grateful to my family members for their patience and support during the preparation of this book.

R. Senthil Kumar
Assistant Professor (SRG)
Department of Textile Technology
Kumaraguru College of Technology
Coimbatore, Tamil Nadu, India

Author

R. Senthil Kumar did his postgraduation in textile engineering from the Indian Institute of Technology (IIT), Delhi, India. He authored a book entitled *Textiles for Industrial Applications*, which was published by CRC Press in August 2013. He also contributed a chapter entitled "Mechanical Finishing Techniques for Technical Textiles" for the book *Advances in Dyes, Chemicals and Finishes for Technical Textiles*, which was published by Woodhead Publishing in January 2013. He has so far published more than 30 technical and management-related review papers in various international and national journals and has prepared manuals on "nonwoven technologies" for international and local textile industries. Senthil Kumar has rich experience in the yarn manufacturing industry, having served for over six years in areas of production and R&D. He also serves as a consultant to various textile industries. He has published nearly 200 technical and management-related textile study materials prepared by him on the online portal (http://www.scribd.com/sen29iit), which has more than 331,857 readers. His areas of specialization are fiber science, textile composites, technical textiles, nonwoven technologies, and textile testing. He has been serving as assistant professor (senior grade) in the Department of Textile Technology, Kumaraguru College of Technology, India, since 2009.

1

Process Control in Mixing

1.1 Significance of Process Control in Mixing

In cotton fiber spinning, raw cotton is the prime factor that influences the quality of yarn. The main technological challenge in any textile process is to convert the high variability in the characteristics of input fibers to a uniform end product. This critical task is mainly achieved in the mixing process, provided three basic requirements are met—accurate information about fiber properties, capable blending machinery, and consistent input fiber profiles. Mixing has a significant impact on the end-product cost and quality. Mixing could be thought of combining fibers together in somewhat haphazard proportions whose physical properties are only partially known so that the resultant mixture has only generally known average physical properties that are not easily reproducible; the term "mixing" is generally meant as the intermingling of different classes of fibers of the same or similar grades, which are nearly alike in staple length. Fiber processors seek to acquire the highest quality cotton at the lowest price and attempt to meet processing requirements by blending bales with different average fiber properties. Blends that fail to meet processing specifications show marked increases in processing disruptions and product defects that cut into the profits of the yarn and textile manufacturers.

Mixing department in the spinning mill plays a crucial role in the formulation of appropriate mix of fibers. Cotton cost forms the largest single component (60%–70%) of the total cost of the yarn (as seen in Figure 1.1). Even a small saving in the cotton cost, therefore, means a considerable increase in the gross profits of the mill. Process control in mixing department starts with the correct selection of fiber with respect to the end product. The physical, chemical, and related characteristics of cotton, including the type and trash present and "fiber configuration" (preparation, neps, etc.), determine its textile processing performance and behavior, in terms of processing waste and efficiency and yarn quality. If the spinning mill plans to produce 40^{s}Ne, then the selection of fiber should be appropriate with the yarn to be produced. The wrong selection of fiber for a particular yarn count will adversely affect the process, quality of the end product, and productivity. The underutilization of fiber properties also significantly affects the profit of the spinning mill.

A mill can control the total cost of cotton by selecting cottons of the right quality by buying them at the most appropriate time since cotton prices fluctuate substantially over the year and by efficient formulation of mixings for the various counts of yarn that are being spun (as seen in Figure 1.2).

The type and proportions of cottons that are to be used in making a mixing are decided upon by taking into account the cotton prices, as well as the previous experience of the mill with the working of different cottons. Mixing quality can be controlled by specifying the

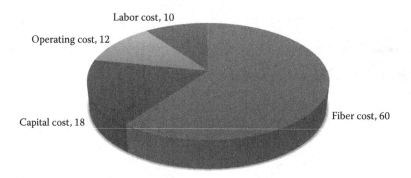

FIGURE 1.1
Cost break down in a typical textile mill.

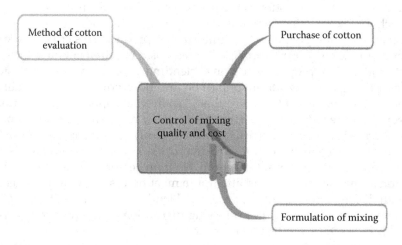

FIGURE 1.2
Factors influencing mixing quality and cost.

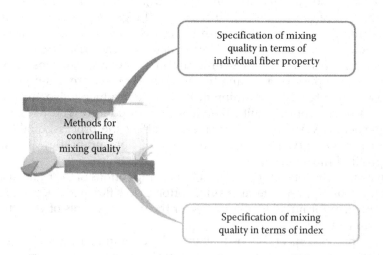

FIGURE 1.3
Methods for controlling mixing quality.

mixing quality in terms of individual property and quality index like fiber quality index (FQI) specifying mixing quality (as seen in Figure 1.3). Linear programming and artificial neural network (ANN) techniques are widely used for optimizing the mixing quality and cost.

1.2 Fiber Quality Index

The quality requirements of yarns mostly depend on its specific end-use applications. The warp yarns meant for shirting and trouser products are expected to be strong and extensible, whereas the yarn meant for hosiery applications should be uniform and reasonably free from imperfections. The different cotton quality characteristics are synthesized into a single index called fiber quality index[1] defined as follows:

$$FQI = \frac{LSm}{f}$$

where

L is the 50% span length in mm measured by digital fibrograph

S is the fiber bundle strength tested on stelometer at 3.2 mm gauge and expressed in g/tex

m is the maturity coefficient measured by NaOH method

f is the fiber fineness (micronaire value)

The achievable yarn CSP for a given FQI of cotton under optimum twist factor is given by the expression[1]

$$\text{Lea CSP} = 320(\sqrt{FQI} + 1) - 13C \text{ for Carded counts}$$

$$= [320(\sqrt{FQI} + 1) - 13C]\left(1 + \frac{W}{100}\right) \text{ for Combed counts}$$

where

C is the count spun

W is the %waste extracted during combing

1.3 Essential Properties of Fiber

Each fiber has particular properties that help to decide which particular fiber should be used to suite a particular requirement. Certain fiber properties increase its value and desirability in its intended end use but are not necessary properties essential to make a fiber. Fiber property requirement may differ with respect to end-use application. Basically, utility of fibers can be categorized into apparel and industrial applications. The property requirements in the fiber for apparel applications are listed as follows:

- Tenacity
- Elongation at break
- Recovery from elongation

- Modulus of elasticity
- Moisture absorbency
- Zero strength temperature (excessive creep and softening point)
- High abrasion resistance (varies with type fabric structure)
- Dyeable
- Low flammability
- Insoluble with low swelling in water, in moderately strong acids and bases and conventional organic solvents from room temperature to 100°C
- Ease of care

The property requirements in the fiber for industrial applications are listed as follows:

- Tenacity
- Elongation at break
- Modulus of elasticity
- Specific strength
- High-temperature resistance
- Impact strength
- High acid/alkali resistance
- Abrasion resistance
- Insulation properties (thermal, sound)
- Other special properties like UV resistance, bacterial resistance, etc.

The spinning mill will run properly only when the proper selection of raw material is done within the acceptable range of selection. The mistake may happen if there is uneven distribution of different cotton fiber properties such as length, strength, fineness, etc. The selection of raw material during mixing is done based on essential properties of fibers such as length, strength, fineness, and trash content. The secondary properties include moisture absorption characteristics, fiber resiliency, abrasion resistance, density, luster, chemical resistance, thermal characteristics, and flammability. The honey dew content especially in the imported cotton will create processing problems like roller lapping that affect productivity and quality of the process. The fiber friction is also a property requirement in some specific end uses. Slenderness ratio (fiber stiffness) plays a significant role, mainly when rolling, revolving, and twisting movements are involved. A fiber that is too stiff has difficulty in adapting to these movements. Fibers that are not stiff enough have too little resilience that do not return to shape after deformation. In most cases, this leads to the formation of neps.

Fiber properties, for example, length, strength, fineness, and color, have always been important in determining the value of cotton fiber.[2] These properties and others are measured soon after the cotton is ginned to determine the market value of the cotton. Klein[3] reported that the fiber property requirements also vary for different spinning system (ring/rotor/air-jet) as shown in Table 1.1.

The important properties that influence the spinning process and the quality of the yarn with respect to cotton spinning process are discussed later elaborately.

TABLE 1.1

Fiber Property Requirements of Different Spinning Systems

Importance Rank	Ring	Rotor (Open-End)	Air Jet
1	Length	Strength	Length
2	Strength	Fineness	Trash
3	Fineness	Length	Fineness
4		Trash	Strength

1.3.1 Fiber Length

Fiber length is universally accepted as the most important fiber property because it greatly affects process efficiency and yarn quality. Cotton buyers and processors used the term staple length long before the development of quantitative methods for measuring fiber properties. Fiber length is usually defined as "the average length of the longer one-half of the fibers (upper half mean length)" obtained from high volume instrument (HVI) or "2.5% span length" obtained from fibrogram. Fiber length is the significant factor influencing the uniformity and tenacity of the yarn. Shorter fiber will tend to create processing problems as well as quality problems. Variations of length are unique to specific varieties of cotton. Length uniformity is defined in terms of uniformity index or uniformity ratio. Uniformity index obtained from HVI is defined as the ratio between mean length and upper half mean length. Uniformity ratio obtained from fibrograph is defined as the ratio between 2.5% span length and 50% span length. Balasubramanian and Iyengar[4] showed that cotton yarn irregularity is considerably influenced by the properties of the fibers from which it is spun and they found that fiber length uniformity was the most important property. Coefficients of fiber length variation, which also vary significantly from sample to sample, are on the order of 40% for upland cotton.[5] The effect of fiber length on various yarn quality parameters and spinning process is given in Table 1.2.

Fiber length influences the productivity in terms of end-breakage rate, quantity of waste, and dust level in the spinning environment. The short-fiber limit has not been standardized but may settle at around 12 or 12.5 mm. Short fiber content (SFC) predominantly influences the hairiness and propensity of neps.[6] SFC is a better indicator of the floating fibers. An increase in SFC increases spinning end breaks, processing waste, fly and optimum roving twist, and causes deterioration in yarn and fabric properties, notably yarn strength and evenness. Excessive fiber length variation (e.g., CV of fiber length, uniformity ratio, or

TABLE 1.2

Effect of Fiber Length on Yarn Quality and Process

Spinning limit
Yarn strength
Yarn evenness
Handle of the product
Luster of the product
Yarn hairiness
Productivity

uniformity index) tends to increase process waste and to adversely affect process performance, including spinning performance and yarn quality. A uniformity index of above 83% and uniformity ratio above 48% are desirable, although it depends upon the spinning system and yarn count. Fiber length characteristic is the important property criterion deciding ring spinning performance. The setting between feed roller and beater, beater to grid bar, and setting between drafting rollers are influenced greatly by 2.5% span length. The drafting roller setting is crucial in deciding the quality and trouble-free running in ring spinning process. Spinning limit for the given fiber is decided by the fiber length parameter. An increase of 1 mm in fiber length increases yarn strength by some 0.4 cN/tex or more. Fiber length can be determined by Baer sorter or Fibrograph or AFIS®.

1.3.2 Fiber Strength and Elongation

The inherent breaking strength of individual cotton fibers is considered to be the most important factor in determining the strength of the yarn spun from those fibers. A yarn should possess adequate breaking force and elongation in order to withstand the stress and strain of subsequent processing. The required force and elongation values depend on the end use of yarn. The ability of cotton to withstand tensile force is fundamentally important in the processing of cotton. Yarn strength correlates highly with fiber strength. Most of the fibers are not usable for textiles because of inadequate strength. The minimum strength for a textile fiber is approximately 6 cN/tex. Cottons with good strength usually give fewer problems and neps during processing than weaker cottons. Except for polyester and polypropylene fiber, fiber strength is moisture-dependent. At optimum yarn twist, fiber tenacity has a greater effect on yarn tenacity than any other fiber property, strength utilization being typically 50%–60% for rotor yarns and 60%–70% for ring yarns. Elongation enables all the fibers to share in contributing to yarn strength. Cotton fibers range from 5.0% elongation, which is very low to 7.6% elongation, which is high. Yarn elongation significantly affects weaving efficiency. An increase in fiber elongation can sometimes reduce spinning end-breakage and yarn strength.

Tensile properties are the result of a tensile test that is defined as "a test in which the resistance of a material to stretching in one direction is measured." Breaking force and elongation are the most widely used tensile properties for assessing the quality of fiber. Historically, two instruments have been used to measure fiber tensile strength, the Pressley apparatus and the Stelometer. Bundle tests measure the tensile properties of the weakest and least extensible fibers included in the bundle. Pressley tester consists of a set of jaws and an inclined lever system in which an ever-increasing load is applied to the specimen until the bundle breaks. Stelometer uses a somewhat different loading concept but still requires clamped and weighed bundles very much like the Pressley. Today, fiber tenacity and breaking extension are rapidly measured through HVI. Tensile tester can work on any one of the following principles:

- Constant rate of loading (CRL)
- Constant rate of extension (CRE)
- Constant rate of traverse (CRT)
- Principle of moments

The tensile properties measured also provide a basis for optimization of the process parameters by identifying deviations apart from helping in selection of raw material suitable for expected end use. Some of the terms[7] relevant to tensile testing of fibers are given with their definition:

- *Breaking force*: It is the maximum force value that is registered when carrying out a tensile test on a sample.
- *Breaking elongation*: Breaking elongation is the difference between the length of a stretched sample at breaking force and its initial length usually expressed as a percentage.
- *Tenacity*: The breaking force of the material under test divided by the linear density of the unstrained material is called tenacity. Tenacity is usually expressed as g/tex, lbf/tex, kgf/tex, cN/tex.

$$\text{Tenacity} = \frac{\text{Breaking force}}{\text{tex}}$$

- *Rkm*: Resistance kilometer (Rkm) is the length of the yarn in kilometers when the yarn breaks by its own weight.
- *CSP*: CSP is count strength product. This factor, used mainly for expressing performance of lea, is a product of count in Ne and strength in lbs.
- *Modulus*: Modulus is defined as the ratio between stress and strain encountered by the test specimen when loaded.

Strength values in g/tex obtained from HVI are given with grading as follows:

- 32 and above = very strong
- 30–32 = strong
- 26–29 = base
- 21–25 = weak
- 20 and below = very weak

It is well known that fiber with high moisture content has a higher strength than "dry" fiber. It is for this reason that fiber moisture is equilibrated to standard conditions (20°C and 65% relative humidity) before testing. Fiber tenacity can be increased in excess of 10% by increasing fiber moisture from 5% to 6.5%.

1.3.3 Fiber Fineness and Maturity

Fiber fineness is basically a measure of its thickness in the transverse direction. The fineness of the perfectly circular fiber is indicated by its diameter. The fineness of noncircular fibers is measured by indirect method using airflow principle. The linear density or weight per unit length of the fiber is the more commonly used index of fineness. Fiber fineness has long been recognized as an important factor in yarn strength and uniformity, properties that depend largely on the average number of fibers in the yarn cross section. Fiber fineness also influences spinning limit, luster, and drape of the fabric. Thirty fibers are needed at the minimum in the yarn cross section, but there are usually over 100.

$$\text{Number of fibers in the yarn cross-section} = \frac{15,000}{\text{Count in Ne} \times \text{Micronaire}}$$

$$CV_{\text{lim}} = \frac{100}{\sqrt{n}} \times \sqrt{\left(1 + 0.0004 CV_d^2\right)}$$

The irregularity of the yarn depends on the coefficient of variation of the number of fibers along the length of a yarn. As the yarn becomes thinner, the number of fibers in its cross section decreases and the yarn becomes increasingly uneven because the presence or absence of a single fiber has a greater effect on the yarn diameter (as seen in Figure 1.4). The spinning limit, that is, the point at which the fibers can no longer be twisted into a yarn, is reached earlier with a coarser fiber. The finer the fiber, the greater the total surface area available for interfiber contact and also less twist is needed to provide the necessary cohesion or interfiber friction.

Spinning larger numbers of finer fibers together results in stronger, more uniform yarns than if they had been made up of fewer, thicker fibers (as seen in Figure 1.5). With the increase in the number of fibers, greater will be the surface area of the contact among the fibers and higher will be their resistance to slippage. Fiber fineness also influences the twist for maximum strength. The twist multiplier for optimum strength is lower for fine fibers due to greater cohesion among the fibers as a result of high-specific surface area of the fine fibers. Barre is a defect occurring in the fabric mostly due to the variation in color or micronaire as mentioned in Table 1.3. Barre is defined as "unintentional, repetitive visual pattern of continuous bars or stripes usually parallel to the filling of woven fabric or to the courses of circular knit fabric."

The cotton fiber consists of cell wall and lumen. The growth of cell wall thickness in the fiber determines the degree of maturity. Schenek[8] suggests that a fiber is to be considered

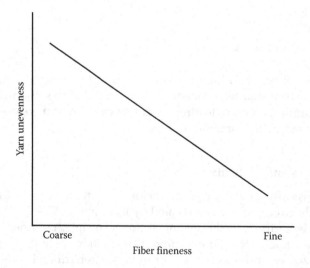

FIGURE 1.4
Effect of fiber fineness on yarn unevenness.

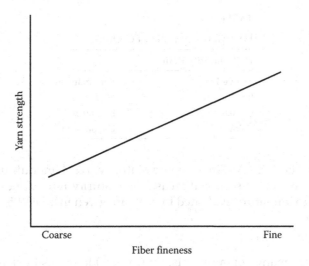

FIGURE 1.5
Effect of fiber fineness on yarn strength.

TABLE 1.3

Measures to Control Barre Defect due to Micronaire Variation

Difference in micronaire average of mixings of the entire lot should not exceed 0.2.
C.V% of the micronaire of individual bales within the mixing should be less than 12%.

as mature when the cell wall of the moisture-swollen fiber represents 50%–80% of the round cross section, as immature when it represents 30%–45%, and as dead when it represents less than 25%. Fibers with very low levels of maturity are sometimes called dead fibers or unripe fibers and do not take dye in the normal way. Immature fibers are usually of lower strength. The presence of immature fibers creates processing problems and several quality problems as given in Table 1.4. Maturity can be determined by differential dyeing method caustic soda swelling method.

Micronaire is the widely used method to estimate fiber fineness and maturity. The fineness factor in micronaire is considered more important in spinning, and fiber maturity is thought to have more effect on dye-uptake success. Airflow instruments are widely used for the estimation of fiber fineness.[9] These instruments are based on the principle that, for equal weights of fiber samples, the rate of airflow across the sample would be less for finer fibers than the coarser fibers due to the relatively more surface area in the case of finer fibers that offer a drag on the flow of air. High-volume instruments measure fineness and

TABLE 1.4

Problems Associated with Immature Fibers

White spots in dyed fabrics
Variations in dye shade
Neppiness
Fiber fly due to fiber breakage
Processing difficulties at the card
End breakage and lappings

TABLE 1.5

HVI Maturity Ratio of Cotton

HVI Maturity Ratio	
Above 1.0	Very mature
0.8–1.0	Mature
0.7–0.8	Immature
Below 0.7	Uncommon

maturity of cotton. The AFIS-A2 Fineness and Maturity (F&M) module uses scattered light to measure single-fiber cross-sectional areas. The maturity ratio values of fibers ranging from very mature to immature evaluated by HVI are given in Table 1.5.

1.3.4 Trash

Cotton bale contains many contaminants, which include pieces of stem, bark, seed-coat fragments (SCFs), motes, leaf and remains of weeds, sand and dust from the fields and dust carried by the wind, and foreign matter such as metal, bale wrapping material, plastic bags blown into the cotton fields. The amount of vegetable matter, sand, and dust in the cotton is referred to as "trash" and can vary from region to region. Even under ideal field conditions, cotton lint becomes contaminated with leaf residues and other trash. Trash accounts for 1%–5% by weight of baled cotton. Pepper trash significantly lowers the value of the cotton to the manufacturer and is more difficult and expensive to remove than the larger pieces of trash. Crop environment, harvest system, genotype, and second-order interactions between those factors all had significant effects on trash grade.

The need to evaluate the amount and type of trash in a particular cotton fiber is based on the effort required to clean it, the adverse effects on yarn quality and process efficiency and the yarn realization, in particular raw cotton. Trash is the nonvalue-added component in the cotton. Broken trash particles are difficult to remove and lead to black spots in the final fabric, which spoils the fabric appearance. The rotor spinning is very sensitive to trash and dust present in the feed sliver, which obstructs the yarn formation in the rotor groove and leads to end breaks. Air-jet and friction spinning require even lower levels of trash content than rotor spinning. The cotton with higher trash content will be prone to harsh treatment during opening and cleaning, which leads to fiber rupture. The problem is more severe when the different cottons used for a mixing have varying degrees of trash. This is because trashy cotton requires severe beating in blow room and cards, which may damage the fibers in the cleaner cotton, mixed with it.

Conventionally, trash content in the cotton sample is determined by using trash analyzer, which separates the lint, trash, and microdust. The result is an expression of trash as a percentage of the combined weight of trash, lint, and microdust of a sample. In HVI cotton classing, a video scanner measures trash in raw cotton, and the trash data are reported in terms of the total trash area and trash particle counts. AFIS—Trash module measures the dust, trash count per gram, trash size (μm) distribution, and visible foreign matter content.

1.3.5 Color

Color is the basic criterion of cotton classification into cotton grade according to the Universal Cotton Standards. It can be affected by many factors connected with cotton cultivation: rainfall, freezes, insects, fungi, staining through contact with soil, grass, etc., as well as by the condition of cotton storage: moisture and temperature. Color deterioration also affects the ability of fibers to absorb and hold dyes and finishes. Cotton color is objectively measured in terms of Nickerson–Hunter color diagram. Raw fiber stock color measurements are used in controlling the color of manufactured gray, bleached, or dyed yarns and fabrics. Color measurements also are correlated with overall fiber quality so that bright (reflective, high Rd), creamy-white fibers are more mature and of higher quality than the dull, gray, or yellowish fibers associated with field weathering and generally lower fiber quality. In the HVI classing system, color is quantified as the degrees of reflectance (Rd) and yellowness (+b), two of the three tri-stimulus color scales of the Nickerson–Hunter colorimeter. The degree of reflectance (Rd) indicates how bright or dull a sample is, whereas the yellowness (+b) indicates the degree of color pigmentation.

1.3.6 Neps

A nep can be defined as a small knot (or cluster) of entangled fibers consisting either entirely of fibers (i.e., a fiber nep) or of foreign matter (e.g., a seed-coat fragment) entangled with fibers.[10] Neps are created when fibers become tangled in the process of harvesting, ginning, and other operations. The nepping propensity of cotton fibres is dependent upon its fiber properties, particularly its fineness and maturity, and the level of biological contamination, for example, seed coat fragments, bark, and stickiness. Many fiber properties such as elongation, fineness, length, maturity, strength, and short fiber content, along with contaminants such as stickiness and seed coat fragments, have been cited as possible predictors or as related to nep formation. Neps can be classified as mechanical neps, biological or shiny neps, and SCF neps.

Nep formation is related to fiber buckling coefficient,[10] defined as the ratio between 2.5% span length and micronaire. The longer, finer, and more flexible the fiber, the more it is prone to nepping during processing. The cotton with higher immature fiber content causes a nep during processing due to higher buckling coefficient (propensity to buckle). Mechanical neps are those that contain only fibers and have their origin in the manipulation of the fibers during processing. Biological neps are neps that contain foreign material, whether the material is seed coat fragments, leaf, or stem material. In unginned cotton, biological neps are typically associated with motes, while in ginned cotton (i.e., cotton lint), they typically contain SCF. Neps in raw cotton can be determined by using AFIS-Nep Module. The nep module gives the parameters like nep count, nep size, and nep removal efficiency.

1.4 Properties of Cotton: Global Scenario

Cotton is one of the most important and widely cultivated cash crops across the world. The cotton cultivated in different countries has different properties. It is cultivated in tropical and subtropical regions of more than 80 countries of the world. The major cotton producing countries are United States, China, India, Pakistan, Uzbekistan, Turkey, Brazil, Greece, Argentina, and Egypt.[11] The properties of some imported cotton varieties

TABLE 1.6

Typical Properties of Some of the Imported Cotton Varieties

S. No.	Cotton Variety	2.5% SL (mm)	U.R%	Tenacity at 3.2 mm G.L (g/tex)	Elongation%	Micronaire
1	GIZA 70	34.3	49	30.52	6.63	4.02
2	PIMA	33.77	48.77	29.05	7.2	3.82
3	SUDAN	33	47.9	25.47	6.61	3.9
4	CIS ELS	32.3	48.15	26.58	7.2	4.1
5	GIZA 86	32.28	50.58	30.6	6.84	4.43
6	CAG	30.7	47.6	23.4	6.5	4.2
7	MANBO	28.8	49.1	22.5	6.03	3.8
8	BOLA-S	28.32	46.93	21.89	5.84	3.8
9	TANZANIA	27.2	48.3	20.8	5.6	3.4
10	MEMPHIS	27.02	46.5	19.1	6.16	4.32

TABLE 1.7

Typical Properties of Indian Cotton Varieties

S. No.	Cotton Variety	2.5% SL (mm)	U.R%	Tenacity at 3.2 mm G.L (g/tex)	Elongation%	Micronaire
1	SUVIN	35.6	47.2	27.8	6.6	3.3
2	DCH-32	33.29	46.38	25.15	6.79	3.1
3	MCU-5	31.32	47.42	23.22	6.11	3.89
4	JULI-S	28.8	47.78	22	6.24	4.05
5	MECH-1	28.74	47.17	21.36	5.9	3.77
6	LK	28.7	46.44	21.29	5.83	3.45
7	S4	28.51	47.67	21.4	6.14	3.96
8	BRAHMA	28.36	46.4	21.07	5.96	3.56
9	S6	27.74	47.85	20.9	5.8	3.84
10	JYOTI	27.2	48.07	21.02	5.8	4.75
11	LRA 5166	26.74	47.66	19.86	5.4	3.8
12	H-4	26.5	47.5	20.1	5.3	3.8
13	J-34	26.17	49.62	20	5.23	4.8
14	NHH-44	25.15	46.73	18.75	4.98	3.71
15	V 797	23.72	52.38	15.57	5.07	5.22

are given in Table 1.6. These countries contribute about 85% to the global cotton production. India stands first in area, third in production, and last in productivity among these countries. The properties of some Indian cotton varieties are given in Table 1.7. Cotton has around 59% share in the raw material consumption basket of the Indian textile industry. Table 1.8 gives the information about the cotton property requirements for the production of various end-use fabrics.

1.5 Process Sequence: Short Staple Spinning

The sequence of process in a short staple spinning is shown in Figure 1.6.

TABLE 1.8

Cotton Property Requirements for the Production of Various End-Use Fabrics

	Yarn Count	Upper Half Mean Length, UHML (in.)	Strength (g/tex)	Micronaire	Maturity Ratio
		Woven Fabric			
Denim	4/1 to 20/1	0.92–1.10	24–30	3.0–5.0	0.80–0.90
Toweling	8/1 to 22/1	0.93–1.10	24–30	3.0–5.5	0.80–0.90
Twill	15/1 to 30/1	1.03–1.12	24–32	3.0–4.9	0.85–0.95
Corduroy	15/1 to 30/1	1.06–1.14	24–32	3.8–5.5	0.90–1.00
Velvets	20/1 to 40/1	1.06–1.16	24–32	3.7–4.9	0.90–1.00
Sheeting	20/1 to 60/1	1.07–1.16	24–32	3.8–4.6	0.90–1.00
Shirting	20/1 to 60/1	1.10–1.18	24–32	3.7–4.4	0.90–1.00
Rugs	3/1 to 6/1	0.95–1.08	24–30	5.0 and higher	0.80–1.00
		Home Furnishings			
Sheer	15/1 to 60/1	1.06–1.16	24–32	3.5–4.9	0.90–1.00
Heavy	3/1 to 12/1	0.95–1.10	24–30	3.2–4.0	0.80–0.90
		Knit Fabric (18–28 Cut)			
Single	16/1 to 40/1	1.04–1.14	24–32	3.5–4.9	0.85–1.00
Double	20/1 to 60/1	1.06–1.16	24–32	3.4–4.6	0.90–1.00

1.6 Bale Packing and Dimensions

Cotton is packed, stored, and transported in units called bales. Bales are formed at the end of the ginning process by accumulating cotton fibers in a chamber called a press box. Bulk cotton fiber is compressed by hydraulic rams in the press box, typically creating forces up to 4 million N (Newton). Bale bands are added at the press box to contain cotton fibers to form the bale. Packaging and labeling requirements also have changed over the past century. Major changes have been made from heavy steel bands and buckles and heavy jute fabrics toward more technically advanced bands, fabrics, and films. Practically all wrapping and strapping materials have realized significant improvements in performance while decreasing shipping weights.

Figure 1.7 from the ISO standard depicts the external dimensions of the cotton bale. *L* is the overall length of the banded bale, *W* is the overall width of the banded bale, and *H* is the overall height of the bale. The inside dimensions of the bale press determine the cross-sectional dimension (length and width) of the bale. Press design is decided by the bale press manufacturers, so once the press is installed, the ginner can control only one dimension: the height. The height is determined by the degree of compaction and the length of the bands. There are numerous weights, sizes, dimensions, and densities of cotton bales produced around the world. Bale weights may be as great as 330 kg as in some Egyptian bales and as low as 100 kg as in old-type bales observed in China. However, recent advances in standardization are rapidly reducing the variation among cotton bales. Imported bales weigh about 225 kg and local Indian bales weigh around 165 kg. Today most bales are compliant with the International Standard ISO-1986 (E). The nominal dimensions and density of the ISO-compliant bales are shown in Table 1.9.

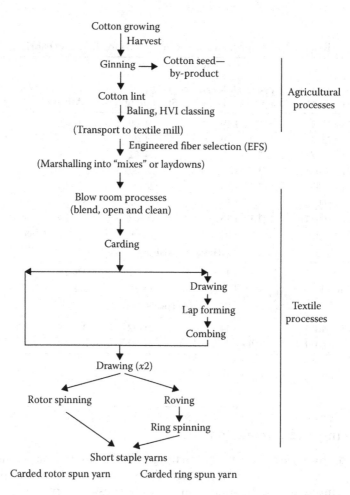

FIGURE 1.6
Process sequence: short staple spinning.

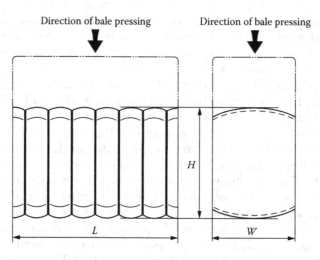

FIGURE 1.7
Bale dimensions.

TABLE 1.9

Dimensions and Density of Cotton Bales (ISO)

Length (mm)	Width (mm)	Height (mm)	Density (kg/m^3)
1060	530	780–950	360–450
1400	530	700–900	

For most stable stacking, bales are normally stacked with their height horizontal, that is, lying down; however, selection from warehouse inventories often makes it most efficient to stack bales on their heads, with the length dimension vertical.

Cotton bales are pressed under high compression; therefore high-strength bands must be used to restrain bales at the desired dimension. Typically, bands may have a strength capacity of up to 9000 N per band. While the average static bale forces may be as low as 4000–5000 Newton per band, an additional safety margin is required to compensate for dynamic forces created during storage, handling, and transport. Heavy bales can create much higher bale strapping forces than bales of average weight. Temperature and humidity changes also affect internal bale forces. Moisture conditions for the cotton fiber during compression are a significant factor: the lower the moisture content, the higher the force on the bale bands.

Bale banding materials are typically constructed of steel bands, high tensile steel wire ties or plastic (polyethylene terephthalate) bands. It is especially critical that bale banding materials be strong enough to withstand the static loads containing the fibers in bales as well as impact forces of handling. Broken bands can represent a significant risk because of probable loss of fiber weight, inefficient handling, and contamination potential. Optimum banding specifications, like bale size and density, represent a compromise of attributes. Steel bands having high load-carrying capacity are less likely to break than plastic bands under a given load. On the other hand, plastic bands allow elongation, relieving compressive static forces from the fibers, which in some cases actually decreases breakage.

Woven polypropylene bagging is the toughest and has the highest tensile and tear resistance of all bale bagging materials. Woven polypropylene is usually the product preferred by ware housers and handlers of cotton bales as they perceive it as protecting the fiber better than other materials. Textile millers do not universally agree on the attributes of woven polypropylene because of the fear of a strand of plastic yarn becoming entrained in raw cotton lint and causing yarn and fabric defects. Because of those concerns, woven polypropylene specifications for US cotton bales mandate that all woven polypropylene fabrics be stabilized with a laminate coating to minimize yarn and fabric fraying.

1.7 Bale Management

Bale management refers to the process of inventory control and selection of fiber according to its properties and also to mix fiber homogeneously to get acceptable spinning performance, consistent production and quality of yarn. The function of bale management is very much unique to the spinning industry. The module should have a function in such a way that the system should automatically generate the issues for mix or a count, where all the lay down or issues should have a consistent quality parameter both in terms of average and SD.

1.7.1 Selection of Cottons for Mixing

The quality of a yarn is maintained uniform for a long period if the raw cotton of specified quality is in stock for that period. It is practically quite difficult to procure a single cotton variety of desired quality for long period and at different times. To overcome this difficulty, mixing of different varieties as well as lots is adopted in the industry. The selection of cottons for a mixing should be done by ensuring that the different fiber varieties or lots are compatible with regard to processing and quality requirements of the yarn. The major fiber properties such as fiber length, fiber fineness, trash, strength, and color must be checked for compatibility before mixing. The blended cottons varying in fiber length even by 5 mm create serious processing problems.

Factors considered for efficient bale management system

Origin (country of origin)
Length and length uniformity
Fiber fineness (μg/in.)
Strength and elongation
SCI (spinning consistency index)
Maturity
Color
Trash%, etc.

A homogenous blend starts with the selection of an appropriate number of bales from large lots in the cotton warehouse. Cotton bales are usually segregated lot wise to provide compatible content and a bale taken from a lot should have similar attributes to the rest of the bales in the lot (as seen in Figure 1.8). If high variation exists within bales in a lot,

FIGURE 1.8
Bale management: lot-wise segregation of bale.

the assignment of bales should be arranged in such a way that it minimizes the variation within the lot. Bales are transported to the mill in the compressed state and they are protected by a bale cover cloth and iron strap. The bales that are going to be processed on the particular day are moved to the mixing department and coverings are removed carefully. Bale straps should be removed using proper equipment to avoid unnecessary injury to the worker. If proper care is not taken during bale strap and bale cover cloth removal, there may be a chance of inclusion into the cotton that will deteriorate the yarn and fabric quality.

The bales are allowed to condition properly. The bale conditioning allows the moisture content and temperature of the fiber to approach stability. After the removal of bale strap and cover cloth, bales try to relax and expand.

1.7.2 Points to Follow for Effective Bale Management

- If the cotton received is from different ginners, it is better to maintain the percentage of cotton from different ginners throughout the lot, even though the type of cotton is same.
- It is not advisable to mix the yarn made of out of two different shipments of same cotton. For example, the first shipment of Sudan cotton is in January and the second shipment is in March, it is not advisable to mix the yarn made out of these two different shipments. If there is no shade variation after dyeing, then it can be mixed.
- Stack mixing is the best way of doing the mixing compared to using automatic bale openers that picks up the material from 40 to 70 bales depending on the length of the machine and bale size, provided stack mixing is done perfectly.
- Improper stack mixing will lead to barre problem. Stack mixing with bale opener takes care of short-term blending and two mixers in series takes care of long-term blending.
- Tuft sizes can be as low as 10 g and it is the best way of opening the material (nep creation will be less, care has to be taken to reduce recycling in the inclined lattice).
- The raw material gets acclimatized to the required temperature and R.H% since it is allowed to stay in the room for more than 24 h, and if the fiber is opened, the fiber gets conditioned well.

1.8 Bale Lay Down Planning

After conditioning, cotton bales are laid in two or more parallel rows on the floor for the bale plucking operation and it is termed as "bale lay down." When a uniform bale lay down is prepared, the average properties are approximately equal to a weighted mean of the component bales. The number of bales to be laid will be in accordance with production planning requirements. Bales should be kept inside markings drawn in the bale lay down area so that the traversing carriage of the bale plucker has pluck from the entire area of the bales kept. Lay down planning for single-variety cotton having two lots will be done in alternate way. If a mill is planning a bale lay down for 40 bales totally consists of 20 bales of MCU-5

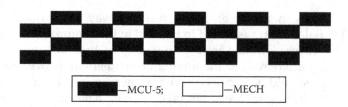

FIGURE 1.9
Mixing: bale lay down plan.

and 20 bales of MECH variety, then alternatively lay the cotton variety in the bale lay down to get homogenous mix. Figure 1.9 illustrates the bale lay down plan of the mixing plan.

With the advent of single-fiber testing, a measure of variation, such as standard deviation, becomes, like the average, a property to be combined. If components, such as bales, have masses w and length, micronaire, linear density, or maturity values x and standard deviations s, the mean value X of the mix is

$$X = \frac{\sum w_i}{\sum (w_i/x_i)} \tag{1.1}$$

Only if all component values are "similar," this reduces to the weighted arithmetic mean

$$X = \frac{\sum w_i x_i}{\sum w_i} \tag{1.1a}$$

and the standard deviation is given by

$$S = \sqrt{X \frac{\left(\sum w_i s_i^2 / x_i + \sum x_i w_i\right)}{\sum w_i - X^2}}$$

If all component means are similar, this reduces to

$$S = \sqrt{\frac{\sum w_i s_i^2}{\sum w_i}}$$

If, for instance, a bale with an average linear density of 200 mtex and standard deviation between fibers of 80 mtex is mixed with one averaging 150, with $s = 60$, the mean of the mix is not 175, but 171 mtex, and the standard deviation of the mix, due partly to mixing different average values, is not 70, but 73 mtex.

1.8.1 Procedure for Bale Lay Down Planning

1. Plan the mixing formulation of the process before taking bales from bale godown. After deciding the cotton varieties and mixing ratio, calculate the number of bales required in each variety in accordance with production requirements.

2. Bales in each variety is taken from bale godown on previous day itself and kept in the mixing department. Lay down bales for 24 h production if space is available.

3. Bales of different varieties are marked with different colors for easy identification.

4. Bale straps are removed using special equipment with utmost care so that any metal contaminants will not get mixed in mixing. Also remove the bale cover cloth with care.

5. Bales should be allowed to "Bloom" and condition for 24 h if possible.

6. Make sure that the floor space where the bales are going to be placed is clean and free from any contaminants.

7. After conditioning, the bales are placed alternatively according to mixing formulations.

8. Bales should be assembled so that they are in close proximity to their neighbors in the lay down and similarly oriented to create a compact mass of fiber suitable for the bale plucker operation. Care should be taken to keep the height of the bales similar.

9. Care should be taken that no two bales of same variety placed near to enhance mixing homogeneity.

10. Ensure that all the bales are kept inside the marked bale lay down area where the bale plucker traversing carriage travels.

11. Bale plucker plucks the cotton with a pair of saw toothed roller and conveys it to the next machine pneumatically.

12. At the run-out of the bales at one side, new bales should be moved into place as soon as possible.

13. The fiber remaining in the previous lay down should be picked up and packed as pieces between the bales of the next lay down.

1.9 Linear Programming Technique for Cotton Mixing

The development of linear programming technique (LPT) is one among the most important scientific advances of the mid-twentieth century. Linear programming uses a mathematical model to describe the problem of concern. Linear programming involves the planning of activities to obtain an optimal result, that is, a result that reaches the specified goal best (according to the mathematical model) among all feasible alternatives.

1.9.1 Formulation of LPT Model

Let[12]

$C_1, C_2, C_3, \ldots C_n$ be the costs of n cottons

$P_1, P_2, P_3, \ldots P_n$ be the percentages of each cotton to be mixed

$L_1, L_2, L_3, \ldots L_n$ be the lengths of each cotton

$S_1, S_2, S_3, \ldots S_n$ be the strengths of each cotton

$M_1, M_2, M_3, \ldots M_n$ be the maturity coefficients of each cotton

$F_1, F_2, F_3, \ldots F_n$ be the micronaire value of each cotton

Objective function:

$$\text{Min } Z = (C1 * P1) + (C2 * P2) + (C3 * P3) + \cdots + (Cn * Pn)$$

S.T. constraints:

$$L1 * P1 + L2 * P2 + L3 * P3 + \cdots + Ln * Pn \geq Lr$$

$$S1 * P1 + S2 * P2 + S3 * P3 + \cdots + Sn * Pn \geq Sr$$

$$M1 * P1 + M2 * P2 + M3 * P3 + \cdots + Mn * Pn \geq Mr$$

$$F1 * P1 + F2 * P2 + F3 * P3 + \cdots + Fn * Pn \geq Fr$$

$$P1 + P2 + P3 + \cdots + Pn = 1$$

$$P1, P2, P3, \ldots Pn \geq 0$$

- Right-hand-side values are obtained from the given set of norms.

1.9.2 LPT in the Optimization of Cotton Mixing

Aim:
To manufacture 10 tex cotton yarn

Required properties for the raw material:

- Length: 31.5–34 mm
- Strength: 20–23 gpt
- Maturity coefficient: 80–83
- Micronaire value: 3.6–3.9

Properties of cottons available and their costs (as seen in Table 1.10):

TABLE 1.10

Properties of Cottons Available and Their Costs

Properties	Cottons			
	1	2	3	Norms
Length (mm)	33	31	30	32
Strength (g/tex)	24	20.5	19	21.5
Maturity coefficient	83	80.2	79.8	82
Micronaire	3.5	3.85	3.9	3.7
Cost per lb (US $)	2.05	1.70	1.66	

Objective function:

$$\text{Min } Z = (2.05 * P1) + (1.70 * P2) + (1.66 * P3)$$

S.T. constraints:

$$33 * P1 + 31 * P2 + 30 * P3 \geq 32$$

$$24 * P1 + 20.5 * P2 + 19 * P3 \geq 21.5$$

$$0.83 * P1 + 0.802 * P2 + 0.798 * P3 \geq 0.82$$

$$3.5 * P1 + 3.85 * P2 + 3.9 * P3 \leq 3.7$$

$$P1 + P2 + P3 = 1$$

Non-negativity constraints: $P1, P2, P3 \geq 0$
$P1, P2, P3$ values are obtained by solving this LP Model using SIMPLEX method (Microsoft Excel can be used).

Results:
Objective function value: Min $Z = 1.925$.

Cotton	Percentage to Be Mixed (%)
1	64.3
2	35.7
3	0

1.10 Cotton Property Evaluation Using HVI and AFIS

The conventional methods used for evaluating fiber properties tend to be tedious and time consuming. Moreover, a certain amount of expertise is required for the results to be reproducible and accurate. The two commonly used instruments, namely, high volume instrument (HVI) and advanced fiber information system (AFIS), are fully automatic machines that test a number of fiber parameters at the same time, which, in turn, intend to save a lot of time and effort spent in testing through the conventional methods.[13]

1.10.1 High Volume Instruments

Fiber samples are now tested on the high volume instrument (HVI)[14] for a range of fiber properties, including strength, elongation, length, uniformity, micronaire, color, and trash. The HVI system was developed in the late 1960s. Before the introduction of the HVI system, cotton in a bale was graded subjectively by experienced cotton classers for properties such as staple length, color, and trash content. HVI systems are based on the fiber bundle testing, that is, many fibers are checked at the same time and their average values determined. In HVI, the bundle testing method is automated. Here, the time for testing is less and so the number of samples that could be processed is increased, quite considerably. The influence of operator is reduced.

The HVI testing is attractive due to the classing of cotton and the laying down of a mix in the spinning mill. The time for testing per sample is 0.3 min. It is best applied to instituting optimum condition for raw material. About 180 samples per hour can be tested and that too with only 2 operators.

1.10.1.1 Modules of HVI

1. *HVI: Length module*

   HVI uses an optical principle for the determination of fiber length. A narrow rectangular beam of light is allowed to fall on the specimen beard. The attenuation of light through the specimen at different areas of the beard is measured and used to obtain the different span length values. In the HVI, the tip of the beard is scanned first and scanning gradually proceeds toward the clamp.

2. *HVI: Strength module*

   HVI uses the "constant rate of elongation" principle while testing the fiber sample. Strength is obtained by measuring the force required to break a sample of known mass. Elongation, the average length of distance to which the fibers extend before breaking, is calculated.

3. *HVI: Fineness module*

   The micronaire module of HVI tester uses the airflow method to estimate the fineness value of cotton. A sample of known weight is compressed in a cylinder to a known volume and subjected to an air current at a known pressure. The rate of airflow through this porous plug of fiber is taken to be a measure of the fineness of cotton. If air is blown through these samples, the plug containing finer fibers will be found to offer a greater resistance than the plug with coarser fibers. The fineness is expressed in the form of a parameter called the micronaire value, which is defined as the weight of 1 in. of the fiber in micrograms. Maturity of cotton also influences the micronaire value.

4. *HVI: Color module*

   The HVI color module utilizes optical measuring principles to define color. The color module has a photodiode, which collects the reflected light from the sample. The photodiode output is converted into meaningful signals using signal conditioners. The illumination of the sample is done with the help of two lamps connected in parallel. Light from the lamps is reflected from the surface of a cotton sample on the test window. The reflected light is diffused and transmitted to the Rd and $+b$ photodiode. These two signals are conditioned to provide two output voltages, which are proportional to the intensity of light falling on the respective photodiodes. These voltages are converted to digital signals from which the computer derives Rd and $+b$ readings to be displayed on the screen.

5. *HVI: Trash module*

   The trash module is an automated video image processor that measures the amount of visible leaf or trash in the sample. The following parameters are obtained from trash module:

   a. Trash area—the percent of sample viewing area occupied by trash

   b. Trash count—no. of trash particles approximately 0.01" in diameter or larger

1.10.2 Advanced Fiber Information System

AFIS[14] is based on the single fiber testing. With the introduction of AFIS, it is possible to determine the average properties of a sample, and also the variation from fiber to fiber.[15] The AFIS method is based on aeromechanical fiber processing, similar to opening and carding, followed by electro-optical sensing and then by high-speed microprocessor-based computing and data reporting.

Fiber length by number is the length of the individual fibers. AFIS length module measures the length of each fiber and places them into length categories. These categories are added together to obtain the length measurement for short fiber and average or mean length. AFIS nep module is able to distinguish between neps and seed coat neps. Each event (fiber, nep, SCN) has its own distinct electrical waveform. Each sample waveform is compared to a standard waveform to determine which classification it most resembles.

AFIS-trash module measures dust and trash particles per gram, in accordance with ITMF recommendations. Trash and dust particles are foreign particles that are mostly part of the cotton plant (leaf or stem fragments, etc.). These particles need to be extracted during the ginning and spinning process.

1.11 Cotton Stickiness

Cotton stickiness[16] caused by excess sugars on the lint, either from the plant itself or from insects, is a very serious problem for the textile industry—for cotton growers, cotton ginners, and spinners. It affects the processing efficiency as well as the quality of the product. The contaminants are mainly sugar deposits produced either by the cotton plant itself (physiological sugars) or by feeding insects (entomological sugars), with the latter being the most common source of stickiness. The main honeydew-producing insects that infest cotton plants are cotton whitefly *Bemisia tabaci* (Gennadius) and the cotton aphid *Aphis gossypii* (Glover).

Whiteflies and aphids are both sap-sucking insects that feed by inserting their long and slim stylets into the leaf tissues. The sap is digested and the excreta discharged as honeydew droplets. The honeydew attaches itself to the leaves and the fibers of opened bolls. The presence of these sugars on the lint reveals that the contamination is coming, at least partially, from insect honeydew. A high percentage of melezitose along with a low percentage of trehalulose reveals the presence of aphid honeydew. When both melezitose and trehalulose are present and trehalulose is dominant, whitefly honeydew contamination is indicated. The other sugars are generally found on both noncontaminated and honeydew-contaminated cottons. It was reported that glucose and fructose contained in the honeydew are synthesized from sucrose by the insect.

During yarn formation the cotton fibers are exposed to friction forces that elevate the temperature of some mechanical parts, which affects the temperature-dependent properties of the sugars present. If one or more of the sugars melt, stickiness results. Obviously moisture will cause sugars to change from a crystalline state (nonsticky) to an amorphous state (sticky). In particular, the relative humidity in the manufacturing environment may affect the moisture-dependent properties of the sugars present.

1.11.1 Effect of Stickiness on Various Processes

1.11.1.1 Ginning

Sticky cotton tends to clog/choke the ginning machines. Stickiness reduces roller gin production by 10–15 lb of lint per hour. Financial losses due to frequent replacement of blades/saws are in addition.

1.11.1.2 Spinning

Stickiness will cause lint to stick to card clothing and draft rollers in subsequent processes. The sticky deposits noticed in the creel calendar roller and drafting zone of draw frame machine are shown in Figures 1.10 and 1.11.

Sticky fibers even if they pass through the spinning back process will create extra centrifugal forces during ballooning causing the yarn to break. The sticky deposit in the drafting zone of ring frame machine is shown in Figure 1.12.

In the OE frames, stickiness will clog the turbine. No matter how we look at stickiness, it will reduce efficiency and production to a considerable extent during spinning.

FIGURE 1.10
Sticky deposits on the draw frame creel.

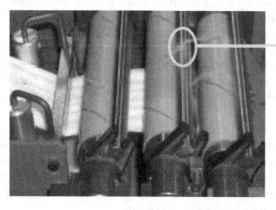

FIGURE 1.11
Sticky deposits on the draw frame drafting zone.

FIGURE 1.12
Sticky deposits on the ring spinning frame.

Low humidity will dry the sugars and they will cease to be sticky. If, however, humidity is allowed to rise, sugars will become sticky again.

1.11.1.3 Effect of Stickiness on Weaving

Stickiness has minimal effect on warp as it is usually sized and the sugar present gets either dissolved in the hot size mix or is covered by it. However, in weft, sugar starts building up in shuttle, gripper, or air jet and weaving efficiency drops to a level where it becomes uneconomic to continue weaving. Frequent cleaning of wefts passage would, therefore, be required. This is time consuming and expensive.

1.11.2 Stickiness Detection and Measurement

The degree of stickiness depends on chemical identity, quantity, and distribution of the sugars, the ambient conditions during processing—especially humidity—and the machinery itself. Stickiness is therefore difficult to measure. Nonetheless, methods for measuring sugars on fiber have been and are being developed. These measurements may be correlated with sticking of contaminated lint to moving machine parts. The physical and chemical attributes of the lint and sugars that are correlated with stickiness have been measured in many ways, each with differing efficiency and precision.

Some of the measurement methods are given as follows:

- Reducing sugar method
- High-performance liquid chromatography
- Minicard method
- Sticky cotton thermo detector
- High-speed stickiness detector
- Fiber contamination tester

1.12 Contamination and Its Impact

Mixing of foreign material/matter with main product at any stage of collection, production, handling, storage, processing in the yarn manufacturing process is termed as contamination.[17] Contamination of raw cotton can take place at every step, that is, from the farm picking to the ginning stage. Contamination, even if it is a single foreign fiber, can lead to the downgrading of yarn, fabric, or garments or even the total rejection of an entire batch and can cause irreparable harm to the relationship between growers, ginners, merchants, and textile and clothing mills. The International Textile Manufacturers Federation (ITMF)[18] reported that claims due to contamination amounted to between 1.4% and 3.2% of total sales of 100% cotton and cotton-blended yarns.

Major source of contamination in all bales continues to be organic matter such as leaves, feathers, paper, leather, etc., which has steadily increased. The next most prevalent contaminant is fabric and string made from cotton, woven plastic, plastic film, and jute/hessian, followed by sand and dust. The incidence of oily substances/chemicals and inorganic matter such as rust and metal has remained fairly consistent.

1.12.1 Effects of Contamination

1. Contamination of cotton causes it to become sticky that creates obstruction in rollers.
2. It causes wastage of dying material and requires extra efforts at cleaning process that unnecessarily inflates cost.
3. Even after cleaning leftover embedded pieces of contamination in yarn affect its quality and value.
4. Contaminants such as stones, metal pieces, etc., cause disturbance to material flow especially in spinning preparatory process which affects production as well as quality of the process.
5. Metal pieces tend to cause fire accident that leads to severe machine and material loss.
6. Fabric appearance produced with contaminated yarn will be poor and prone to rejection (as seen in Figure 1.13).
7. Dyeing affinity of contamination is different from dyeing affinity of fabric that leads to uneven fabric coloration.

1.12.2 Measures to Reduce Contamination

1. Introduction of standardized picking storage and marketing of raw cotton.
2. Dissemination of awareness through mass media to the targeted segment.
3. Cloth bags instead of jute and fabric must be provided by farmers and ginning factory owners to pickers.
4. Cotton should be stored on clean and proper floors.
5. Metal body open trolleys should be used for quick transportation of cotton from field to factories.
6. Sheds and platforms should be built properly in the market.

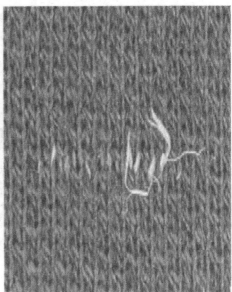

FIGURE 1.13
Polypropylene contaminants in knitted fabric.

7. Bags should be opened by unsewing instead of cutting twine into small pieces.
8. Bags should not be beaten on heap. Instead it should be done separately and obtained cotton should be cleaned properly to be added in heap.
9. Conveyers can greatly facilitate.
10. Plastic strips should be used for strapping bales to avoid contamination by rust. Bale packing should be graded and awareness created to improve bale packing.

1.12.3 Contamination Cleaning Methods

1.12.3.1 Hand Picking Method

A small number of spinning mills are able to manually check and remove contamination from every bale of cotton before it is repacked and released for processing in the mill. This manual sorting is either done directly from the bale or the bale is first opened using a bale opener with a spiked lattice to open the cotton prior to manual sorting (as seen in Figure 1.14).

Spinning mills situated in countries where labor costs are comparatively low employ large numbers of people to patrol the bale lay down and remove contamination from bales before cotton is fed into the blow room line by the bale opener (as seen in Figure 1.15).

1.12.3.2 Blow Room Equipped with Contamination Detection and Ejecting Units

Although manual intervention is helpful, even low labor cost spinning mills have come to realize that it is not always sufficient as generally only the bigger contaminants are removed.

FIGURE 1.14
Hand picking method.

FIGURE 1.15
Patrolling in the bale lay down.

Hence, they are equipping blow room with systems for detection, separation and measurement of foreign material (as seen in Figure 1.16). These systems detect contaminants using acoustic, optical, and color sensors that monitor the material as it flows (is processed) through the machinery. When a sensor is activated by a contaminant it is measured (registered) and, depending upon the system, mechanically removed via an alternate material flow outlet. These systems are normally installed at the beginning of the blow room line before the final cleaning stage.

FIGURE 1.16
Contamination clearer attached in blow room line.

1.13 Soft Waste Addition in Mixing

Spinning machineries produce two types of waste such as soft waste and hard waste.[19] Soft waste is reprocessable whereas hard waste is not. Addition of soft waste to the raw cotton in the mixing process is a common practice in the spinning mill. Soft waste or useable waste can be classified as lap bits, pneumafil waste, rove ends, sliver bits, etc., usually generated in the various departments of the spinning process. Soft waste is inevitable in the spinning process but the amount of soft waste generated can be controlled. The soft waste added to the cotton mixing should be as low as possible. The soft waste consists of good fibers that already get processed experiencing lot of stress in each and every machine. The properties of these waste fibers which have undergone both mechanical and surface degradation vary considerably and their compatibility with virgin fibers might be expected to be poor. Reprocessing of soft waste again in the spinning process will tend to create fiber rupture in blow room and carding, neps and hairiness propensity in the yarn. Further, it will create unacceptable level of variation in fiber properties. The yarn spun from such a mixing will obviously have properties inferior to the yarns from mixing without waste.

Soft waste generated in each shift to be weighed machine-wise and segregated waste type-wise in the waste godown. During mixing lay down planning, recommended percentage of soft waste should be spread over the top of the cotton bale uniformly. If mixing is laid manually, soft waste should be added uniformly to each layer of stack mixing. It must be ensured that soft waste should not have any entanglement or contaminants which will get jammed between the spiked rollers in the bale plucker. Roving ends are harder to

TABLE 1.11

Recommendation for the Addition of Soft Waste in the Mixing

Ring-Spinning Process	Rotor-Spinning Process
• Carded up to 5%	• Coarse up to 20%
• Combed up to 2.5%	• Medium up to 10%
	• Fine up to 5%

open due to certain amount of twist in the rove. Roving is opened into fibers with the help of roving end opener prior to addition in the mixing.

The percentage of soft waste addition will vary depending upon the yarn count whether it is fine or coarse count. For fine counts, the amount of soft waste added should be low to control the process and quality of the yarn. The recommendation for soft waste addition in the mixing is given in Table 1.11.

Comber noil is normally utilized in rotor spinning for producing very coarse counts.

References

1. Ratnam, T.V., Seshan, K.N., Chellamani, K.P., and Karthikeyan, S., *Quality Control in Spinning*, SITRA Publications, Coimbatore, India, 1994.
2. Gordon, S. and Hsieh, Y.L., *Cotton: Science and Technology*, Woodhead Publications, Cambridge, U.K., 2007.
3. Klein, W., *Technology of Short Staple Spinning*, Vol. 1, The Textile Institute, Manchester, U.K., 1987.
4. Balasubramanian, N. and Iyengar, R.L.N., Relationship between yarn irregularity, draft, and fibre properties. *Journal of the Textile Institute Transactions* 55(7): T377–T379, January 1964.
5. Bradow, J.M. and Davidonis, G.H., Quantitation of fiber quality and the cotton production-processing interface: A physiologist's perspective. *The Journal of Cotton Science* 4: 34–64, 2000.
6. Zeidman, M.I. and Batra, S.K., Determining short fiber in cotton. Part I: Some theoretical fundamentals. *Textile Research Journal* 61: 21–31, 1991.
7. Textile Institute, Manchester, *Textile Terms and Definitions*, 5th edn., The Textile Institute, Manchester, U.K., 1963.
8. Schenek, A., Massnahmen zur Vermeidung von Reklamationen bei der Verarbeitung von Baumwolle. *Textil-Praxis* 39: 559–563, June 1984.
9. Anderson, S.L., The measurement of fibre fineness and length: The present position. *Journal of the Textile Institute* 68: 175–180, 1976.
10. Bel-Berger, P. and Roberts, G., Neps devaluate cotton, August 2000, http://www.insidecotton.com/xmlui/bitstream/handle/1/843/img-302114335.pdf?sequence=1, Accessed Date: November 12, 2013.
11. Singh, P. and Kairon, M.S., Cotton varieties and hybrids. *CICR Technical Bulletin* 13, http://www.cicr.org.in/pdf/cotton_varieties_hybrids.pdf, Accessed Date: November 12, 2013.
12. Pavani, H. and Naidu, J., Linear programming for cotton mixing, November 2002, http://webcache.googleusercontent.com/search?q=cache:4fniaB- fZnIJ:faculty.philau.edu/NaiduJ/lpt.ppt, Accessed Date: November 12, 2013.
13. Nair, A.U., Nachane, R.P., and Patwardhan, P.A., Comparative study of different test methods used for the measurement of physical properties of cotton. *IJFTR* 34: 352–358, December 2009.
14. Parthasarathi, V., Evolution in cotton testing instruments. *The Indian Textile Journal* April 2008, http://www.indiantextilejournal.com/articles/FAdetails.asp?id=1065, Accessed Date: November 12, 2013.
15. Dönmez Kretzschmar, S. and Furter, R., A new single fiber testing system for the process control in spinning mills. USTER AFIS PRO 2 Application Report, July 2008.
16. Senthil Kumar, R., Cotton stickiness: Causes, consequences & remedies. *Indian Textile Journal*, June 2008, http://www.indiantextilejournal.com/articles/FAdetails.asp?id=1239, Accessed Date: November 12, 2013.
17. Senthil Kumar, R., Impact of contamination on yarn and fabric quality. *Pakistan Textile Journal*, August 2010, http://www.ptj.com.pk/Web-2010/08-10/R.Senthil-Kumar.htm, Accessed Date: November 12, 2013.
18. Schindler, C., The ITMF cotton contamination of 2005, *Proceedings of the International Cotton Conference*, Bremen, Germany, 2006, pp. 57–60.
19. Senthil Kumar, R., Process control in spinning and weaving—Class notes, May 2012, http://www.docstoc.com/docs/143676084/, Accessed Date: November 12, 2013.

2

Process Control in Blow Room

2.1 Significance of Blow Room Process

The main tasks of blow room process are opening, cleaning, and blending of cotton fiber tufts without overstressing of fibers. One important function of the blow room is to disintegrate the fiber bales into a flow of very small clumps of fiber, which are sufficiently small in size to be digested by the cards. The intensity of fiber treatment is different because the tufts continually become smaller as they pass from stage to stage. Cotton bale has nonlint content apart from cotton that has to be removed in the initial stages of the spinning process itself. About 40%–70% trash is removed in the blow room section. Blender equipped in the blow room sequence reduces the lot-to-lot variability existing in the raw material by homogenous blending. A blow room line is a sequence of different machines arranged in series and connected by pneumatic transport ducts. The sequence of machinery arrangement in the blow room process for a particular process depends on the following factors:

- Fiber type
- Fiber characteristics
- Trash content
- Material throughput
- Mixing formulation

Long staple cotton with low trash would require lesser beating and more opening than short staple, trashy cotton. For processing 100% synthetic fiber, we can bypass several openers and cleaners for gentle treatment of fibers and obviously synthetic fibers do not contain any trash content to remove. Though not preferred, sawtooth wires can be used for opening polyester or nylon fibers. However, they can cause several problems if used for opening soft fibers like viscose since such fibers have the tendency to disintegrate under stress.

Process control in blow room encompasses all the areas such as opening, cleaning, and blending. Opening and cleaning shares nearly about 5%–10% of the manufacturing cost in the spinning mill. A good blow room line ensures gentle opening, effective cleaning, and homogenous blending. Opening means to increase the specific volume (cc/g) of the feed material and is adjudged from the tuft size. The tuft size of 2–3 mg is considered to be optimum, but as low as 0.1 mg is achievable. Increase in the opening/cleaning intensity increases waste removal but also leads to fiber damage, fiber loss, and an increase in neps level. Blow room process has tremendous impact on yarn quality as well as spinning process. The advent of modern openers and cleaners facilitate in achieving the process requirements of the blow room. Choice of beaters and sequence of opening depend on the nature of fiber and the process requirements. Conventional blow room line for the cleaning of cotton consists of four to six and, in some cases, even seven beating points. The underlying idea was to open and clean the material slowly and gradually. In the last decade, the trend has shifted to minimize the number of machines in the blow room line.

Fiber opening is the key to good yarn spinning. Good, gentle opening ensures maximum retention of fiber strength by minimizing fiber rupture, reducing the level of neps, effective thrash removal, and minimal amounts of microdust and lint. In modern blow rooms, four types of beaters are primarily used—disk beaters, peg beaters, pinned beaters, and sawtooth beaters, each having its own function and suitability to certain requirements and conditions. The sawtooth opener tends to cause fiber rupture and lint generation. This tends to increase the percentage of short fibers and the level of neps. The trash contained in the fiber supply also tends to disintegrate into microdust due to the sawtooth action. In comparison, the pin has a smooth round surface and a spherical tip, which opens the fibers through a gentle untangling action. The round profile of pins also has another significant advantage—that of higher performing life and more consistent quality of opening.

2.2 Intensity of Fiber Opening

Opening is the breaking up of the fiber mass into tuft. Opening in the spinning process has two stages: opening to flocks done in the blow room process and opening to fibers done in carding. Finer trash particles are more difficult to remove and require more intensive treatment. The various factors affecting the intensity of fiber opening are given in Table 2.1.

Cotton bales consist of a large number of tightly packed tufts of fibers along with a large number of trash particles embedded among the fibers. The fiber tufts are processed through a series of opening machines, and as a result the larger tufts are broken into smaller tufts and the fibers lose their tightness of packing, thereby reducing tuft density or increasing its specific volume. The effective opening of fiber flocks gives a better chance of trash removal with minimum fiber loss in waste. Fiber openness generally means the

TABLE 2.1

Factors Influencing Intensity of Fiber Opening in the Blow Room

Raw Material	Machine	Process
Thickness of the feed	Type of feed—loose or clamped	Speed of the beater, feed roller
Density of the feed	Form of feeding device	Throughput speed of the material
Fiber coherence	Type of opening device	Ambient conditions: humidity, temperature
Fiber alignment	Type and point density of clothing	
Size of flocks in the feed	Arrangement of pins, needles, teeth, etc., on the surface	
	Spacing of the clamping device from the opening device	

reduction in tuft density or the increase in its specific volume. Openness value is a measure of how fluffed up the fiber mass has become on passing through a beater system (i.e., degree of opening).

Openness value = Specific volume of fiber mass × Specific gravity of the constituent fiber

Specific volume of the tuft can be determined by the beaker test method. The higher the openness index of a tuft, the higher the openness of fibers and vice versa. Opening action of a beater is also termed as the intensity of opening. It can be defined as the amount of fibrous mass in milligrams per one striker of a beater for a preset production rate and beater speed.[1]

$$\text{Intensity of opening } I = \frac{P \times 10^6}{60 \times n_b \times N}$$

where
  I is the intensity of opening (mg)
  P is the production rate (kg/h)
  n_b is the beater speed (rpm)
  N is the number of strikers

The intensity of opening is an estimate of the tuft size produced by a given beater. I value for Kirschner beater and sawtooth beater falls in the range of 5 and 0.88 mg, respectively.[2] From the I value, we can get the number of fibers (n_f) constituting a tuft produced by a given beater.[2]

$$n_f = \frac{I \times 10^5}{L_f \times T_f}$$

where
  n_f is the number of fibers in the tuft
  L_f is the average fiber length (cm)
  T_f is the average fiber linear density (mtex)

Fiber openness increases with the opener speed. The fiber openness can be increased by increasing the angle of grid bars present underneath the bale opener or underneath the fine opener or underneath both of them. The length of the fibers is decreased very slowly with an increase in the openness of fiber due to fiber breakage that occurs at higher fiber opening resulting from higher opening roller speed and higher grid bar angle. Consequently, the short fiber content is found to increase slowly with the increase in fiber openness. Higher fiber openness results in less imperfection in yarn due to the fact that higher fiber openness results in better fiber individualization in card sliver, which in turn produces quality yarn.

Trützschler[3] adopts the technology of opening in such a way that gradual reduction of tuft size can be achieved after every beater (as seen in Figure 2.1). The wire point density of the beater also gradually increases from first beater in blow room to cylinder in carding machine to facilitate gradual and gentle opening at every stage of the process. The intensity of opening achieved in the Trützschler blow room is 0.1 mg, and combined intensity of opening (blow room and carding) is 0.001 mg.

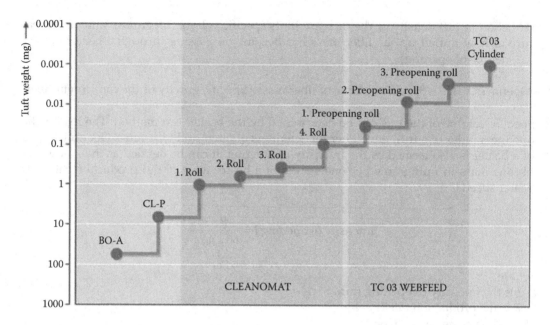

FIGURE 2.1
Effect of opening and cleaning on tuft weight in the Trützschler blow room and card. (Courtesy of Trützschler Spinning, Mönchengladbach, Germany, http://www.truetzschler.com.)

2.3 Cleaning Efficiency

Cleaning the nonlint content is one of the important tasks done in blow room process. The term cleaning means removing the trash content from the fiber tuft. Cleaning can be accomplished by the action of beater against grid bar. The well-opened fiber tuft will be easily cleaned. Trash particles may be in different sizes, ranging from larger seeds to leaf fragments. The larger size trash particles are easy to remove as they fall with their own weight by gravity while opening the tuft. Seed coat fragments are infinitely small particles of fractional trash that stick to the fiber tuft that is difficult to remove. The seed coat fragments are virtually impossible to extract from the bulk of raw cotton because of the tuft of fibers attached generally incorporated into the yarn as a neps. Cleaning efficiency depends largely on smaller fragments removal. An increase of seed coat fragments in raw material is also associated with the cotton cleaning difficulty. The smooth leaf needs only gentle cleaning at the mill, but the hairy-leaf cotton needs much more aggressive cleaning to remove the hairy-leaf particles, which tend to attach to the cotton fibers. Seed coat fragments were the main reason for end breaks, deposits in rotors, increased neps, and other problems.

The efficiency of the blow room process is judged with the term cleaning efficiency. Poor cleaning efficiency in blow room and card may lead to many quality and process problems. Black spots (kitties) in the fabric, fabric barre, and obstructing yarn formation in rotor spinning are the few problems associated with poor cleaning efficiency. The numerical evaluation of the cleaning effect of a machine in spinning preparation is generally effected by detailing the cleaning efficiency, which indicates in percentage terms the quantity of trash removal relative to the trash content present in the feed material.

A typical blow room line has approximately 40%–70% cleaning efficiency. The cleaning efficiency of the blow room process depends on the type of beater used, number of beater, beater-grid bar setting, feed roller to beater setting, speed of beater, environmental conditions, etc. The intensity of cleaning depends on the spacing of the grid from the opening device, the setting angle of the bars relative to the opening device, and the width of the gaps between the bars.

2.3.1 Classification of Cleaning Efficiency

It is clear from Table 2.2 that cleaning efficiency should not be the same for various trash levels in the cotton. It is easier to achieve higher cleaning efficiency with the high trash content raw material. In that case, cleaning intensity will be more while removing more trash from cotton, which leads to fiber loss as well as nepping propensity. The fiber loss and nepping tendency have a positive correlation with cleaning efficiency (as seen in Figure 2.2). Normally, fibers represent about 40%–60% of blow room waste termed as "lint loss." With increasing micronaire values, there is a rise in the cleaning efficiency of the cotton, which suggests that the finer cotton fibers have low rigidity and high buckling coefficient, and are thus more easily entangled into neps and attached to the trash during cleaning. Thus, their cleaning and spinning efficiencies will be affected.

TABLE 2.2

Classification of Cleaning Efficiency

Class (%)	Interpretation
>40	Very good
30–40	Good
20–30	Average
10–20	Bad
<10	Very bad

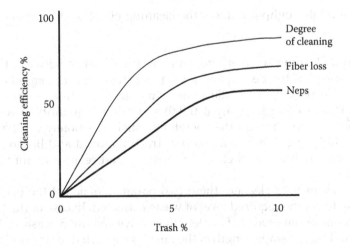

FIGURE 2.2
Effect of trash removal on the degree of cleaning, neps, and lint loss.

2.3.2 Determination of Cleaning Efficiency

The cleaning efficiency of the blow room can be evaluated by comparing the trash present in the mixing and trash in the blow room lap:

$$\text{Cleaning efficiency (\%)} = \frac{(\text{Trash in feed material} - \text{Trash in delivery material})}{\text{Trash in feed material}} \times 100$$

Using Trash analyzer, the trash in the input material and output material is determined. The cotton sample of 100 g is taken and fed through the feed roller. Taker-in separates the lint and nonlint content by intensive opening. The separated lint and trash are weighed, and the data are applied in the aforementioned formula to attain cleaning efficiency of the process or a particular beater.

2.3.3 Points Considered for Attaining Better Cleaning Efficiency

- Larger trash particles are easier to remove than smaller fragments.
- Eliminate trash in the initial stage of blow room process to avoid shattering of trash particles.
- Opening should be followed immediately by cleaning. The higher the degree of opening, the higher the degree of cleaning.
- Better cleaning efficiency can be achieved always with some amount of fiber loss.
- Cotton containing more seed coat fragments should be opened well in the blow room so that effective cleaning is done in carding to attain a higher combined cleaning efficiency.
- Higher roller speeds result in a better cleaning effect, but also more stress on the fibers.
- Cleaning is made more difficult if the impurities of dirty cotton are distributed through a larger quantity of material by mixing with clean cotton.
- Damp stock cannot be cleaned as well as dry.
- High material throughput reduces the cleaning effect, and so does a thick feed sheet.

The opening and cleaning intensity depends on the distance between the beater and the feed roller, speed of the beater, and grid bars setting apart from other various factors. The Rieter VarioSet System[4] adjusts these parameters while the machine is in the running state. The cleaning intensity (0.0–1.0) and relative quantity of waste (1–10) are entered into the VarioSet through the operator's panel or remotely via the ABC-Control system (as seen in Figure 2.3). The amount of trash removed and lint loss experienced with respect to different levels of cleaning intensity and relative quantity of waste are shown in Figure 2.4.

On Rieter UNIclean B 12 cleaner,[4] these two parameters adjust the beater speed and grid bars setting to get the required level of waste extracted. In case of the Rieter UNIflex cleaner,[4] the parameters entered are fiber length, relative amount of waste (1–10), and cleaning intensity (0.0–1.0). The staple length of the fibers is converted for the basic setting of the feed trough nip. The relative amount of waste primarily adjusts the grid bars setting, and the cleaning intensity adjusts the rotational speed of the beater and the feed trough nip.

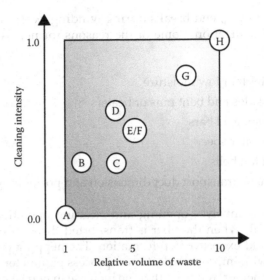

FIGURE 2.3
Cleaning intensity and relative quantity of waste. (Courtesy of Rieter Blow Room Systems, Rieter, Winterthur, Switzerland.)

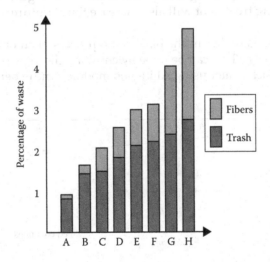

FIGURE 2.4
Fiber and trash proportion in waste removed. (Courtesy of Rieter Blow Room Systems, Rieter, Winterthur, Switzerland.)

2.4 Neps in Blow Room

Neps are small knot-like aggregates of entangled fibers generated in the blow room due to overstressing of fibers and wrong selection of beaters. Basically immature fibers in the cotton tend to create neps while processing. High incidence of neps is responsible for the poor appearance of yarns and fabrics, formation of spotty and streaky materials during dyeing and printing, end breaks during winding, warping, weaving and knitting, and lower price realization. Some of the reasons for nep generation in the blow room department are

1. Cotton with too high or low moisture
2. Rough or blunt blades and bent pins or beaters
3. Damaged and rusty grid bars
4. Too high or low beater speed
5. Slack or too tight fan belts
6. Improper pneumatic transport duct dimension and positioning

The blow room process involves opening and cleaning operation that combines nep generation and removal. When the fiber is transported through pneumatic duct, there can be fiber damage and excessive nep formation. The nepping propensity in the blow room process is high with improper selection of process parameters, machinery, and nep proneness of fiber. Schneider[5] reported that the increase in nep level by 100% or less from bale to card input is generally considered to be acceptable. If the increase is greater than 150% or 200%, extreme caution is required. The gradual reduction of neps irrespective of its type by increasing the cleaning treatments is shown in Figure 2.5. It is obvious that the increase in cleaning treatment will also increase fiber rupture due to repeated stress on fibers.

If the nep generation is higher in the blow room process, then one should identify the machine responsible for it. This can be done by checking the neps per gram of the cotton tuft before and after each beater using AFIS-Nep module. One experimental study shows

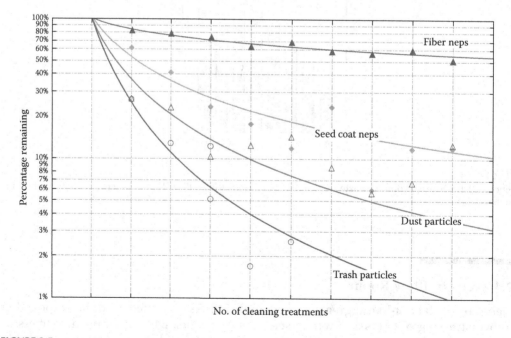

FIGURE 2.5
Effect of the number of cleaning treatments on trash and neps.

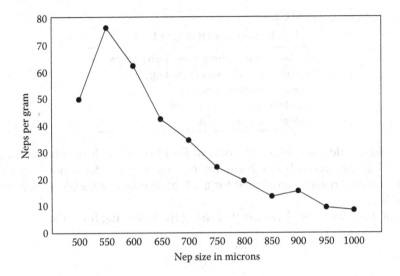

FIGURE 2.6
Nep size distribution and its quantity in the blow room lap.

nep size of 500–700 μm found higher in the blow room lap. The nep size distribution and its quantity in the blow room lap are shown in Figure 2.6.

$$\text{Nep increasing}\,(\%) = \frac{\text{Neps/g in the output material} - \text{Neps/g in the input material}}{\text{Neps/g in the input material}} \times 100$$

Bogdan[6] reported that reprocessing increases tangled fibers due to excessive mechanical treatment. Kistler[7] reported that the addition of waste of about 4% did not significantly affect the nep levels of 72 and 36 tex-carded yarns, whereas a significant increase in nep levels was found for 20 tex-carded yarn. One study shows that the nep levels in cotton lint can vary from below 100 to far over 1000 neps/g. Schneider[5] reported that too intensive beating, increasing number of beaters, too high a feed rate all encourage the formation of neps. Leifeld[8] reported that incorrect cleaning and overaggressive cleaning with sawtooth clothing may cause an increase in seed coat neps. Schneider[5] reported that transport fans running at excessive speed and conveyors moving too slowly can influence nep levels.

2.5 Lint Loss

For an annual production of 10,000 tons, 0.5% raw material savings mean 50 tons or approximately 200 bales of material saved. The costs for the raw material are the largest position in yarn price calculation. Raw material savings thus are the best opportunity to realize cost savings. Fibers should be part of the yarn and not of the waste. The amount of waste extracted in the blow room is mostly determined by the trash level in cotton. In modern blow room lines, emphasis is given to opening of cotton than cleaning. For good cleaning efficiency, the waste extracted in the blow room should be about the same as the trash in mixing. If the cleaning efficiency achieved is less than 50%–60%, then the total

TABLE 2.3

Various Reasons for High Lint Loss

Too close grid bar setting with a high beater speed
Too wide feed roller to beater setting
Improper feed roller weighing
Angle between grids is too open
Waste plate setting is high

waste extracted should also be low. The overall fiber loss or lint loss in waste should not be more than 40% in cottons with a high amount of trash and 50% for cottons with a low level of trash. The various reasons attributed for the high lint loss in the blow room process are listed in Table 2.3.

The expected lint loss[10] can be estimated using the following formula:

$$W_b = \frac{(t - t_L)100}{(100 - L)}$$

$$L = 1 - \frac{(t - t_L)}{W_b}$$

where
   t is the trash in mixing (%)
   t_L is the trash in lap (%)
   W_b is the waste extracted in blow room (%)
   L is the lint% in waste

Trützschler's[3] waste sensor "WASTECONTROL BR-WCT" is an optical measuring system that monitors the share of good fibers in the waste. The sensor detects how many fibers and how many trash particles are contained in the waste.

Using special optimization software, the deflector blades as shown in Figure 2.7 in the cleaner are adjusted until an optimum cleaning is obtained with as few good fibers as

FIGURE 2.7
Trützschler WASTECONTROL BR-WCT. (Courtesy of Trützschler Spinning, Mönchengladbach, Germany, http://www.truetzschler.com.)

possible in the waste. Measuring of the good fiber share is carried out at exactly defined points in the suction hood. The system makes a distinction between dark trash particles and light good fibers. The ideal setting of the deflector blades is the compromise between minimum good fiber loss and maximum cleaning.

2.6 Blending Homogeneity

Homogeneity is an important aspect to be considered in connection with blending. In most cases, the homogeneity of mixing, leading to the homogeneity of physicomechanical characteristics of blended yarns, is required. The blending of different quality fibers of the same type is a well-established technique for achieving quality and economic advantages. Cotton fiber is inherently heterogeneous in its characteristics. The characteristics of cotton are governed by various factors such as seed, place of cultivation, climatic conditions, soils, irrigation, and cultivation techniques. Blending is the process of combining two or more fibers to get all their desirable properties and minimize the blend cost. Blending fibers of the same type from different sources can be used to produce a uniform and consistent product economically. Blending different fiber types can improve aesthetic or functional properties of the end product. In polyester/cotton blends, the crease resistance of polyester helps to retain fabric shape without losing fabric comfort provided by cotton. The type and the quality of the blending process can have a significant effect on yarn quality and characteristics.

Yarn structure is influenced by the radial disposition of the fibers along the yarn length. With sufficient mixing, it is possible to create an almost perfect blend of the materials. Fiber arrangement in the yarn is highly influenced by blending. The extent of intermingling of components within a cross section is an equally important parameter of degree of mixing. Grouping of fibers of one type could also lead to streaks and uneven appearance in dyed fabric. A perfect blend involves laying fibers next to each other in an orderly three-dimensional structure (as seen in Figure 2.8). An increase in the number of fibers per cross section by using finer denier fiber is obviously one of the ways of reducing blend variability.

Blending uniformity[9] can be defined by the following ways:

- Arrangement of fibers in the cross section (radial homogeneity)
- Variation of mixing degree or composition between cross section (axial homogeneity)
- Variation of number of fibers between cross sections (mass unevenness)

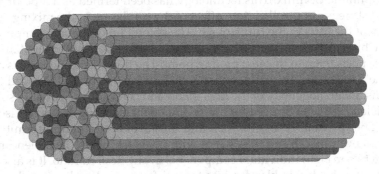

FIGURE 2.8
Blending homogeneity. (Courtesy of Rieter Blow Room Systems, Rieter, Winterthur, Switzerland.)

Intimacy of mixing and blend variation are important parameters to be controlled in blends in view of their critical influence on the appearance, fault incidence, grade, and sale value of fabric. Blend irregularity also influences variability in strength, feel, and abrasion resistance of fabric, streakiness, shade variations, weft bars, and weft way defects. Blend irregularity is broadly of two types: (1) variations in the blend proportion of the component fibers in each cross section along the length of yarn. The variation can be short-term or long-term type. (2) Inadequate intermixing of the components within a cross section. This is termed as degree of lateral mixing.

Bale mixing offers good blend in transverse direction. Flock blending carried out in weighing hopper feeders and blenders has satisfactory blend in both longitudinal and transverse directions due to metering. Flock blending is widely used in many countries. In the conventional blow room line, different lots or varieties of cottons laid horizontally layer by layer and withdrawn vertically by manually are termed as stack mixing. The basic principle of stack mixing is to fill, sequentially, a series of vertical compartments in a storage bin (providing stacks of tufts) and then to remove the layers from consecutive stacks in a manner that sandwiches the layers, thereby dispersing and mixing tufts, say, from the first traverse of the bale lay down with tufts from subsequent traverses. The factors influencing the various types of blending are

- Capital cost
- Labor intensiveness
- Blending precision
- Liability to error and simplicity

Trützschler[3] Multimixer MPM can have 6 or 10 hoppers depending upon the application. Multimixer consists of series of hoppers arranged side by side. The rotating flaps forward the material in sequence to the individual trunks, but these trunks are emptied at the same time, resulting in the homogeneity of the mix. Once all the trunks are filled, the transport air is routed past the trunks to prevent material compaction. As soon as the trunks are empty, refilling starts immediately so that a loose layer of tufts is formed on the conveyor belt from the individual trunks. At the end of the conveyor belt, material from all the trunks lies in layers in a sandwich form that ensures ideal feeding for the CLEANOMAT Cleaner. For better mixing and homogeneity, two mixers can also be set in tandem.

The Rieter[4] UNImix B 70 blending machine claims to have random distribution of the tufts to the eight mixing chambers and then the fiber blending takes place at three different points within the UNImix. This technology has been termed as a 3-point mixing process. The first stage is a controlled time offset of the tuft layers in the mixing chambers by 90° deflection of the tufts in the tuft storage. The second mixing point comprises a spiked feed lattice, which picks small tufts at random out of the layers of the eight mixing chambers and transports them to the next mixing point. The third blending level is achieved by mixing of the fine tufts in the active mixing chamber above the spiked feed lattice. This results in constant homogeneity of the fiber blend and constant yarn quality subsequently. Rieter's UNIblend A 81 can be used for multicomponent blending. It can precisely measure deviation in fiber percentage less than 1% and thus avoid color nonuniformity in the end product. It can even mix/blend 98% white with 2% black fibers. This blending machine can mix from two to eight individual components in any desired ratio. It is also possible to split the line, after the dosing blender, into four different carding lines, each of which can contain a different blend ratio of the same components.

For optimum utilization and reasonable prediction of the properties of yarn or fabric from the fiber properties, it is essential to have constituent fibers blended as homogeneously as possible. The blend homogeneity could be expressed as index of blend irregularity[10]:

$$\text{Index of blend irregularity (I.B.I.)} = \sqrt{\left(\frac{1}{M}\right)\sum \frac{(T_i p - W_i)}{(T_i pq)}}$$

where
T_i is the total number of fibers in yarn cross section
M is the number of sections examined
W_i is the number of fibers of component W
p is the average fraction of component W for all the sections
$q = (1 - p)$

When index is equal to unity, it means that the arrangement of fibers is random. A value of zero would mean that the arrangement of fibers is perfect.

2.7 Fiber Rupture and Its Measurement

Fiber rupture refers to the reduction in the fiber length due to break of fibers during processing. During the blow room and carding process, fiber experiences too much mechanical stress while undergoing opening and cleaning operation. A weaker fiber will be more prone to fiber rupture while a beater acts on it. If the number of beating points is more, then even a stronger fiber could not withstand the stress beyond the limit and tends to break. Some of the causes of the fiber rupture in the blow room department are listed as follows:

- More number of beating points
- Closer setting between beater and feed roller
- Very high beater and fan speed
- High speed and closer setting in sawtooth beaters

Fiber rupture is critical that significantly affects yarn quality and processability. Fiber rupture will have a huge impact on imperfections and irregularity in the yarn. The choice of beaters and sequence of opening depends on the nature of fiber and the process requirements. The various beater factors influencing fiber rupture are listed in Table 2.4. Good, gentle opening ensures maximum retention of fiber strength by minimizing fiber rupture and reducing the nep level. Gripped beating creates more fiber rupture than loose beating.

TABLE 2.4

Beater Design Factors Influencing Fiber Rupture

Beater wire point density
Projection and angle of points
Wire point tip profile

In this case, the setting between feed roller nip and the beater is crucial in determining the extent of opening and cleaning and simultaneously the extent of fiber rupture. Opening action of sawtooth wire is characteristically different from pins that tend to cut open the fibers causing fiber rupture and lint generation.[11] In comparison, the pin has a smooth round surface and a spherical tip, which opens the fibers through a gentle untangling action. Fiber rupture would be minimized as well as the consequent generation of microdust and lint would also be reduced considerably with the use of pins. The beater speed and setting also significantly influence the fiber rupture. If the mill is processing long staple cotton, then the process parameters should be in accordance with the fiber used. Synthetic fibers need gentle opening and no harsh treatment.

After setting the process parameters in the blow room, mills have to ensure that any occurrence of fiber rupture in any opening or cleaning machine. A measure of fiber rupture is determined by the reduction in span length of the fiber after processing in relation to its value before processing. Based on the digital fibrograph estimates, a difference of 4% in 2.5% span length between the length of feed and delivery material is an indication of fiber rupture.

Fiber damage is another general term used in some countries, which is defined as a substantial change in one or more of the basic fiber characteristics that can result in a loss of fiber contribution to yarn or fabric performance. Fiber damage is classified into many forms such as fiber breakage, loss of elasticity, surface deterioration, and fiber neps. The extent of fiber damage (EFD)[12] is a method to determine the occurrence of fiber damage before and after the process with respect to change in short fiber content:

$$\text{Extent of fiber damage (EFD\%)} = \frac{\left(\text{SFC}_{\text{out}} - \text{SFC}_{\text{in}}\right)}{\text{SFC}_{\text{in}}} + \text{CF}_{\text{w}}$$

where
  SFC_{out} is the short fiber content in the output material
  SFC_{in} is the short fiber content in the input material
  CF_{w} is the correction factor that accounts for short fiber content extracted with the waste

The correction factor can be determined using the equation

$$\text{CF}_{\text{w}} = W \frac{\left(\text{SFC}_{\text{w}} - \text{SFC}_{\text{out}}\right)}{\text{SFC}_{\text{in}}}$$

where
  SFC_{out} is the % short fiber content in the output material
  SFC_{in} is the % short fiber content in the input material
  SFC_{w} is the % short fiber content in the waste material
  W is the waste%

2.8 Microdust

Cotton after cultivation contains very little dust. During the opening in the ginning and blow room process, a large amount of microdust is liberated. Even though the dust is removed, again the dust is being created through shattering of impurities. The microdust comprises 50%–80% fiber fragments, leaf and husk fragments, 10%–25% sand and earth,

TABLE 2.5

Classification of Dust

Type	Size of the Particle (μm)
Trash	Above 500
Dust	50–500
Micro dust	15–50
Breathable dust	Below 15

and 10%–25% water-soluble materials. The high proportion of fiber fragments indicates that a large part of the microdust arises in the course of processing. Nearly about 40% of the microdust is free between the fibers and flocks, 20%–30% is loosely bound, and the remaining 20%–30% bound to the fibers. Microdust has to be removed effectively to safeguard the health of the operative and to reduce the microdust in the cleaned cotton.

2.8.1 Classification of Dust

The dust collected in the spinning mill can be classified according to its size as listed in Table 2.5.

2.8.2 Problems Associated with Microdust

Workers exposed to cotton dust–laden environment generally become patients of byssinosis.[13] It is a breathing disorder that occurs in some individuals with exposure to raw cotton dust. Characteristically, workers exhibit shortness of breath and/or the feeling of chest tightness when returning to work after being in the mill for a day or more. There may be increased cough and phlegm production.

2.8.3 Microdust Extraction

Dust that separates easily from fibers is removed as far as possible in the blow room. The transport fans are used in the scutcher to transfer the cotton from one machine to another. The dust-laden air exhausted by the fan is blown into the cellar (dust room) that is positioned under the scutcher. In the cellar-less blow room, material from the preceding machine is thrown with a great force by the transport fan on the screen. The oscillating dampers help in distributing the material evenly across the width of the screen. This results in smaller tufts coming in contact with the perforated screen. The liberated microdust is removed by the dust transport fan that is placed on the opposite side of the screen. After the material slides down, it is sucked away by either a condenser or a material transport fan. It is recommended to use the microdust remover after the last beating point where the cotton is in open form so that the microdust can be removed easily.

Dedusting is the term often used to tell about microdust extraction in the modern blow room line.[4] Effective dedusting of the material is paramount to running efficiencies in downstream machines specifically in rotor spinning applications. A good dedusting leads to considerably higher efficiencies in the winding process. Dedusting in the blow room happens by air suctioning only either between the machines, for example, by dust cages, dust extractors, etc., or within the machine by normal air separation. Usually two filter stages are used because a great deal of fly is carried along in the removal of dust by suction. The stages are preliminary filtering and fine filtering as shown in Figure 2.9. These operations can be performed with individual filters or a central filter.

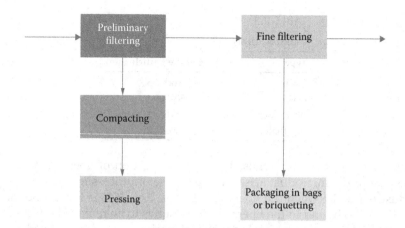

FIGURE 2.9
Dedusting operation. (Courtesy of Rieter Blow Room Systems, Rieter, Winterthur, Switzerland.)

In new installations, the dust-laden air flows against a slowly rotating filter drum (as seen in Figure 2.10). The air rotary filter has rotating drum made of steel grid mesh with suitable mounting frames, driven by a geared motor at a low rpm. Air rotary filter is used to clean exhaust air properly that afterward can be utilized in the department by passing through air-conditioning washer chambers.

The unique design of the Crosrol Dust Remover shown in Figure 2.11 ensures that unsurpassed cleaning efficiencies are achieved. In this dust remover, positive pressures and negative pressures are used to dedust the material. It has a rotating perforated disk in which tufts of fiber are blown onto the upper sector of the rotating disk's surface under positive pressure. Negative pressure is applied on the rear side of the disk that holds the material onto the disk assisting in the removal of the fine dust particles. As the disk rotates, tufts of fiber move from the upper to the lower side of the disk. The lower sector is isolated from the negative pressure, thus allowing the fiber to be freely released from the disk surface.

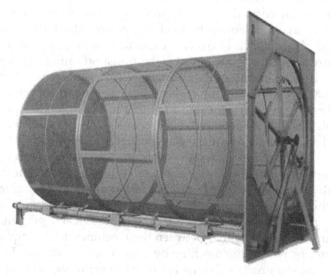

FIGURE 2.10
Air rotary drum filter.

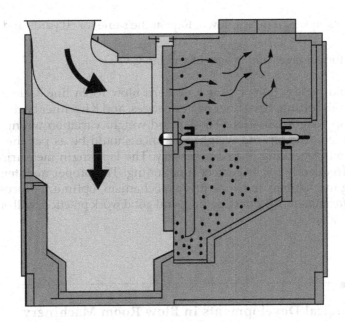

FIGURE 2.11
Crosrol dust remover.

2.9 Lap Uniformity

In the conventional blow room line, lap is the end product of the blow room and intermediate product of the spinning process. Lap uniformity plays a predominant role in deciding the card sliver quality. The conversion of opened and cleaned cotton into a continuous uniform sheet called lap is performed in the lap-forming unit. The lap-forming unit in the blow room process is also called a scutcher. Cotton tufts opened and cleaned in the fine opener are allowed to deposit over perforated cage condenser that is under a negative pressure. Piano feed regulating motion and cone drive mechanism ensure the uniform feeding of cotton tuft to the fine opener. A good opening of cotton reduces the size of the tuft. As the tuft size reduces, lap uniformity can be improved. Pedals in the piano feed regulating motion sense the difference in batt thickness and convey it to cone drive mechanism that adjusts the speed of the feed roller according to the batt thickness. Lap weight should be checked and recorded for the set length of the lap. If lap weight deviates from the set standard lap length, it is the sign of deviation of lap hank due to improper opening or malfunction in piano feed regulating motion.

Lap C.V% is the critical factor that influences the uniformity of the lap. The coefficient of variation of the lap should be checked at frequent intervals by the quality department. The procedure to check the lap C.V% is given next:

1. The lap should be taken from the blow room department to the testing place.

2. A square wooden scale of 1 yard (length) × 1 yard (width) is taken for testing lap C.V%.

3. Unroll the lap to a certain extent and place the wooden scale on the sheet of the lap and using scissors cut the sample of size 1 yard (length) × 1 yard (width) dimension.

4. Weigh the sample and note it down. Repeat the same for 50 yards and generate 50 readings.
5. Calculate the average sample weight and lap C.V%.

Lap C.V% should be less than 1% for a modern blow room line. Calendar rolls pressure, stripping rail setting, Kirschner beater gauges, and Kirschner beater speed mainly influence the lap weight variation. The lap rod weight variation wrongly projects the lap hank variation. The lap rod weight tolerance should be as per the recommendations, and periodic checking is also necessary. The lap length measuring motion has to be checked frequently for its correct functioning. The proper maintenance of piano feed regulating mechanism and cone drive mechanism, optimum process parameters, proper mixing formulation, proper opening, and good work practices will ensure a proper control of lap C.V%.

2.10 Technological Developments in Blow Room Machinery

Opening, cleaning, and blending are the main tasks of any blow room line. Conventional blow room lines have a main focus on achieving higher cleaning efficiency in the blow room itself so that the fiber tuft may experience intensive and harsh treatment that leads to fiber rupture and nep propensity. The number of openers and cleaners in conventional blow room line is more due to the low cleaning efficiency achieved by individual beater. The modern blow room line emphasizes more on opening of tufts than cleaning to ensure a gentle treatment to the fibers. The concept of better opening is accomplished in a gradual manner without overstressing of fibers. In this section, the developments of blow room machineries in the recent decades are discussed.

2.10.1 Automatic Bale Openers

The action of breaking the baled fiber mass down into initially large and then smaller size tufts is termed opening.[3,4,14] The manual bale opening process relies more on the sincerity and efficacy of the worker and has more variation in the tuft size fed to the bale opener. The automatic bale opener gives a smaller tuft size, thus resulting in a better opening and cleaning efficiency of the subsequent machines. Small tufts are essential for efficient cleaning because they offer a large surface area. Efficient bale opening must prepare the material properly for the subsequent cleaning stages.

Rieter's UNIfloc A 11 as shown in Figure 2.12 uses single plucking roller, called take-off roller. The take-off roller along with a narrow grid results in a small tuft size. The same roller can be used for processing cotton as well as man-made fibers. The roller teeth can also be replaced individually. The automatic bale opener regularly measures the profile of the lay down material for gradually leveling out the bales. Trützschler BLENDOMAT BO-A as shown in Figure 2.13 is fitted with two plucking rollers rotating in the opposite direction. It can process three different cotton/blend lots simultaneously that can be fed into three separate cleaning lines. The large working width of the BLENDOMAT of 1720 or 2300 mm allows lay down of up to 150 or 200 and more bales (machine length of 50 m). Marzoli automatic bale opening Super Blender B12SB has two plucking rollers

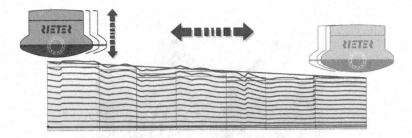

FIGURE 2.12
Rieter UNIfloc A 11. (From Rieter, Rieter Blow Room Machineries Brochure, Rieter Company, Winterthur, Switzerland, 2011. With permission.)

FIGURE 2.13
Trützschler BLENDOMAT BO-A. (From Trützschler GmbH & Co KG, Fibre+Sliver Technology, Trützschler GmbH & Co KG Textilmaschinenfabrik, Information brochure, Trützschler GmbH & Co KG, Mönchengladbach, Germany, 2011. With permission.)

with 254 blades on each roller, which ensures a small flock size. Automatic bale opener can process four different mixings. The production, with a working width of 2250 mm, is stated to be 1600 kg/h. Lakshmi Automatic Bale Plucker LA 17/LA 28 is fitted with twin plucking rollers. The grills of the plucking rollers are selectable according to the required tuft size.

2.10.2 Openers and Cleaners

Rieter's UNIclean B 11 as shown in Figure 2.14 consists of specially designed pin roller, grid bars with electrocylinder control, and VarioSet program for optimizing cleaning intensity and waste rate (as seen in Figure 2.15).[3,4,14] The cleaning intensity and waste rate can be adjusted by changing the rotational speed of the drum and grid bar angle, respectively, while the machine is running. Material transport inside the UNIclean B 11 takes place mechanically by means of special pins, independently of the conveying air. The material is circulated seven times inside the machine by the hooks on the drum. The raw material is forced over the cleaning grid bars where trash is removed.

Rieter's UNIflex B 60 as shown in Figure 2.16a is a fine cleaner for processing natural fibers that works on single cylinder concept. A single opening and cleaning cylinder as shown in Figure 2.16b minimizes the negative side effects such as higher nep generation,

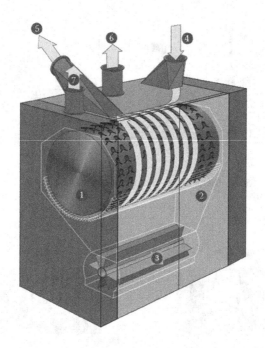

FIGURE 2.14
Rieter UNIclean B 11. 1, cleaning cylinder; 2, cleaning grid; 3, airlock cylinder; 4, material feed; 5, material outlet; 6, filter exhaust air; 7, waste removal. (From Rieter, Rieter blow room machineries brochure, Rieter Company, Winterthur, Switzerland, 2011. With permission.)

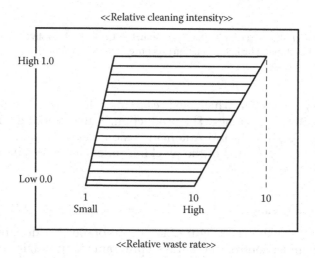

FIGURE 2.15
Cleaning characteristic diagram. (From Rieter, Rieter blow room machineries brochure, Rieter Company, Winterthur, Switzerland, 2011. With permission.)

fiber rupture, and lint loss. Unidirectional feeding is performed by a feed roller and a feed trough. The modular grid with the carding element and the cleaning cylinder work together in the removal of fine trash particles. The speed of the cleaning cylinder and grid bar can be adjusted by means of VarioSet. With only two settings, fiber processing and cleaning can be adjusted in a range from gentle to intensive.

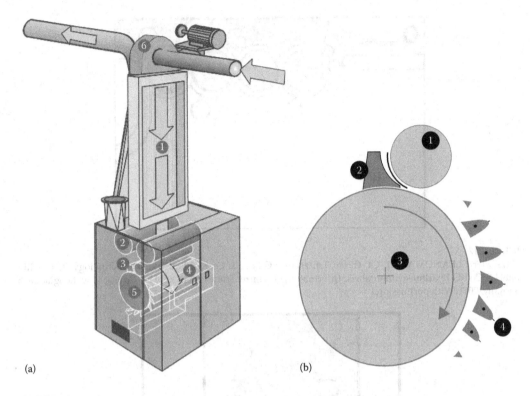

(a)   (b)

FIGURE 2.16
(a) Rieter's UNIflex B 60. 1, lamellar chute; 2, perforated drum; 3, adjustable feed trough; 4, adjustable grid; 5, opening and cleaning cylinder; 6, fan with divider element. (From Rieter, Rieter blow room machineries brochure, Rieter Company, Winterthur, Switzerland, 2011. With permission.) (b) UNIflex B 60—opening and cleaning cylinder. 1, feed roller; 2, feed trough; 3, opening and cleaning cylinder; 4, adjustable grid.

Trützschler's CLEANOMAT CL-C4 as shown in Figure 2.17 consists of four beaters in series in the same machine rather than two or four machines in tandem in line that gives the same cleaning. Thus, the modern blow room line becomes shorter without any compromise on the quality of the material.

Trützschler's waste sensor WASTECONTROL BR-WCT is attached to a Cleaner CLEANOMAT and optically measures good fibers in the waste and amount of suction for fibers.

Trützschler's Universal Cleaner CL-U (as seen in Figure 2.18) ensures gentle fiber processing due to the four-roll feed unit that is free from tight clamping. Initial dust removal takes place already during feeding. Trützschler's Universal Opener—TO-U (as seen in Figure 2.19) is a high-performance opener that can process all fibers up to 130 mm. It offers three different rollers for each material and application.

2.10.3 Blenders or Mixers

Blending of dissimilar fibers should also be done properly at blow room stage.[3,4,14] Mix homogeneity depends on the methods of mixing and type of mixing/blending machine used. When mixing the different bales in tuft form, the smaller the tuft size the better would be the homogeneity. Rieter UNImix B 70 blending machine claims to have random distribution of the tufts to the eight mixing chambers and then the fiber blending takes

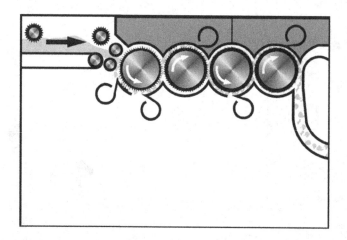

FIGURE 2.17
Trützschler CLEANOMAT CL-C4. (From Trützschler GmbH & Co KG, Fibre+Sliver Technology, Trützschler GmbH & Co KG Textilmaschinenfabrik, Information brochure, Trützschler GmbH & Co KG, Mönchengladbach, Germany, 2011. With permission.)

FIGURE 2.18
Trützschler Universal Opener TO-U. (From Trützschler GmbH & Co KG, Fibre+Sliver Technology, Trützschler GmbH & Co KG Textilmaschinenfabrik, Information brochure, Trützschler GmbH & Co KG, Mönchengladbach, Germany, 2011. With permission.)

place at three different points within the UNImix. This technology has been termed as the "3-point mixing process." Rieter "UNIblend A 81" can be used for multicomponent blending. It can precisely measure deviation in fiber percentage less than 1%, and thus avoid color nonuniformity in the end product. It can even mix/blend 98% white with 2% black fibers. Trützschler mixer can have 6 or 10 trunks depending upon the application.

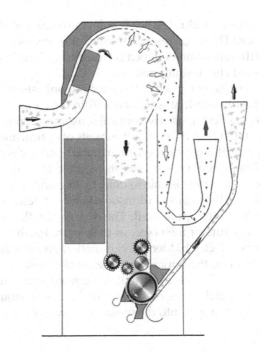

FIGURE 2.19
Trützschler Universal Cleaner CL-U. (From Trützschler GmbH & Co KG, Fibre+Sliver Technology, Trützschler GmbH & Co KG Textilmaschinenfabrik, Information brochure, Trützschler GmbH & Co KG, Mönchengladbach, Germany, 2011. With permission.)

2.10.4 Contamination Sorting

Contamination, even if it is a single foreign fiber, can lead to the downgrading of yarn, fabric, or garments or even the total rejection of an entire batch and can cause irreparable harm to the relationship between growers, ginners, merchants, and textile and clothing mills.[14] Most contamination arises from impurities being incorporated into the bale as a result of human interaction during harvesting, ginning, and baling. Detection of foreign fibers in spinning preparation is an indispensable part of the spinning process today. Trützschler's SECUROPROP SP-FPU[15] foreign parts eliminations system consists of matrix containing 15 broad types of contaminants and an assessment of the negative impact of each on product and process quality. The following are the modules present in SP-FPU:

- Color module—using multiple cameras to detect colored fibers and yarns as well as white PP
- P module—using polarized light to detect transparent and semiopaque PP
- UV module—utilizing ultraviolet (UV) light to isolate brightened fibers such as PP and polyester

A color camera scans the web on the surface of the opening roll, and on the detection of a foreign particle, the compressed air impulse of a nozzle blows it into a waste suction device. The timing of the ejection jets, which blow out the impurities, is obviously critical and is efficiently achieved by having sensors that determine the speed of tufts in the duct.

Jossi's Vision shield[16] uses two ultrafast CCD color cameras and detects contamination as being different in color. The high resolution and photorealistic real-color processing guarantees the utmost differentiation between cotton and contaminations. The pneumatic ejection eliminates detected contaminations.

The critical point, however, is that an optical sensor can only see what is visible, meaning that it cannot detect contamination that is hidden within the cotton tufts. To overcome this, most systems, like the Loptex Sorter, use two optical sensors each positioned at the opposite sides of the pipe. Loptex's OPTOSONIC system[14] consists of optical and sonic sensor. Optical sensor measures average brightness of the raw material and recognize colored contamination as being darker. The optical sensor consists of standard fluorescent light tubes and photo sensor arrays. The sonic sensor is able to detect white and transparent contamination including all types of plastics. By means of ultrasonic waves, it detects all the contamination with major density than the cotton processed. The degree of reflectance of acoustic waves depends on the surface structure of the object in their path. Furthermore, since the acoustic waves penetrate the fiber flocks, hidden contamination are also detected. In case of the detection of a contamination by the optical or the acoustical system, the electronic control will activate pneumatic valves. The air blow will be targeted since only the valves are activated that are located in front of the passing contamination. The contamination will be deviated through an opening in the pipes into the waste container of the machine.

2.11 Chute Feed System and Its Control Mechanism

Modern blow room lines are equipped with chute feed system to supply tuft to carding machine. The chute consists of reserve top chamber, feed roller, beater, bottom chamber, delivery rollers, and in-built fan. The material feed is regulated through a pressure transducer and a VFD drive, which is fitted at the last feeding point to ensure constant and uninterrupted feed to cards. The reserve top chamber of chute that receives the fiber tufts from the fine opener of blow room and separates it from the air. The chute contains photo cell sensor to measure the tuft position in the chute. The package density of the fiber tufts in the chute column is often influenced by vibrating front plates, air flow, and proper designing of chute elements. Vibrating front plates or compressed air in continuous manner is responsible for uniform and firm fiber batt. The control of fiber tuft density and fiber level in the chute reserve box will decide the good carding operation.

2.12 Process Parameters in Blow Room

The selection of appropriate process parameters plays a significant role in any manufacturing process that influences quality and productivity. The basic purpose of the blow room process is to supply smaller tuft and cleaner tuft to the carding machine. The quality of ginning decides the quality of cotton in the bale. Harsh treatment in the ginning process breaks the larger trash particles into smaller ones, which is difficult to remove in blow room. The technologist in the spinning mill should take utmost care while deciding the process parameters in the blow room process. An efficient blow room process accomplishes the basic objectives such as opening, cleaning, mixing of cotton without increasing fiber rupture, fiber nep and broken seed particles, and extracting more amounts of trash without good fiber loss. Efficient blow room process is possible with the right selection of process parameters. The process parameters in the blow room process can be listed as follows:

- Speed of the beater
- Speed of the feed roller, feed type
- Number of beating points, type of beater
- Feed roller to beater setting
- Beater to grid bar setting, grid bar angle
- Pneumatic air velocity
- Calendar roller pressure (in lap feed system)
- Piano feed regulating motion
- Ambient conditions

The efficient opening at blow room stage not only improves fiber cleaning but also yarn properties such as yarn tenacity and total imperfections.[17] Effective preopening results in smaller tuft sizes, thus creating a larger surface area that facilitates easy and efficient trash removal by subsequent openers and cleaners. It must be ensured that during preopening, larger trash particles should not be broken, which will be difficult to remove from the tuft. Optimum speed of the beater in any opener or cleaner ensures smaller tuft with less fiber damage and neps. Kumar et al.[17] reported that the increase in imperfections at higher openness is due to over beating than that which is necessary; the fibers are stressed and damaged, which then buckle and tend to form neps. The speed of the opener or cleaner in a blow room process is predominantly influenced by factors such as fiber fineness, trash, fiber type (natural or synthetic), and fiber length. Beater speed has a positive correlation with fiber quality, that is, fiber rupture and fiber neps. The yarn hairiness remains almost unchanged initially with the increase in openness at blow room, but at a higher level of openness it increases sharply.[17] Ishtiaque et al.[18] attributed that the overstressing of fibers at higher openness with staple shortening and generation of short fibers is the reason for increase in hairiness tendency in yarn. For processing synthetic fibers and/or fine cotton varieties, a lower beater speed is normally preferred to minimize fiber damage and nepping tendency. The influence of beater type was discussed enough in previous topics. The number of beating point in a blow room process is decided by the trash% in cotton and fiber type used.

Conventionally, terms such as bite and blow were widely used with respect to feeding of cotton tuft. "Bite" refers to the point at which the feed rollers hold the cotton and "blow" refers to point at which the blades of the beater strike the cotton that is held by the feed rollers. The distance between the bite and blow is highly influenced by the staple length of the fiber used. The feeding type (loose or gripped) of cotton tuft to beater influences the opening intensity as well as fiber damage. Gripped feeding of cotton tuft facilitates intensive opening and cleaning but fiber flocks get harsh treatment prone to fiber rupture. Gripped feeding is normally employed at later stages of the blow room process. The combination of beater and feed roller in conventional beaters such as Kirschner beater or three-bladed beater or ERM cleaner provides an intensive opening of bigger tuft into smaller tuft. Modern beaters such as CVT cleaner, Uniclean, also have beater feed roller system for an intensive opening of fiber tuft. An intensive opening of fiber tuft helps to effectively release the fine trash particles adhered to the fiber tuft. Bhaduri[19] reported that the degree of opening greatly affects the cleaning and lint loss at carding and does not appear to have a significant effect on yarn quality, especially yarn strength, evenness, and performance. Loose feeding of cotton tuft to beater gives gentle opening, but not a very intensive opening, which reduces the cleaning efficiency. The coarse beaters of the blow room process such as axiflow cleaner, step cleaner, and monocylinder adopt the loose feeding technique to ensure gentle treatment of cotton tuft. Step cleaner opens the larger tufts by the actions of opposing spikes with the grid bars. A high degree of opening out in the blow room reduces shortening of staple at the cards.[20] A closer setting between feed roller and beater enhances the degree of opening, but it also stresses the fiber tuft.

The angle of grid bars and the space between them can be adjusted so as to optimize the cleaning efficiency and lint loss. The grid bar settings in modern cleaners can be normally adopted to handle any degree of impurities and opening in the cotton to be processed by means of the lever provided with the scales. With reference to the grid bar setting, it must be remembered that at first the cotton requires to be held quickly so that it receives a shock that frees the impurities, after which the bars should subject the cotton more to a scrapping action. The grid bar angle should be kept less acute so that the cotton can roll over the bars, thus removing many of the trash particles carried over with the cotton. The cleaning intensity controls both the speed of beater and compensatory setting of grid bars in the modern beaters.

Air velocity in a blow room process plays a significant role in the effective transport of fiber tuft from one machine to another machine. Improper air velocity generates processing problems as well as quality defects. The ratio of fan speed to beater speed for a typical cotton spinning process is 2:1. Optimum calendar roller pressure is prerequisite for a uniformly built compact lap in the lap feed system. The proper maintenance of piano feed regulating motion is essential to control lap uniformity or tuft uniformity. Ambient conditions such as temperature and RH% influence the process performance significantly.

2.13 Defects Associated with the Blow Room Process

2.13.1 Lap Licking

The sticking of fibers between lap layers is termed as lap licking. The problem of lap licking arise mainly due to the factors such as excessive addition of soft wastes in mixing, higher rack pressures, too high fan speed, excessive beating, higher honey dew content in cotton, lower compacting of laps, and excessive dampness in cotton. In case of synthetic fibers, especially polyester, this problem shall be mainly due to static charges and higher bulk of fibers. The tendency of lap licking can be reduced by increasing the calendar roller pressure, reducing the pressure on racks, increasing the quantity of antistatic, proper selection of fan speed and beating points, use of roving ends or lap fingers behind the calendar roller nip, blocking of top cage, and reducing the lap length.

2.13.2 Conical Lap

Conical laps occur mainly due to either higher quantity of cottons coming on one side of the lap or due to unequal calendar and rack pressures in scutcher. The occurrence of conical laps can be controlled by maintaining several factors such as

- Ensure equal opening of air inlets under grid bars
- Replace torn leather lining at the cage
- Clean the cage thoroughly with emery paper

- Make pressure on lap spindle uniform on both the sides
- Remove the pedals and clean thoroughly, and check the pedals where it rests on fulcrum and also pedal fulcrum bar

2.13.3 Soft Lap

Inadequate calendar roller pressure due to wear and tear of weighting mechanism produces the soft lap.

2.13.4 Curly Cotton

The very important step in avoiding curly cotton is to avoid chocking of materials in beaters. Excessive use of cotton-spray oil, water, etc., during mixing may be prone to chocking in pneumatic duct or beater, which leads to curly cotton. The other various factors influencing the generation of curly cotton are too closer grid bar setting, hooked beater spike or pin, wide setting of stripping rail, too much bent conveyor ducts, and lower fan speed.

2.13.5 Nep Formation in Blow Room

Nep is a fiber entanglement. Nep can be classified as fiber neps and seed coat neps. Most opening and cleaning machines in the blow room process double the number of neps that were in the bale material. All cleaning machines generate neps, but the actual increase depends greatly on the aggressiveness of the machine's components. The major reasons for nep generation in blow room process are

- Blunt beaters, higher beater speed
- Lower fan speed, inappropriate ratio of fan to beater speed
- Cotton with high or low moisture content
- Too much reprocessing of laps and sliver in blow room
- Higher soft waste addition in mixing
- Closer setting between feed roller and beater
- Presence of immature fibers
- Long and too much bends in pneumatic transport duct

2.13.6 High Lap C.V% or Tuft Size Variation

Modern blow room lines can produce lap or tuft with high uniformity compared with conventional blow room line. High lap C.V% or tuft size variation will in turn affect the card sliver C.V% and yarn C.V%. Modern blow room line assures output tuft size in the range of 5 mg. The reasons for high C.V% in the lap are as follows:

- Improper action of feed regulators, viz., cone drums, pedals, photocells, direct driving gear motors, etc.
- Improper mixing especially with the manual process
- Insufficient preopening
- Improper levels in the hopper

Salhotra[21] reported that the heavier lap causes excessive beating at the card causing lap-ups in the card cylinder and formation of too many neps.

2.13.7 Other Defects

- *Patchy lap*: The unopened tuft results patchy lap. The various reasons for patchy lap are insufficient opening, improper setting between feed roller and beater, obstruction in cage, and poor suction in cage.
- *Holes in lap*: The reason for this is due to damage in cages and higher tension draft.

2.14 Work Practices in Blow Room

- Mixing compartment (conventional blow room line) in the blow room department should have proper information cards consisting of variety name, lot, count, and date of mixing laid.
- Bale trolley should be used to transport bales from bale godown to blow room.
- Contamination sorter should be checked frequently for proper functioning.
- Roving waste should be preopened before adding to the mixing.
- Compressed air should not be used while machines are in running condition.
- Proper R.H% should be noted frequently.
- Lap should be weighed for each lap produced and registered in the log register.
- If there is high variation found between laps, front attendant must check the proper functioning of piano feed system.
- Lap should be covered with polypropylene sheet to prevent from dust accumulation.
- Lap rod weight variation to be avoided.
- Front and back attendant of blow room department to be trained for proper material handling, machine operations, process details, meeting emergency conditions such as fire hazard, etc.

2.15 General Considerations in Blow Room Process

- Fan speed should be optimum for smooth transporting of fiber tufts. A higher fan speed will generate turbulence in the bends, which results in curly fibers and neps.
- The reserve chamber for the feeding machine should have adequate volume to avoid long-term variations, especially for the blow room running at a higher production rate.
- The selection of fan speeds and layout of machines should assure the material chocking in pneumatic duct, beater jamming, etc., does not happen. The layout should be in such a way that pneumatic ducts should have minimum bends to avoid curly fibers.
- The feed roller speed of the machine should be selected in such a way that it functions at least 90% of the running time of the next machine to attain maximum blow room efficiency.

- Metal fragments, bale iron straps, etc., that contaminate the cotton should be removed using magnetic extractors before they damage machine parts.
- Fire eliminators that detect any spark occurrence and eliminate should be a part of the blow room line to prevent fire accidents.
- The number of opening points is decided on the fiber type. The number of cleaners adopted in a blow room line is influenced by ginning type and trash% in mixing.
- The blow room machinery should be selected based on fiber type, yarn count spun, and production requirements.
- The storage chamber should always be in filled condition and the material level should not fall below 25% of its capacity.
- Periodic monitoring of conditions of grid bars should be done. Damaged grid bar should be immediately replaced.
- Cotton with pronounced stickiness creates deposition on machine parts that obstructs the fiber movement. It should be cleaned frequently.
- The occurrence of fiber rupture must be checked for each opener and cleaner by measuring drop in 2.5% span length. 2.5% span length should not drop by more than 3%.
- Blow room tremendously increase neps, but it should not be more than 100%.
- With the help of AFIS, nep increase in each blow room machines can be determined. Optimization of beater speeds and settings to be done to control neps.
- R.H% to be maintained as per recommendations.

References

1. Szaloki, S.Z., *Opening, Cleaning and Picking*, Vol. 1, Institute of Textile Technology, Charlottesville, VA, 1976, 126p.
2. Lawrence, C.A., *Fundamentals of Spun Yarn Technology*, CRC Press, Boca Raton, FL, 2003.
3. Fibre+Sliver Technology, Trützschler GmbH & Co KG Textilmaschinenfabrik, Information Brochure.
4. Blow room system: Variations on success, Rieter Textile System Information Brochure.
5. Schneider, U., Varioset—A blowroom concept for maximum flexibility and efficiency. *Melliand International* 89(2): 89–91, June 1995.
6. Bogdan, J.F., Measurement of the nepping potential of cotton. *Textile Research Journal* 24: 491, 1954.
7. Kistler, W., Improvement of raw material utilization by by-pass cleaning. *Textil-Praxis International* 779, 1982.
8. Leifeld, F., The influence C of cotton in the cleaning process. *Melliand Textilber* 69: 309, 1988; English, 5: E162, 1988.
9. N. Balasubramanian, *Intimacy of Mixing and Blend Variation*, http://www.cottonyarnmarket.net/books/article.html, Accessed Date: November 12, 2013.
10. Ratnam, T.V., Seshan, K.N., Chellamani, K.P., and Karthikeyan, S., *Quality Control in Spinning*, SITRA Publications, Coimbatore, India, 1994.
11. Leach, B. and Khaitan, K., Selection of beater design in fibre preparation. *Indian Textile Journal*, December 2008, http://www.indiantextilejournal.com/articles/FAdetails.asp?id=1738, Accessed Date: November 12, 2013.
12. Elmogahzy, Y. and Farag, R., Minimizing fibre damage caused by spinning (Chapter 15), in: Lawrence, C.A. (ed.), *Advances in Yarn Spinning Technology*, Wood Head Publications Ltd., Cambridge, U.K., September 2010.

13. Senthil Kumar, R., Cotton dust—Impact on human health and environment in the textile industry. *Textile Magazine*, 45–47, January 2008.
14. Singh, R.P. and Kothari, V.K., Developments in blow room, card & draw frame. *Indian Textile Journal*, April 2009, http://www.indiantextilejournal.com/articles/FAdetails.asp?id=2010, Accessed Date: November 12, 2013.
15. Blow room—Technical Performance, Trützschler-Spinning brochure, http://www.truetzschler-spinning.com/en/downloads/brochures/blow-room/english-en/, Accessed Date: November 12, 2013.
16. JOSSI systems, Detection and elimination of foreign materials, http://www.ptj.com.pk/Web%202004/03-2004/joosi.html, Accessed Date: November 12, 2013.
17. Kumar, A., Ishtiaque, S.M., and Salhotra, K.R., Impact of different stages of spinning process on fibre orientation and properties of ring, rotor and air-jet yarns: Part 1—Measurements of fibre orientation parameters and effect of preparatory processes on fibre orientation and properties. *Indian Journal of Fibre & Textile Research* 33: 451–467, December 2008.
18. Ishtiaque, S.M., Chaudhuri, S., and Das, A., Influence of fibre openness on processability of cotton and yarn quality: Part-I—Effect of blow room parameters. *Indian Journal of Fibre & Textile Research* 28: 399–404, 2003.
19. Bhaduri, S.N., Effect of openness of cotton on subsequent processing, Resume of papers, *First Joint Technological Conference*, BTRA, Mumbai, India, 1959.
20. Klein, W., *Technology of Short Staple Spinning*, Vol. 1, Textile Institute, Manchester, U.K., 1987.
21. Salhotra, K.R., *Spinning of Manmade Fibres and Blends on Cotton Systems*, Textile Association, Mumbai, India, 1989.
22. Rieter, Rieter blow room machineries brochure, Rieter Company, Winterthur, Switzerland, 2011.

3

Process Control in Carding

3.1 Significance of the Carding Process

"Well carded is half spun" is a famous adage widely used among the spinning technologists around the world. Carding is considered as the heart of the spinning process. Carding is a mechanical action of reducing tufts of entangled fibers into a continuous web of individual fibers suitable for subsequent processing. The disentangling of fiber tufts facilitates effective trash removal. This is achieved by passing the fibers between closely spaced surfaces clothed with opposing sharp points. The carding process has a major impact on the final product, that is, yarn, in terms of uniformity, imperfections, and cleanliness. It is of more importance particularly for carded route, because it is the last process in which the opening and cleaning of cotton is effected. This process converts the fiber in the lap form to the sliver form. The neps generated tremendously by the blow room machineries are disentangled and/or removed to a maximum extent in the carding process. The basic objectives of a carding operation are as follows:

- Individualization of fibers
- Cleaning of fine trash particles that were left in the blow room process
- Disentangling and/or removal of neps
- Short-fiber and microdust removal
- Fiber blending
- Drafting and orienting of fibers
- Transformation of the lap into a sliver, therefore into a regular mass of untwisted fiber

The carding machine consists of cylinder, licker-in, doffer, flats, feed plate and feed roller, web-forming zone, calender rollers, coiler zone, and other auxiliary components (as seen in Figure 3.1). The main parts such as cylinder, licker-in, and doffer are clothed with rigid wire clothing, whereas flats and stripping rollers are clothed with flexible wire points. The licker-in facilitates better opening and cleaning, which runs at around 1100 rpm. The licker-in zone has a grid bar with a mote knife, which effectively removes the major trash present in the feed material. Some of the modern cards have three-licker-in systems, which are recommended for coarser cottons. The carding zone plays a predominant role in fiber individualization in which fibers are picked apart by the cylinder from flats. Stationary flats are employed in the entry and exit of cylinder to facilitate effective fiber opening. The function of the doffer is to deliver the fibrous web to the web-forming zone. The web is condensed in the web-forming zone and converted into sliver with the help of the coiler zone. The sliver then gets deposited in the can, which is required for the subsequent process.

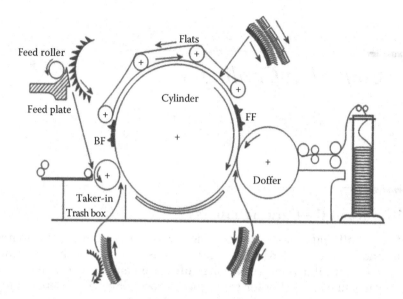

FIGURE 3.1
Parts of a carding machine. (From Lawrence, C.A., *Fundamentals of Spun Yarn Technology*, CRC Press LLC, 2003, p. 136. With permission.)

Lewis Paul invented a hand-driven carding machine in 1748. Richard Arkwright made improvements in this machine and in 1775 took out a patent for a new carding engine. The carding machine has tremendously developed since its inception. Since 1965, production rate has been increased from about 5 kg/h to about 180 kg/h. The current-generation carding machines are highly sophisticated with many features such as autoleveler, nep monitoring, microprocessor-controlled operations, servomotor-controlled settings, integrated grinding systems (IGSs), triple licker-in, stationary flats, increased carding area and automatic waste collection, fluff extraction suction points, and maintenance-friendly machine design. These features of latest cards help to achieve higher production rate, quality sliver, and minimum machine downtime for maintenance operations.

The quality of the carding process can be judged by the factors such as transfer efficiency, nep removal efficiency (NRE), and fiber orientation in sliver. Transfer efficiency is defined as the amount of fiber transferred from the cylinder to the doffer per rotation of the cylinder. The transfer efficiency of a card is important in determining the level of loading in the cylinder. Poor transfer efficiency leads to excessive loading of fibers on fibers, which restricts the quality and production of card.

3.2 Neps in Carding

Carding both generates and removes neps. Furter and Frey[1] reported that carding is the most significant process where significant decrease in the level of neps can take place, by a factor of between about 4 and 10. Nep removal is accomplished in carding machine by disentangling and/or removing fiber entanglements. A report from Uster[2] stated that a 70% nep reduction by the card is low, 80% is average, and 90% is high. Nep formation is related to fiber buckling coefficient, defined as the ratio between 2.5% span length and micronaire.

The longer, finer, and more flexible the fiber, the more prone it is to nepping during processing. The neps generated by the blow room beaters are mostly eliminated by the carding action. Large fragments breaking into small pieces increase the number of seed coat fragments, which are not removed by carding. Leifeld[3] reported that the average nep reduction in carding lies between 60% and 70%. Kaufmann[4] indicates that of the remaining neps, 30%–33% pass on with the sliver, 5%–6% is removed with the flats strips, and 2%–4% is eliminated with the waste. Flat settings do not have an effect on the level of neps. The more the flat strip waste, the less the neps present in sliver. Pearson[5] reported that a high nep count and a cloudy web are often indicative of poor carding due to damp cotton, damaged or worn wire, poor maintenance and setting, grinding, and stripping. The other parameters influencing nep removal are licker-in speed and setting, the number of licker-in, wire point geometry, and stationary flat settings. Bogdan and Feng[6] claimed that increasing licker-in speed for processing immature cotton will result in an increase in yarn neps, a licker-in speed of 800 rpm giving the best results. Hayhurst and Radrod[7] reported that the use of stationary flats, instead of the conventional revolving flats, produced a more even yarn with fewer neps. The maintenance of carding components, especially wire point grinding, predominantly influences the nep removal in carding. Bogdan[8] reported that the level of neps in the card web increases with the increase in card production rate.

The removal of one grid bar in licker-in undercasing helps in removing the short fiber and trash particles, which reduces the neps. Similarly, licker-in undercasing with large-diameter perforation helped in reducing the nep level. Neps were found to be minimum immediately after stripping and were found to increase gradually after 6–8 h of working, even though no loading was noticed on the cylinder surface. Neps were also found to reduce after grinding of the wires. Long-nose feed plates are recommended for long-staple fibers to avoid fiber rupture and subsequently neps. Low front angle with too low cylinder speed and with high frictional force will result in bad quality, because the fiber transfers from cylinder to doffer will be less. Hence, recycling of fibers will take place, which results in more neps and entanglements. Cripps[9] reported that higher cylinder speeds and flat speeds are advantageous for removing neps. Artzt[10] reported that cylinder and flat wire in terms of points/inch in conjunction with the front angle of the wire has a definite effect on neps in yarn. Improper feed roller loading and the setting between feed roller and feed plate will affect the quality, especially C.V% and neps. Modifications in licker-in zone are being tried more for reducing neps. The loading developments include triple-licker-in system, additional cleaning roller on licker-in, and fiber retriever.

3.2.1 Nep Monitoring and Control

Nep monitoring and control are the important aspects of process and quality control in the spinning process. Nep formation occurs mostly in blow room and carding process. Traditionally, nep counting templates consisting of large holes were used to count the neps in the carded web. This method gives the number of neps per 100 square inches. The latest testing equipment such as USTER AFIS®-nep module will provide the information about the quantity and size of the neps in intermediate fiber forms such as cotton tuft, web, or sliver. NRE is an efficient tool to check the efficiency of carding in the removal of neps. The NRE is calculated as follows:

$$\text{Nep removal efficiency (NRE\%)} = \frac{\text{Neps in feed batt} - \text{Neps in card silver}}{\text{Neps in feed batt}} \times 100$$

TABLE 3.1

AFIS-Nep Module: Parameters

Term	Definition	Application
Neps	Fiber entanglements	Causes imperfections in yarn and fabric
Nep count	No. of neps in 1 g of material	Analysis of processing equipment
Nep size	Average nep diameter in microns	Determines impact on yarn and fabric
NRE	(In − Out)/In × 100%	Equipment evaluation and comparison

A hundred percent would be the perfect removal efficiency from card feed batt to card sliver or comber lap to comber sliver. The card is designed to remove neps and trash and align the fibers.

Neps can be monitored online as well as offline in the carding process. The AFIS-nep module gives information about neps such as nep count, nep size, and NRE (as seen in Table 3.1).

Neps can be monitored in the delivery web through online monitoring system attached in the modern carding machines. Trutzschler's NEPCONTROL® (NCT)[11] provides a device that ensures that these requirements placed on high production cards are met. In this connection, a camera located below the stripper roll traverses in a hollow profile and detects the size and number of interfering particles by constantly providing sample images over the width and length of the web produced. A computer installed at the profile classifies the type of interfering elements as neps, seed coat fragments, and trash and transmits the result to the card control. Afterward, the particle counts per gram can be shown on the card display. In addition to plotting neps over time, the nep distribution can also be established over the working width and automatically monitored for limiting values. By utilizing the NCT, the setting of the card to a constant nep level can, for the first time, be carried out in a quick and accurate way.

3.3 Influence of Licker-In Zone on the Carding Process

Licker-in zone in the carding machine is mainly responsible for effective feeding, opening, and cleaning of the fiber tuft. The licker-in zone comprises feed plate, feed roller, licker-in, undercasing, and mote knives (as seen in Figure 3.2).

3.3.1 Feeding

The object of perfect feeding is to comb out the fibers, clean them, and feed them, as far as possible, individually to the surface of the cylinder. Lap or fiber batt is held between the fluted feed roller and the smooth-curved surface of the feed plate under a certain load and is fed to the licker-in part by the revolutions of the feed roller. A gap is defined between the feed roll and the feed plate between which the fibers are fed. The feed roller in conjunction with a feed plate feeds the material to the licker-in roller at a constant rate. The front of the fiber tuft is combed downward by the tips of the teeth of the metallic wire mounted on the licker-in roller, which are running downward. In modern cards, the feed plate has sensors to monitor the thickness of the feed batt and the feed roller speed is altered to ensure consistent feed thickness. The feeding to the licker-in can be done in two ways, that is, counterfeeding and concurrent feeding, as shown in Figure 3.3. Counterfeeding is the feed material gripped by the feed roller and is supported at the feed plate nose while the licker-in combs away the material. Concurrent feeding, which is used in some modern cards,[12] is to operate the feed roller so that

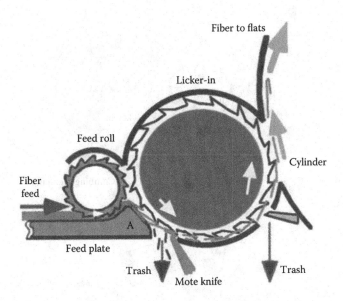

FIGURE 3.2
Feeding systems. (From Lord, P.R., *Hand Book of Yarn Technology: Technology, Science and Economics*, Woodhead Publishing Ltd., Cambridge, U.K., 2003. With permission.)

its surface moves in the same direction as the licker-in. The latter feeding method is claimed to offer a gentle opening action and thus a potential reduction in fiber damage.

The processing of the fiber tuft in the licker-in zone is significantly influenced by the following factors:

- Licker-in speed
- Number of licker-in
- Setting between licker-in and mote knife
- Licker-in wire specifications
- Feeding type—counter or concurrent
- Feed fiber batt homogeneity and its parameters
- Feed roller speed
- Setting between the feed plate and the licker-in

Although the surface of the feed plate is smooth, the fiber batt is subjected to a shearing force due to friction on its undersurface, because the feed plate is fixed. Thick laps increase the friction in the lap and the fiber cohesion due to compression. The position of the fibers in the fiber batt and the position where they are contacted by the licker-in wire point play a crucial role in determining the fiber stress. The setting between the feed plate and the licker-in for short-staple cotton variety should be closer than that for long-staple cotton. The setting between the feed plate and the licker-in is around 0.45–0.7 mm, depending upon the feed weight and fiber type. Fiber rupture often results in medium- and long-staple cotton, when closer setting is kept between the feed plate and the licker-in.[13] The length of the feed plate nose also has a significant effect on the opening of the fiber tuft. The shape of the feed plate nose is constructed so as to suit the quality of the cotton being used. Since the lap bends sharply near the point where it is gripped, there is also a layer-by-layer

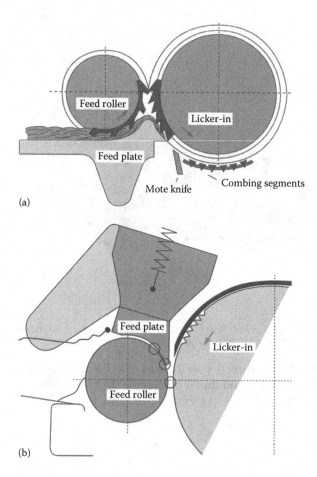

(a)

(b)

FIGURE 3.3
Feeding system—(a) conventional type and (b) unidirectional feed. (Courtesy of Rieter, Winterthur, Switzerland, http://www.rieter.com/en/rikipedia/articles/fiber-preparation/the-card/the-operating-zones-of-the-card/feed-device-to-the-licker-in/)

difference in the combing action of licker-in wires. The reduction in lap or fiber batt feeding rate increases the fiber openness due to increased residence time with the licker-in wire points, which leads to uniform sliver production. The impurities and the neps increase in the sliver significantly as the lap feeding rate increases. If the feed lap linear density is coarser, the fiber tuft will not be properly cleaned, and there will be great wear and tear, whereas tendency of web breakage will occur with the finer lap. The feeding rate is normally slow to allow exposition of a small portion of the fiber mat to the action of the high-speed licker-in, typically 700–1200 rpm for cotton and 400–600 rpm for man-made fibers.

3.3.2 Licker-In: Opening and Cleaning

Licker-in is where fibers are first opened up into individual fibers. Good opening action in the licker-in part gives good cleaning action in the licker-in part as well as good carding and cleaning actions between the cylinder and the flats, resulting in good sliver quality. Greater licker-in surface with higher surface speed assists complete separation of trash from lint. The lap fed by the feed roller is opened by the rigid metallic wire wound on the licker-in near the nose of the feed plate, and the trash and a part of the fibers contained in the lap get a chance to leave the licker-in surface. The licker-in, usually measuring about 250 mm in diameter, runs at a much higher surface speed (>10 m/s) than the feed roller (about 0.05 m/s max). The aggressive combing action of its teeth detaches the fibers and tufts from the feed lap. Licker-in along with mote knives removes the significant amount of trash particles along with some lint. The mote knives along with the licker-in can scrape off the leaf and dirt from the fibers. The closer the setting of mote knives with the licker-in, the greater will be their scraping action and the smaller the fly liberation. The setting between the licker-in and the first mote knife is around 0.35–0.5 mm, which helps to remove the heavier trash particles and dust. The setting between the licker-in and combing segments is around 0.45–0.6 mm, which helps to enhance fiber tuft opening. Trash and fibers that have left here move along their respective paths in the induced flow around the licker-in. The mote knives can be adjusted in conjunction with the feed plate. The disposition angle of the mote knives is important, which heavily influences the cleaning intensity and lint loss.

The higher licker-in speed increases the fiber openness due to higher centrifugal force, but at the cost of fiber damage and lint loss. This is mainly due to the increase in the degree of combing with the increase in licker-in speed. The increase of licker-in speed combined with constant lap feeding rate significantly improves the opening and cleaning. The effective breakdown of fiber mass feed into the tufts with minimal fiber breakage is accomplished with the help of coarser licker-in wire, low number of points per unit area, and a not too acute angle of rake. The intense reduction of fiber mass is mainly achieved in the zone between the feed roller and the licker-in. The draft ratio (the ratio of surface speed between the licker-in and the feed roller) is typically around 1000.

The number of licker-in in a card has significantly influenced the opening, cleaning, and nepping tendency.[14] Conventional cards have single licker-in, which carries out 90% of the cleaning work in a card. Modern cards equipped with three licker-in, as shown in Figure 3.4, are suitable for processing coarser and trashy cotton. The throughput rate of single licker-in has increased several times since the five decades, but the speed of licker-in has not increased in the same proportion. Therefore, without any modifications, the licker-in would deliver mostly flocks, to the main cylinder. These tufts are compact and relatively poorly distributed across the surface of the licker-in and would result in considerable loading of the cylinder and flats clothing. This ought to result in greater stress on the carding elements and on the fibers when the large flocks are being carded. The danger of fiber shortening also increases.

The importance of producing small size fiber tufts is evident from the various components fitted in the fiber opening zone on modern short-staple cards. Sawtooth wire–covered

FIGURE 3.4
Triple-licker-in system. (Courtesy of Rieter Company, Rieter Card C60, Rieter, Winterthur, Switzerland, 2013.)

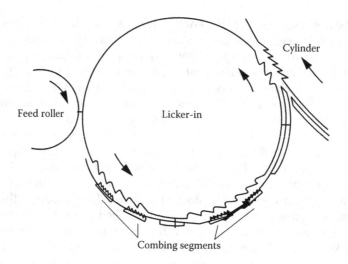

FIGURE 3.5
Combing segment under licker-in.

plates, termed combing segments, fitted below the licker-in or built into the licker-in screen are claimed to give improved trash removal and smaller fiber tufts (as seen in Figure 3.5). The purpose of equipping three licker-in in the modern card is to facilitate better opening and cleaning.[12] The better opening of fiber tufts achieved with three licker-in minimizes the cylinder loading, which helps to improve productivity.[12] The triple-licker-in configuration allows a gradual acceleration of the fibers toward the high surface speed of the main cylinder by simultaneously orientating the fibers properly.

3.3.3 Quality of Blow Room Lap (in Lap Feeding System)

The condition of blow room lap significantly affects the cleaning as well as parallelizing of the fibers at the card. The weight of the fiber mat may typically range from 400 to 1000 g/m (ktex). The weight per yard or C.V% of the blow room lap should be under control to ensure better sliver quality and productivity. Irregularities in the feed will be reproduced in the sliver delivered by the card. The laps should not be dirty or too thick, as these characteristics tend to increase the percentage of waste generated in the carding machine. Lap licking should be carefully prevented as it deteriorates the quality of sliver and smooth functioning of process. Lap transportation from the blow room department to the carding machine should be carried out with proper material-handling equipment. The lap should be wrapped with polypropylene sheets to prevent dust accumulation and surface damage while handling.

3.4 Influence of the Carding Zone on the Carding Process

3.4.1 Cylinder

The cylinder takes the fiber tuft from the licker-in with the aid of its higher speed and carries it under the action of flats, whose work is to comb, parallelize, and remove short fibers. In the carding zone, it is the interaction of the fiber mass and the wire-teeth clothing of the

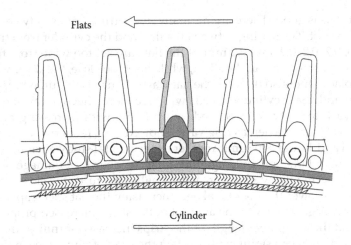

Flats

Cylinder

FIGURE 3.6
Cylinder–flat zone.

cylinder and flats that fully individualizes the fibers and gives parallelism to the fiber mass flow (as seen in Figure 3.6). The term "carding action" literally means separating one fiber from another. The carding zone essentially consists of the cylinder, flats, stationary flats (entry and exit), and cylinder undercasing. The cylinder, running around 450 rpm, measuring a diameter of about 1000 or 1290 mm, and having a greater density of metallic wire clothing than that of the licker-in, strips the fibers from the licker-in surface and transfers them to the carding zone. The stripping of fibers from the licker-in by cylinder wire points is possible only when the surface speed of the cylinder (1000–2400 m/min) is higher than the surface speed of the licker-in (700–950 m/min). The draft given between licker-in and cylinder will be in the range of 1.5–2.5. The setting between licker-in and cylinder will be in the range of 0.178–0.127 mm. Krylov's[15] experimental data show that an increase in cylinder speed reduces the load on the cylinder. Van Alphen[16] reports that increasing cylinder speed causes more fiber breakage than increasing licker-in speed and that this is reflected in the yarn properties.

3.4.2 Flats

In the nineteenth century, cards were designed with the top flats carried on an endless chain so that they could move continually forward as the card worked. This type of card is called a "revolving flat top" (R.F.T.) card or more commonly a "revolving flat" card. Flats are the flexible wire–clothed bars rotating at a very slow speed (8–20 cm/min) compared with the high-speed cylinder. Flats move slowly either in the same direction or in the opposite direction of the cylinder, but the wire points are always backward, which is opposite to the direction of the cylinder wire points. The number of flat bars mounted over the cylinder frame is in the range of 80–110 of which 40 flat bars are in the carding action. The optimum flat speed is influenced by the staple length of the fiber, the amount of trash in the fiber, the weight of the lap fed, and the waste% desired to be removed. The short fibers removed between the cylinder and the flats and the trash particle that comes out with them are called "flat strips." Fibers that are deeply embedded in the flats and cannot be reached by the cylinder wires become flat strips. For this reason, the closeness of the flat setting to the cylinder is important. An increase in flat speed results in an increase in card waste%, affecting lint loss and overall yarn quality.

The successful working of card depends on the correct setting between cylinder and flats. While processing cotton, it can be as close as 0.175 mm provided the mechanical accuracy of flat tops is good. There are normally five setting points between cylinder and flats from entry to exit. The setting between the flats and the cylinder from the entry to the exit can be 0.25, 0.2, 0.2, 0.2, and 0.2 mm. If the flats are set too wide from the cylinder, it will be prone to nep formation in the sliver, while if set too close, the sliver will have many raw uncarded places and also increase the flat waste. It may be assumed that a closer flat/ cylinder setting and faster cylinder speeds will give more effective carding and combing actions. Rotor yarn tenacity was reduced by up to 5% with increasing cylinder speeds between 480 and 600 rpm, whereas ring yarns showed a 5% reduction for speeds between 260 and 380 rpm and 10% at 600 rpm. Another experimental study attributes fiber damage to the cylinder/flat interaction and suggested that the degree of damage depends on the size of the tufts entering the working area; the smaller the tufts, the closer the setting that can be used and the lower the fiber breakage. Increasing the flat speed appears to have no effect on fiber breakage. However, the amount of flat strips increased proportionally with the flat speed and the mean fiber length of the strips increased significantly. Van Alphen[16] found flat setting to have no significant effect on the level of impurities in the card sliver.

3.4.3 Precarding and Postcarding Segments

The carding zone essentially consists of precarding and postcarding sections (also called stationary flats) as shown in Figure 3.7, respectively, apart from the cylinder–flats section. Wire points on the cylinder, the revolving flats, and the stationary flats play a vital role in minimizing fiber damage. The carding segments or stationary flats ensure further opening, reducing flock size, and primarily spread out, thereby improving the distribution of flocks over the total surface area. Precarding segments are installed in the feed side and postcarding segments are installed in the delivery side. Both segments are comprised of identical components such as control flat, control plate, trash knife, and stationary flats. These components are arranged in ascending order in the precarding and descending order in the postcarding segment. Both precarding and postcarding segments improve the cleaning efficiency of the card by facilitating the removal of motes, seed coat fragments, and dust. These stationary flats also show highly significant effect on yarn evenness and yarn strength.

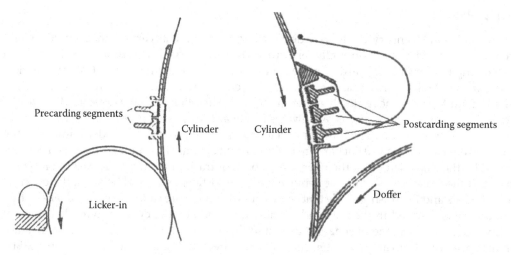

FIGURE 3.7
Precarding and postcarding segments.

The leading edge of the control plate is set closer to the cylinder wires than the backward edge. It regulates the fiber density in conjunction with the stationary flats, allows only a thin layer of fibers to reach the main carding area, and promotes effective carding action between the cylinder and the flats. As the cylinder is rotating at high speed, a strong centrifugal force is developed and strong air current is generated. The leading edge of the control plate being closer to the cylinder back pressure is developed. The air escapes around the control plate, leading to extraction of dust and trash particles under the trash knife. The waste extraction takes place in the precarding as well as postcarding segment.

The stationary flats of the precarding segment have coarser metallic wire of approximately 250 points per square inch (PPSI) as compared with the cylinder wire.[17] The unopened tufts of the material are broken down, and short fibers and neps are removed by these flats. The stationary flats of the postcarding segment are covered with finer metallic wire of approximately 620 PPSI as compared with the cylinder wire. The cotton has already been opened into individual fiber state as it leaves the cylinder/flat area. The stationary flats of the postcarding segment promote fiber alignment and parallelization. Grimshaw[18] reports that the use of postcarding segments just before the cylinder/doffer top transfer zone improves fiber parallelism in the card web, up to 20% reduction in fiber hooks and 25% improvement in fiber parallelism, which results in improved card sliver quality. Well-designed stationary flat applications not only open the fiber tuft properly but also curtail imperfections.

3.5 Doffer Zone

The doffer has a diameter of 686 mm, which is the large component next to the cylinder in the carding machine (as seen in Figure 3.8). Its surface speed is several times lower than the cylinder (10–20 times slower). The primary functions of the doffer are to collect

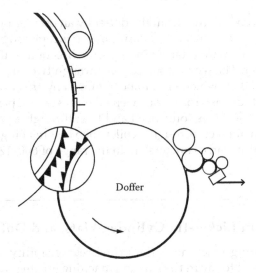

FIGURE 3.8
Doffer zone.

cotton from the cylinder as a uniform sheet and to carry this to the calender roller section. The wire clothing of the doffer is arranged so that it will work point against with that of the cylinder.

Varga[19] reports that the action of fiber mass transfer to the doffer is similar to the transfer at the input to the cylinder–flat zone. The doffer has a finer count of clothing that increases the retaining power of the doffer and contributes to hold the fibers that the cylinder brings around. The doffer diameter, setting of the cylinder, doffer wire population, tooth height, and tooth angle all together have an effect on transfer efficiency and subsequent sliver quality.

3.5.1 Fiber Transfer Efficiency

"Transfer efficiency is defined as the percentage of fiber transferred to the doffer from the cylinder per revolution of the cylinder." The transfer efficiency of card is important from the point of view of determining the level of loading of the cylinder. Fibers make an average of four or five revolutions on the cylinder before being transferred to the doffer. Some fibers get transferred during the first few revolutions, but the rest may take up several revolutions. Krylov's[15] experiment reported that by keeping production rate constant, if doffer speed is enhanced with a proportionate reduction in sliver hank, the load on cylinder decreases and transfer efficiency increases. Ghosh and Bhaduri[20] have shown that with an increase in cylinder speed, the load on the cylinder reduces with a concomitant increase in the transfer coefficient. Ghosh and Bhaduri[21] also show that an increase in production rate through the doffer speed results in an increase in loading and transfer efficiency. Sengupta and Chattopadhyay[22] have shown using fluorescent tracer fibers that transfer efficiency increases with closer setting between cylinder and doffer. Simpson[23] analysis reveals that a higher wire point density on the doffer will reduce the cylinder load.

3.6 Sliver Formation

In conventional cards, web is doffed from the doffer by an oscillating doffer comb. It oscillates up to 2500 strokes per minutes. In modern cards, it is replaced by a roller. A stripping roller clothed with metallic wire rotating in opposite direction to the doffer removes the fibrous web off the doffer. The web is then passes through two smooth steel crush rollers arranged one over the other, which crush and remove any trash particles remaining in the web. The web from the crush rolls, via a web take-off system, passes through a funnel known as a "trumpet" where it is condensed and then through a pair of calender rollers. The resulting sliver is then taken through a coiler tube before being deposited into a can. Coiling is done cycloidally. Can diameters are in the range of 600–1200 mm.

3.7 Wire Geometry in Licker-In, Cylinder, Flats, and Doffer

Control of fiber in carding machine is the prime responsibility of card clothing. The wire clothing widely used today in the carding machine is categorized as flexible, semi-rigid, and rigid clothing. The components of the carding machine such as licker-in, cylinder doffer, and stripping roller are mounted with rigid wire clothing, whereas flats are mounted with flexible and semirigid wire clothing. The metallic card clothing used in the cards plays a major role in ensuring efficient carding in terms of both quality and productivity. Metallic card clothing has two essential functions: pulling fiber and discharging fiber. Both of these actions take place on every single wire tooth. The front or the tip of the tooth pulls the fiber, and the back of the tooth discharges it. If the tooth cannot release the fiber, it will load up and will not be able to take any more fibers, which will cause inefficiency in carding. The various factors influencing the selection of wire clothing are summarized as follows:

- Speed of the cylinder
- Fiber type and characteristics
- Licker-in speed
- Production rate
- Count

The spinning mill technician has to choose clothing according to its process requirements to achieve the required production and quality levels. One of the main features of the metallic wire clothing is its ability to keep fibers on the surface of the clothing more than flexible wires. The loading capacity of the metallic wire is 20%–25% less than the fillet clothing. The geometric parameters relating to the card clothing are as follows:

- Point density
- Tooth pitch (*P*)
- Point profile
- Land
- Tooth depth (*H*)
- Wire angle (α)—front angle and back angle
- Tip width

3.7.1 Point Density

Point density can be defined as the number of points per unit surface area, that is, per square cm (PPSCM) or PPSI. The choice of point density for licker-in, cylinder, and doffer is highly influenced by the fiber type and fiber dimensions. The opening effect in carding can be correlated to the number of wire points per fiber. In the licker-in, this ratio is approximately 0.3 (three fibers per point), and in the main cylinder, it is about 10–15. In general, the higher the point population on the cylinder and the flats, the better is the carding effect. Coarser fibers need less point density, and finer fibers more point density. The retaining capacity of doffer can be increased by increasing the point density of the card clothing on the doffer. Point density can be determined by the pitch (*P*) and the number of tooth points across the card (*R*). The "pitch" (*P*) is the number of points along a 1 in. (25.4 mm) length of the wire. The point density is calculated by multiplying the pitch by the number of rows. The flats use a wire point density in the range of 240–550 PPSI.

3.7.2 Wire Point Profile

Wire point profile influences factors such as point penetration into the fiber and fiber to metal friction. A sharp, symmetrical, uniformly shaped point provides good penetration into the fiber while allowing free release. The area at the tip of the teeth is termed as "land." Increasing the land area behind the effective front edge of the tooth creates support and added strength. However, an increase in the size of the land reduces the carding efficiency. The wire for the cylinder is cut in a special way so that the tip is very sharp and the land area is extremely small.

3.7.3 Wire Angle

The angle of penetration is largely decided by the overall tooth angle. The efficacy of fiber tuft opening in the licker-in zone strongly depends on feed plate setting and proper selection of licker-in clothing. The wire angle plays a significant role in effecting opening as well as carding. The wire angle can be categorized as front angle and back angle. The front angle influences the degree to which a tooth captures fiber from the roller with which it interacts. The higher the front angle, the greater is the hold on the fibers, especially in high-speed carding and fine fiber carding. The front angle could be negative or positive, depending on the application. Licker-in clothing meant for man-made fibers uses a negative or neutral angle, while those meant for cotton use a positive angle:

- Licker-in: + 5° to –10°
- Cylinder: +12° to +27°
- Doffer: +20° to +40°

The cylinder uses a larger front angle with a low wire height. Greater front angle gives greater holding and retaining capacity while the lower wire height keeps the fibers in the active carding zone. A more acute angle of doffer wire pulls fibers deeper into the wire interspaces, which similarly increases the doffer's retaining capacity, leading to better doffing efficiency. The "back angle" of a tooth influences the card wire loading properties as well as the overall strength of the tooth. Cylinder and stripping rollers utilize lower back angles that minimize the trapping of fibers in the mouth of the teeth.

3.7.4 Tooth Depth

The distance from the tip of the point to the bottom of the mouth is termed as "working depth," which determines the holding capacity of the wire. The working depth affects the loading capacity of the roller onto which the wire is wound. The selection of tooth depth is influenced by fiber diameter and fiber length. The recent cylinder wires have a profile called "No Space for Loading Profile" (NSL). With this new profile, tooth depth is shallower than the standard one and the overall wire height is reduced to 2 mm, which eliminates the free blade in the wire. This free blade is responsible for fiber loading.

3.7.5 Basic Maintenance of Card Wire

3.7.5.1 Grinding

Fiber–metal friction results in wearing out of teeth over a period of time. Wire points become round at the top and lose aggressiveness. This results in improper carding, and as a consequence, fiber rolling and nep formation increase. In order to resharpen the teeth, grinding is therefore necessary. Grinding is a process performed on card wire to give the shape and sharpness to the ends of the card wire. A low nep count is directly related to the sharpness and thickness of the points of the cylinder wire, carding segments, and flat wire. The number of neps gets reduced after each grinding. However, as the number of grindings increases, quality drops due to reduction of height and broadening of land of the wire points. Softer metals are gradually exposed and more frequent grinding becomes a necessity. The grinding interval depends on factors such as the amount of fiber processed by the card, type of fiber, clothing quality, and nep level permitted. The usual practice of grinding is given in Table 3.2.

For doffer, the grinding frequency is half of cylinder grinding frequency. Grinding is not done for licker-in clothing, and it is replaced after 100,000–200,000 kg of fiber processing. This is because there is no land in licker-in wires.

3.7.5.2 Stripping

Stripping is often required for flexible card clothing in order to clean the wires from the knee, as over a period of time, wire knees get loaded. In the case of metallic clothing, stripping is not usually required, as there is no knee in the wires. However, if the cylinder gets loaded, then problems appear in the running of the card and then the cylinder should be cleaned. This is often done by a hand scrapper/brush while cylinder is rotated slowly. Brushing must be carried out in the direction of teeth and not against them.

3.7.5.3 Stationary Flats

The work done by the first few stationary flats is very high, and they wear out faster. Fifty percent of the flats are recommended to be changed after 100,000 kg of production and the rest after 150,000 kg of production.

TABLE 3.2

Grinding Frequency of Cylinder and Flat Wire Point

Grinding Frequency	Cylinder (kg)	Flats (for Regrindable Flats) (kg)
First grinding	80–150,000	120–150,000
Each additional grinding	80–120,000	80–120,000

3.8 Autoleveler in Carding

With increasing global competition, the superior and constant sliver and yarn quality become the important requirements of the customer. The high-quality sliver is assured only when the carding machine is provided with an autoleveler. Autoleveler is an auxiliary device attached to the carding machine in order to correct linear density variations in the delivered sliver by changing either the main draft or break draft according to the feed variations. The main objectives of an autoleveler are as follows:

1. To measure sliver thickness variation on a real-time basis
2. To alter the machine draft so that a high consistent sliver thickness is continuously produced

Generally two types of autolevelers are available: open loop and closed loop. In open-loop autolevelers, sensing is done at the feeding end and the correction is done by changing either a break draft or main draft of the drafting system. In a closed-loop system, sensing is at the delivery side and correction is done by changing either a break draft or main draft of the drafting system. The open-loop system is very effective, because the correction length in the open-loop system is many times lower than the closed-loop system. The open-loop system may generally be used for the correction of short-term variations. In the case of an open-loop system, since the delivered material is not checked to know whether the correction has been done or not, sliver monitor is fixed to confirm that the delivered sliver has the required linear density. But in the case of a closed-loop system, it is confirmed that the delivered sliver is of required linear density.

3.8.1 Benefits Associated with Autoleveler

- Count C.V% will be consistent and good.
- Count deviations will be very less in the yarn; hence, off-count cuts will come down drastically in autoconers.
- Thin places in the sliver and, hence, in the yarn will be low.
- Ring frame breaks will come down; hence, pneumafil waste will be less.
- Fluff in the department will be less; therefore, cuts in autoconer will be less.
- Fabric quality will be good because of the lower number of fluffs in the yarn.
- Labor productivity will be more.
- Machine productivity will be more.
- Idle spindles in speed frame and ring frame machine due to better feed material quality will be less.
- RKM C.V% will be low, because of the low number of thin places.
- Workability in warping and weaving will be good, because of the less number of thin places and lower end breaks in spinning and winding.
- Sliver U%, hence yarn U%, will be good.
- Variation in blend percentage will be very less, if both the components are auto-leveled before blending; hence, fabric appearance after dyeing will be excellent.

Most of the carding machine manufacturers combine both medium-term and long-term autoleveling to obtain superior uniformity of sliver.

3.9 Process Parameters in Carding

The selection of proper process parameters is vital for any manufacturing process to attain expected productivity and quality. The carding process has a predominant influence on the final yarn quality in terms of uniformity and imperfections. The carding machine consists of several zones such as feed zone, licker-in zone, carding zone, and doffing zone. In each of these zones, speeds and setting particulars play a crucial role in achieving the objectives of the carding process. Process parameters can vary for different end-use applications and fiber type. The various factors considered while setting the process parameters in the carding process are summarized as follows:

- Fiber type
- Fiber linear density
- Feed lap linear density
- Amount of trash in cotton and lap
- Quality requirements
- Production requirements
- Condition of the machine

For processing synthetic fiber in carding, process parameters should be selected accordingly to avoid the occurrence of fiber rupture and neps. While processing cotton, factors such as linear density and trash content are taken into account for deciding the speeds and settings in various zones. The requirement of fiber cleanability for a particular end use also decides the settings especially in the licker-in zone. Fine cotton requires gentle opening to prevent fiber from rupture and nepping tendency. Coarser and trashy cotton requires intensive opening and cleaning to attain required cleaning efficiency. Carded sliver for producing rotor yarn should have very less trash content in the sliver. The trash present in the card sliver may obstruct the yarn formation in the rotor groove and create end breakage in rotor spinning process. The card sliver meant for producing knitted yarn should have very less trash content in the card sliver. The trash particles present in the card sliver are converted into very tiny trash in the yarn, which form black spots in the fabric that spoil the fabric appearance.

The most important settings in the card are those of the feed plate–licker-in setting, between the cylinder and flats, and the cylinder and the doffer. As far as the speeds are concerned, the speeds of licker-in, cylinder, flats, and doffer are very critical in determining the carding quality. The optimum speed of the licker-in has a significant impact on the openness of the fiber as well as the cleaning intensity of the fibers. The selection of the speed of the licker-in for a particular process is based on factors such as fiber type, amount of trash, feed weight, fiber linear density, and production rate. Synthetic fibers such as viscose rayon are difficult to open compared with cotton. The setting of the feed plate to the licker-in depends on fiber openness requirements. The setting of licker-in to mote knives is highly influenced by the amount of trash present in the feed material.

The force required to individualize a tuft at the cylinder–flat zone is called carding force.[24] The speed of cylinder and flats is set in accordance with the draft and carding force requirements. The increase in carding force per unit load decreases both the nep content and the U% of the sliver.[24] The optimum cylinder speed is dependent on the cleaning propensity of the cotton used. On the other hand, raising the cylinder speed is more detrimental to the staple length than increasing the licker-in speed, and yarn tenacity decreases at higher speeds. The settings between cylinder and flats are crucial in the carding process, which decides the quality of the sliver. The speed and setting between cylinder and doffer plays a significant role in fiber transfer efficiency.

The linear density of sliver is altered either by changing the linear density of feed material for constant card draft or by changing card draft for constant linear density of feed lap. Kumar et al.[24] reported that the increase in sliver weight through either lap hank or card draft increases the relative coefficient of fiber parallelization and projected mean length. The fiber transfer from the cylinder to the doffer decreases and cylinder load increases with the increase in sliver weight.[25] The yarn grade deteriorates with the increase in sliver weight and carding delivery rate.[26] End breakage rate may increase with the increase in sliver weight.[27] Better carding quality at lower card production rate improves the yarn tenacity.[28] The yarn unevenness and imperfections increase with the increase in card production rate.

3.10 Defects Associated with the Carding Process

The defects or substandard quality of card sliver in the carding process will create severe production loss and also subsequently affect yarn quality. The defects can be avoided by frequent monitoring of card sliver quality and correct follow-up of maintenance activities. The carding operator should have conscious on the basic quality aspects which helps to report the abnormalities found in the carding process immediately to the superiors.

The main reasons for the occurrence of defects in the carding process are listed as follows:

- Malfunction of any carding components—damage in wire point or undercasing, etc.
- Wrong selection of process parameters such as speed, settings, and wire point selection
- Poor feed material quality—lap defects, high lap C.V%, etc.
- Wrong work practices—lap surface damage, sliver mending faults, etc.
- Irregular maintenance of machine—delayed grinding or wire clothing, auto-leveler checking, etc.
- Improper ambient conditions—R.H%, etc.

The defects associated with the carding process have several reasons. Each and every defect is a peculiar type and may arise due to mechanical faults and process-related faults. Let us discuss the various defects that occur in the carding process and their causes and remedial measures.

3.10.1 Patchy Web

The various causes of patchy web in the carding process and the remedial measures to control the fault occurrence are given in Table 3.3.

TABLE 3.3

Patchy Web: Causes and Remedies

S. No.	Causes	Remedies
1	Cylinder loading	Ensure effective preopening in licker-in zone.
2	Damaged wire points	Rectification of damaged wire points.
3	Waste accumulation in cylinder undercasing	Proper follow-up of cleaning schedule; polishing of undercasing.

3.10.2 Singles

The various causes of singles in the carding process and the remedial measures to control the fault occurrence are given in Table 3.4.

3.10.3 Sagging Web

The various causes of sagging web in the carding process and the remedial measures to control the fault occurrence are given in Table 3.5.

3.10.4 Higher Card Waste

The various causes of higher card waste in the carding process and the remedial measures to control it are given in Table 3.6.

TABLE 3.4

Singles: Causes and Remedies

S. No.	Causes	Remedies
1	Lap licking	Correct process parameters in the lap forming unit of the blow room
		Proper R.H% in carding; use of roving ends between layers
2	Less feed in chutes	Proper allotment of the number of cards per chute; proper setting of the chute feed system
3	High suction pressure in the waste extractor, which sucks part of the carded web	Optimizes the suction pressure in the waste extractor
4	Damaged doffer wire points	Rectification of wire damage
5	Disturbance of air currents in the web-forming zone	Proper control of air currents

TABLE 3.5

Sagging Web: Causes and Remedies

S. No.	Causes	Remedies
1	Tension draft too low	Set tension draft as per norms
2	High R.H%	Maintain proper R.H% as per recommendations
3	Heavy feed material	Control feed material variation, especially thick laps
4	Inadequate calender roller pressure	Optimum setting of calender roller pressure w.r.t fiber type, process

TABLE 3.6

High Card Waste: Causes and Remedies

S. No.	Causes	Remedies
1	Damaged undercasing (licker-in, cylinder)	Proper maintenance and setting of the undercasing
2	Higher suction pressure in suction points	Optimizes the suction pressure in the waste extractor
3	Flat speed too high	Proper flat speed w.r.t process requirements
4	Closer setting in the licker-in zone and carding zone	Optimum setting in the licker-in and carding zone
5	Occurrence of fiber rupture	Proper setting in the licker-in zone to avoid fiber rupture

3.10.5 Low Nep Removal Efficiency

The various causes of low NRE in the carding process and the remedial measures to control it are given in Table 3.7.

3.10.6 Higher Unevenness of Sliver

The various causes of higher sliver unevenness in the carding process and the remedial measures to control the fault occurrence are given in Table 3.8.

3.10.7 Higher Sliver Breaks

The various causes of higher sliver breaks in the carding process and the remedial measures to control them are given in Table 3.9.

TABLE 3.7

Low Nep Removal Efficiency: Causes and Remedies

S. No.	Causes	Remedies
1	Too wide setting between the feed plate and licker-in	Optimize setting
2	Blunt wire points	Proper follow-up of grinding and wire clothing schedules
3	Improper wire selection	Proper selection of wire w.r.t fiber type, fiber count
4	Improper setting of stationary flats	Proper setting as per recommendations

TABLE 3.8

Higher Unevenness of Sliver: Causes and Remedies

S. No.	Causes	Remedies
1	Uneven feed material	Ensure blow room lap C.V% to be maintained below 1%
2	Improper functioning of autoleveler	Frequent checking and maintenance of autoleveler
3	Stretch during material passage due to worn-out parts	Replacement of worn-out parts
4	Eccentric movement of calender rollers	Rectification of eccentric calender rollers
5	Improper settings	Proper setting w.r.t process requirements
6	Fly accumulation in the feed zone	Ensure proper functioning of suction ducts

TABLE 3.9

Higher Sliver Breaks: Causes and Remedies

S. No.	Causes	Remedies
1	Very small trumphet, worn-out trumpet	Proper selection of trumphet w.r.t hank of the sliver
2	Higher lap C.V%	Ensure lap C.V% is under control
3	Disturbance of air currents in the web delivery zone	Control of air current
4	Damaged clothing	Rectify damaged wire point areas
5	Poor control of ambient conditions	Effective R.H% maintenance
6	Too high tension draft	Tension draft as per recommendations

3.11 Ambient Conditions

Correct ambient conditions are essential to prevent the degradation of textile materials. Maintaining appropriate ambient conditions can lower the energy costs, increase productivity, save labor and maintenance costs, and ensure product quality. Temperature and relative humidity are important considerations for carding cotton in textile manufacturing. Higher relative humidity decreases the stiffness of fibers and increases the moisture content of the fibers. Strang's[29] work suggested that temperature (>10°C) and relative humidity (50%–55%) are important considerations for carding. Sengupta et al.[25] investigated the impact of various variables on carding forces (the force required to individualize fibers from tufts of fibers) and concluded that increased relative humidity resulted in decreased carding forces, which they attributed to the reduction of the flexural rigidity of the cotton fibers as a result of an increase in the moisture content of the fibers. The reduction in the flexural rigidity of fibers as a result of increased moisture also reduced the breakage rate of fibers at carding. The absorption of moisture by cotton fibers changed the mechanical and frictional properties of the cotton, which affected the behavior of the fibers in processing.[30] When cotton is sticky, higher humidity creates sticking of fibers to rollers and other parts of the machine. Higher humidity reduces the static problems. Humidification reduces fly and microdust, giving a healthier and more comfortable working environment.

3.12 Cleaning Efficiency and Lint Loss%

Blow room and carding departments are responsible for cleanliness of the cotton fibers. Around 30% of end breaks in yarn spinning are due to the presence of trash particles in fibers. Modern blow room line focuses more on gentle opening than intensive cleaning to avoid fiber rupture. Well-opened fibrous tuft in the blow room process facilitates effective cleaning in carding. The blow room process removes the larger-sized trash particles, whereas the carding process removes the smaller broken trash particles and pepper trash particles adhering to the fiber tuft. Modern cards are equipped to extract the finest of trash particles. A very high cleaning effect is almost always purchased at the cost of a high fiber loss.

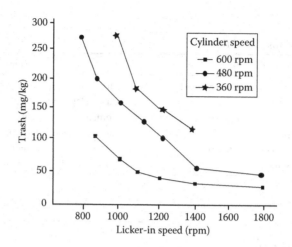

FIGURE 3.9
Effect of licker-in speed on trash removal. (From Lawrence, C.A., *Fundamentals of Spun Yarn Technology*, CRC Press, Boca Raton, FL, 2003. With permission.)

The overall carding system has a cleaning efficiency of 95%, which is much higher than the 45% for the opening and cleaning lines. Taking the carding zones individually, the cleaning efficiency of the licker-in on its own approximates to 30%, the carding segments give 30% and the cylinder/flats give 90%. The cylinder/flat carding action therefore gives the highest cleaning effect. Carding efficiency is better the higher the number of points, the closer the settings and the higher the cylinder speed. The use of higher cylinder surface speeds has helped to improve the cleanliness of the sliver. A mill study reported that the trash removal is higher at combined effect of lower cylinder speed and licker-in speed of 1000 rpm (as shown in Figure 3.9). The use of efficient suctioning arrangements further enhances the removal of trash and dust particles. The level of cleaning efficiency achieved in spinning mills is about 80% for most mixings. However, for fine and superfine mixings for which very high flat speeds and low production rates are employed, a cleaning efficiency of 85% or more is achieved. A modern card can clean over 90% of trash and 65% of microdust present in the infeed fiber material.

Cleaning takes place in three major zones in the card when carding cotton fibers. The first is in the region of the licker-in, the second is in the region of contact between the flats and the cylinder, and the last is under the card. The major amount of trash in the feed material is removed in the licker-in zone, which constitutes nearly about 30%. The optimum setting of licker-in with mote knives and grid bar facilitates the effective removal of trash. Modern cards consisting of three-licker-in system achieve effective cleaning, and those systems are only recommended for trashy cotton. Trash ejection arrangements on the cylinder and the revolving flats remove about 90% of the remaining trash. The carding zone removes the flat strips, which constitute trash, neps, and short fibers. The combined cleaning efficiency will be generally in the range of 90%–98% with modern cards. To illustrate, for 4% trash in cotton and 0.12% trash in sliver, the combined cleaning efficiency is 97%. Cleaning efficiency can be assessed with the help of trash analyzer. The procedure for determining cleaning efficiency was already discussed in the previous chapter.

Lint is a term, often used in blow room and carding, usually referred to good fibers. Lint loss is inevitable during cleaning process in blow room and carding. The amount of lint loss can be kept under control with the proper setting of process parameters. The closer setting in the licker-in and carding zone may be prone to lint loss. The more amount of lint loss affects the yarn utilization as well as output sliver quality.

3.13 Control of Waste

3.13.1 Control of Nonusable Waste

Waste extracted in cards is usually in the range of 4%–7%, depending upon the type of card and mixing. Between the same type of cards and mixing, the waste% should not vary more than ±0.5% from the average. The card waste is also governed by the cleaning efficiency achieved in blow room. Thus, while assessing the waste, combined waste extracted in blow room and cards should be taken into account.

Waste removed from the flats is called "flat strip." Forward motion of flats is commonly employed in which the cylinder motion helps drive the flats and the removal of waste is easier. Modern cards employ backward motion of flats in which flats meet the fiber in a clean condition at the front of the card, but they accumulate short fiber and dirt as they continue to the back for cleaning. Flats embedded with waste do not have any contact with the cleanest fibers. The flat wire retains sufficient amounts of short fiber to clog them fairly quickly unless the fibrous material is removed. Thus, flats should be cleaned often using the stripping roller.

Waste can be controlled by maintaining the correct setting in the licker-in zone and carding zone. The waste% to be removed in the carding process depends on the process and quality requirements of the product. As discussed already, rotor spinning requires sliver with negligible amount of trash as it hinders the yarn formation process. Cylinder-to-cylinder undercasing setting influences air currents and leads to fly generation, and too wide setting causes the loss of fibers. Closer setting of licker-in to licker-in undercasing increases the good fiber loss with waste. If the setting between the feed plate and licker-in is too close, more lint loss will occur, and it will increase the waste%. The speed of the flat is also directly related to the amount of waste removed in the form of flat strips; that is, the higher the flat speed, the heavier the flat strips.

3.13.2 Control of Soft Waste

The waste occurring in the spinning mill can be classified normally as soft waste and hard waste. Soft waste is reusable in the spinning process, whereas hard waste is not reusable. Soft wastes generated in the carding process are classified as lap bits, web waste, and sliver waste. The occurrence of soft waste should be controlled to avoid unnecessary reprocessing of good fibers, which leads to fiber damage. The major reasons attributed for the generation of soft waste in carding are poor feed material quality, poor material handling, higher production speed, wrong settings, air current disturbances in the web delivery zone, poor work practices, wrong selection of sliver trumpet, and improper ambient conditions.

Feed material quality plays a crucial role in reducing soft waste generation, especially in a carding machine installed with autoleveler. In carding, every web or sliver breakage creates a huge amount of soft waste. The reduction in sliver breakage minimizes the soft waste as well as machine downtime. Better lap uniformity or minimum tuft size achieved in blow room significantly reduces the sliver breakage in carding, which minimizes the soft waste. Material handling is crucial especially in lap feed system. Lap should be transported from blow room to carding machine using trolley only. Good material handling will ensure no damage to the surface of the lap.

Speed influences the productivity and quality in the carding process, but in a different manner. The increase of delivery speed in carding is prone to sliver breakage due to uncontrolled web tension, which gives rise to higher soft waste. The selection of optimum speed with respect to fiber parameters and feed material quality is essential for the trouble-free processing. The machinery setting in carding is crucial in the control of breakage rate and soft waste. Work practices in carding should be proper for effective control of soft waste, especially during feeding lap, mending broken web and sliver.

The proper selection of the trumphet is important in controlling the soft waste generation. The smaller trumphet for coarser sliver may give rise to sliver choking problems frequently. Ambient conditions such R.H% should be kept as per norms for the smooth functioning of the card. Poor R.H% in the carding department will tend to increase the web breakage as well as lapping problems. The effective maintenance of R.H% in the carding process is vital in controlling the soft waste generation.

3.13.3 Automatic Waste Evacuation System

Automatic waste evacuation system (AWES) has become an integral part of the spinning mills, especially in the preparatory section. In conventional cards, waste generated in the card was removed manually by stopping the machine. The latest cards are now equipped with the manual or automatic waste removal system. In the manual waste removal, the waste gets deposited on the side filter screen. The carding tenter removes the waste from the filter screen once in a particular time period and puts it into the provided waste box. The collected waste from each carding machine is then weighed at the shift end and sent to the waste room. The carding machine must be stopped during waste removal in the manual system so that productivity does not suffer. The differential pressure switch fitted in the card ensures proper waste removal at a particular time interval; otherwise, it stops the machine until the waste gets removed. With the evolution of AWES, waste generated in the card is automatically evacuated from the card to the waste collection room.

3.14 Productivity and Quality for Different End Uses

Productivity and quality are the two important vital factors for any manufacturing process. Productivity of the carding process depends on delivery speed, sliver hank, efficiency of the machine, and labor allocation. High productivity results in lower cost per unit of output, resulting in higher levels of profit for a business. The delivery speed of the carding machine can be decided based on the quality requirements. The increase in the delivery speed of the card will substantially increase unevenness and imperfections of the sliver.

Yarn quality requirements may differ for various fabric manufacturing technologies such as weaving and knitting. The yarns produced in spinning have a wide range of applications, from hosiery products to shirting fabrics. Each application has different yarn quality requirements. In weaving, there are various types in which feed yarn requirements will vary such as shuttle looms and shuttleless looms. Yarn meant for the weaving process needs higher strength, so higher twist is to be imparted in the spinning process. Hosiery products require yarn with less twist and minimum imperfections compared with yarn meant for woven fabric applications. The delivery speed of the card is kept low for hosiery applications.

3.14.1 Points for Effective Control of Quality in the Carding Process

- Equipment maintenance lays the basis for ensuring quality consistency.
- Process optimization is the means for enhancing quality.
- Equipment configuration also offers a way to improve quality.
- Specified operating procedure can reduce yarn faults, while environment control is closely related to production efficiency.

Taking good control of all these factors can ensure carding machine to reach the ultimate quality goal.

3.15 Technological Developments in Carding

The technological developments in carding machine in the last three decades have provided tremendous improvement in the productivity, quality, and maintenance aspects. The engineering aspects of the card design have undergone extensive changes to achieve higher speeds and productivity rates. Precision engineering has made it possible to achieve extremely close settings, low vibration and noise levels, and reduced card maintenance and care requirements. During the 1950s, increased speeds and improved stability of card settings have been achieved with the adaptation of antifriction bearings to the cards. From 1960 to 2013, the production rate increased rapidly from 8 to 95 kg/h with the help of technological advancements. The technological developments in cards manufactured by various machinery manufacturers are detailed here.

3.15.1 Developments in Rieter Card: C70®

Rieter[12] introduces the latest card C70 with more technological advancements compared with its previous versions. The major developments in the feed zone in the C70 card claimed by Rieter are integrated fine opener for gentle and effective fiber opening in the chute, which ensures smaller fiber tufts and a uniform fiber batt. The active flat area is increased by 45% compared with Rieter's C60 version. The active flat area of the C70 is as much as 60% larger compared with conventional cards (as seen in Table 3.10). The active carding index (ACI) is a measure of the active carding area—the number of flats in contact is multiplied by the working width of the card.

The precarding zone of the Rieter C70 card has six carding units along with guiding elements and mote knife. Flat speed can be infinitely adjusted to output and quality via a

TABLE 3.10

Activating Carding Index of Conventional Card and Rieter Card Models

Card Models	Conventional Card	C60 Card	C70 Card
Total number of flats in rotation	84	79	99
Total number of flats in the contact zone	30	22	32
Card width (m)	1.0	1.5	1.5
ACI = number of active flats × card width	30	33	48
ACI compared to conventional 1 m cards	—	10%	60%

Source: Courtesy of Rieter Company, Rieter Carding Machines, Rieter, Winterthur, Switzerland, 2013.

frequency converter—independently of cylinder speed. IGS-classic features a grindstone that moves automatically over the cylinder clothing during production. This process is performed repeatedly throughout the planned service life of the clothing, and not after every 80–100 tons of production, as is the case with laborious manual grinding. IGS-top grinds the flat clothing fully automatically. Rieter claims that the production performance of the Rieter C70 card with a 1.5 m wide card, compared with that of the C60 card, can be increased by up to 40% with equal or better sliver quality.

3.15.2 Developments in Trutzschler TC11®

High production card from Trutzschler[13] named TC11 is said to be providing 40% higher productivity. This carding machine evolves with a 300 mm wider carding zone. The pre-carding zone of the Trutzschler TC11 card increases the preopening. The large postcarding area takes care of the rest, thus ensuring even cleaner sliver and higher fiber parallelism. The precision knife setting (PMS) system of the Trutzschler card adjusts the distance of the knife to the needle points and clamping point between the feed roll and needle roll to alter the degree of cleaning. The flat measuring system (FLATCONTROL TC-FCT) is used to measure the distance between the cylinder and the flat. For measurements, three regular flats are removed with measuring flat. Trutzschler NEP CONTROL TC-NCT monitors the card web during production and provides information regarding neps in card sliver. An optical electronic camera films the web under the take-off roll approximately 20 times per second. The camera moves about the whole working width of the card in a special, fully closed profile. The computer attached to the profile evaluates the pictures with special analysis software and indicates neps, trash particles, and seed coat fragments in the card web. The Trutzschler MAGNOTOP system uses high-energy "superstrong neodymium magnets" to hold the clothing strips on the flat bars. T-Con calculates the distance of the carding elements objectively, based on various measuring values under production conditions. The various settings displayed on the card monitor include the flat–cylinder gauge, fixed carding segments, and cylinder gauge, and hence, these settings can be optimized. T-Con also registers even the slightest contacts of the clothing and shuts down the card long before damage can occur and protects against clothing damages.

3.15.3 Developments in Marzoli C701®

In the new architecture of the Card C701, the main cylinder has been raised; the licker-in and the doffer have new diameters and have been located underneath the main cylinder. The new carding machine[31] has a carding surface 60% more than the previous model: 3.74 m^2 for the C701 and 2.34 m^2 for the C601. Small dimension of the licker-in that works at high speed results greater centrifugal force that, in combination with the knife and the carding segments positioned underneath the licker-in roller, contributes to an easy elimination of dust and trash.

References

1. Furter, R. and Frey, M., Analysis of the spinning process by counting and sizing neps, Zellweger Uster, Uster, Switzerland, 1990, SE476.
2. Douglas, K. (ed.), Measurement of the quality characteristics of cotton fibers. *Uster News Bulletin* 38: 23–31, July 1991.

3. Leifeld, F., New features of a high-tech card. *Melliand Textilberichte International Textile Reports*, 10: 75, 1994.
4. Kaufmann, D., Investigations of the revolving flat card. *Textil Praxis* 16(11): 1093, November 1961.
5. Pearson, N.L., Neps and similar imperfections in cotton. *USDA Technical Bulletin* 396: 19, November 1933.
6. Bogdan, J.F. and Feng, I.Y.T., Neps and how to control them. *Textile World* 102(5): 91, 1952.
7. Hayhurst, E. and Radrod, J., Carding with stationary flats. *Melliand Textilber* 53: 866, 1972; English 866, 1972.
8. Bogdan, J.F., A review of literature on neps. *Textile India* 114(1): 98, 1950.
9. Cripps, H., High speed revolving flats: An enhancement to card performance, *Proceedings of Beltwide Cotton Conference*, San Antonio, TX, 1995, p. 1389.
10. Artzt, P., Short staple spinning: Quality assurance and increased productivity. *ITB—International Textile Bulletin* 49(6): 10, 2003.
11. Card TC 11-Brochure, Trutzschler Spinning,Trützschler GmbH & Co. KG Textilmaschinenfabrik, Mönchengladbach, Germany.
12. Rieter Card C 60, The Concept for Excellence, Rieter Machine Works Ltd., Winterthur, Switzerland, http://www.rieter.com/en/rikipedia/articles/fiber-preparation/the-card/the-operating-zones-of-the-card/feed-device-to-the-licker-in/.
13. Schlichter, S., Improved raw material utilization with new concepts in Cleaning and Carding. *Pakistan Textile Journal*, February 2001.
14. Vasudevan, P., An investigation into the effect of licker-in design on carding performance, PhD thesis, The University of Leeds, Leeds, U.K., April 2005.
15. Krylov, V.V., Some theoretical and experimental data concerning the design of high speed cotton cards. *Technology of the Textile Industry, USSR* 2: 47–53, 1962.
16. Van Alphen, W.F., The card as a dedusting machine. *Melliand Textilberichte* 12: 1523, 1980 (English edn.).
17. Sheikh, H.R., Impact of carding segments on quality of card sliver. *Pakistan Textile Journal*, September 2009, http://www.ptj.com.pk/Web-2009/09-09/Practical-Hint.htm.
18. Grimshaw, K., Benefits for cotton system from the use of fixed carding flats, *Conference Proceedings: Tomorrow's Yarns*, UMIST, Manchester, U.K., June 26–28, 1984, pp. 166–181.
19. Varga, J.M.J., Technical innovations in carding machines. *Textile Month* 31–36, December 1984.
20. Ghosh, G.C. and Bhaduri, S.N., Transfer of fibres from cylinder to doffer during cotton staple-fibre carding. *Textile Research Journal* 39(4): 390–392, 1969.
21. Ghosh, G.C. and Bhaduri, S.N., Studies on hook formation and cylinder loading on the cotton card. *Textile Research Journal* 38: 535–543, May 1966.
22. Sengupta, A.K. and Chattopadhyay, R., Change in configuration of fibres during transfer from cylinder to doffer in a card. *Textile Research Journal* 52: 178, 1982.
23. Simpson, J., Relation between minority hooks and neps in card web. *Textile Research Journal* 42(10): 590–591, 1972.
24. Kumar, A., Ishtiaque, S.M., and Salhotra, K.R., Impact of different stages of spinning process on fibre orientation and properties of ring, rotor and air-jet yarns: Part 1—Measurements of fibre orientation parameters and effect of preparatory processes on fibre orientation and properties. *Indian Journal of Fibre & Textile Research* 33: 451–467, December 2008.
25. Sengupta, A.K., Vijayaraghavan, N., and Singh, A., Studies on carding force between cylinder and flats in a card: Part II—Effect of fibre and process factors. *Indian Journal of Textile Research* 8(9): 64–67, 1983.
26. Ishtiaque, S.M. and Vijay, A., Optimization of ring frame parameters for coarser preparatory. *Indian Journal of Fibre & Textile Research* 19: 239–246, 1994.
27. Sands, J.E., Little, H.W., and Fiori, L.A., Yarn production and properties. *Textile Progress* 3: 34, 1971.
28. Rakshit, A.K. and Balasubramanian, N., Influence of carding conditions on rotor spinning performance and yarn quality. *Indian Journal of Textile Research* 10: 158, December 1985.
29. Strang, P.M., The theory of carding with some of its implications in textile manufacture and research, Whitin Machine Works, Whitinsville, MA, 1954.

30. Morton, W.E. and Hearle, J.W.S., Equilibrium absorption of water, in: Morton, W.J. and Hearle, J. (eds.), *Physical Properties of Textile Fibers*, Textile Institute, Manchester, U.K., 1997, pp. 159–177.
31. Singh, R.P. and Kothari, V.K., Developments in blow room, card & draw frame. *Indian Textile Journal*, April 2009, http://www.indiantextilejournal.com/articles/FAdetails.asp?id=2010.
32. Lawrence, C.A., *Fundamentals of Spun Yarn Technology*, CRC Press LLC, Boca Raton, FL, 2003, p. 136.
33. Lord, P.R., *Hand Book of Yarn Technology: Technology, Science and Economics*, Woodhead Publishing Ltd., Cambridge, U.K., 2003.

4

Process Control in Drawing

4.1 Significance of the Drawing Process

In the spinning process, the unevenness of the product increases from stage to stage after the draw frame due to the steady decrease in the number of fibers in the section. The uniform arrangement of fibers becomes more difficult because of their smaller number. Drawing is the important process where the card slivers are doubled and drafted in order to level the unevenness present in each sliver. The various parts of a drawing machine are shown in Figure 4.1. Doubling is the combining of several slivers, and drafting is attenuation. Drawing therefore involves the processes of drafting and doubling, with good fiber control being the essence throughout. Doubling refers to the action of combining two or more slivers during a process, such as drawing, doubling taking place at the input to the drawing stage. In principle, every doubling is also a transverse doubling because the feeds are united side by side.[1] Lateral fiber blending and uniformity improvement in the sliver take place during the drawing process. Slivers from different cards vary in evenness and other properties and should be blended to reduce the irregularity. The drawing process especially fitted with autoleveler plays a predominant role in the control of count C.V% and enhances fiber orientation in the sliver. The draw frame primarily improves medium-term and especially long-term sliver evenness through doubling and drafting. The carding process generates fiber hooks, which cause errors in drafting, reduce the strength of yarn, increase the end breakage rate, and lead to a general deterioration in performance. It is well known that the drafting process, in general, improves the fiber parallelization and straightens the hooks present in the card sliver.[2]

The magnitude of the drawing process is huge in terms of quality perspective. A draw frame machine is feeding approximately two speed frame machines, and a speed frame is feeding approximately three ring frames. Thus, the fault generated in the sliver delivered from a draw frame significantly affects the yarns produced in three ring frames from that sliver. The drawing process aims at achieving the following objectives:

- Attenuate the card slivers
- Reduce the fiber hooks and improve fiber alignment
- Blend and mix fibers
- Reduce the irregularity of card slivers by doubling

The draw frame contributes less than 5% to the production costs of the yarn.[3] However, its impact on quality, especially evenness, is all the greater for this. The drawing depends on some factors such as the number of doublings, feed sliver hank, and delivery sliver hank. The number of doublings may be kept as 6 or 8 depending upon the process requirements. Since the total draft achieved is limited in a draw frame, two draw frame passages

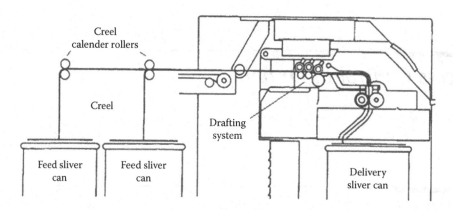

FIGURE 4.1
Parts of a draw frame.

are adopted to effectively enhance the uniformity of the sliver. The amount of draft to be applied immediately after the card cannot be very high as fiber entanglement is very high and the strand is thick. As such, draft has to be increased gradually. Draw frame is a machine where a very high degree of fiber to fiber friction takes place in the drafting zone; this is ideal for separating dust. Many modern draw frames have appropriate suction removal systems; more than 80% of the incoming dust can be extracted.

4.2 Fundamentals of Drafting System

In general, two drafting stages are applied to the card sliver on the draw frame, the first referred to as the "break draft" and the second as the "main draft."[4] The total draft of the draw frame is not the addition, but multiplication, of the drafts in separate zones.

For instance, if a draw frame has three drafting zones as shown in Figure 4.2 with drafts of D1, D2, and D3, respectively, then the total draft of the draw frame should be the product of D1, D2, and D3. The actual draft and mechanical draft are not always equal. The actual

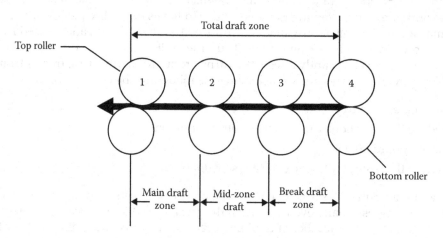

FIGURE 4.2
Distribution of draft in a drafting system.

draft is the real draft. The ratch is also known as the ratch length or ratch setting. It is set according to the length of the longest fibers in order to prevent these fibers from being stretched to break. There are several important definitions with respect to roller drafting:

1. $\text{Actual draft} = \dfrac{\text{Input count (tex)}}{\text{Output count (tex)}}$

2. $\text{Mechanical draft} = \dfrac{\text{Output surface speed}}{\text{Input surface speed}}$

3. *Drafting zone* is the region between the front and back rollers, where drafting occurs.

4. *Ratch* is the distance between the nip points of the front and back rollers.

5. *Doublings* is the number of slivers fed to the drafting system for one output sliver.

In general, the fibers in the card sliver are not very well aligned (i.e., not very parallel), and one of the consequences of the drafting that takes place during the drawing process is that the fibers become better aligned and straighter. Two draw frame passages are commonly applied after carding, the first being referred to as "breaker drawing" and the second as "finisher drawing."

4.3 Fiber Control in Roller Drafting

Drafting is the process of elongating a strand of fibers, with the intentions of orienting the fibers in the direction of the strand and reducing its linear density. In a roller drafting system, the strand is passed through a series of sets of rollers, and each successive set rotates at a surface velocity greater than that of the previous set. In ideal drafting, all fibers would move at the back roller surface speed until the leading end of each fiber was gripped by the nip of the front rollers and instantaneously accelerated to the front roller surface speed.[5] Under these conditions of complete fiber control, the distance between any two fiber leading ends after drafting would equal the distance before drafting, multiplied by the draft. In actual drafting, the fibers held in the back roller nip are known as back beard fibers, while the fibers held in the front roller nip are called front beard fibers. The fibers lie between the back and front rollers, without being held in the nip of either of them, and are termed as "floating fibers" (as seen in Figure 4.3).

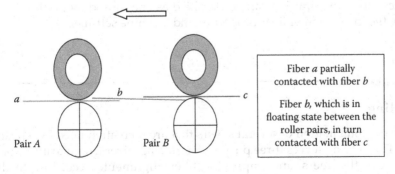

FIGURE 4.3
Floating fibers in the drafting zone.

Floating fibers accelerating out of turn can cause adjacent fibers to also accelerate, creating a thick place. The thick place is then drawn forward by the front roller nip, leaving a thin place behind. The effect of floating fibers is the production of a succession of thick and thin places in the output length. The thick and thin variation has a sinusoidal waveform and is therefore called the drafting wave. Lord and Johnson[6] described the acceleration of adjacent fibers by one floating fiber as "avalanche effect." The drafting wave gives an irregularity, in addition to that of the input irregularity. Foster and Martindale[7] concluded that the drafting waves are characteristics of a mechanism that tends to oscillate with a given wavelength band and amplitude, but the regularity of this oscillation is continually being upset by random disturbances. The wavelength of a drafting wave is about 2.5 times the mean fiber length. McVittie and De Barr[8] contended that the fibers have an intermediate velocity in the floating region of the draft zone. The various factors influencing drafting wave irregularity are as follows:

- The size of the draft
- The count of the input material
- Multiple inputs or doubling
- Roller and drafting zone setting
- The degree of parallelism, length and fineness of fibers in the input material

The main aim of fiber control is to keep the floating fibers at the speed of back rollers until they reach the front roller nip (i.e., to prevent fibers from being accelerated out of turn) while still allowing long fibers to be drafted. Different yarn manufacture systems, and different processes in the same system, often apply different control devices in drafting. The control roller and pressure bar force the fiber assembly (in the drafting zone) to take a curved path, thus increasing the pressure on fibers at the control roller or pressure bar. The increased pressure helps to control fiber movement during drafting. Lamb states that, though an irregularity-causing mechanism does exist in drafting, drafting also actually reduces the strand irregularities by breaking down the fiber groups. If $C.V_{in}$ is the coefficient of variation of the linear density of input strand prior to drafting and $C.V_{added}$ is the additional irregularity due to drafting, the resulting variance of the drafted product is given by the addition of variances:

$$C.V_{out}^2 = C.V_{in}^2 + C.V_{added}^2$$

Irregularities caused by drafting can be classified as one caused by machine defects and the other by the interaction of fiber properties and machine settings.

4.4 Doubling

Doubling is the combination of several slivers that are then attenuated by a draft (as seen in Figure 4.4). Doubling serves three purposes: reducing sliver irregularity, improving the blend or mix of the fibers, and improving fiber alignment. According to the law of

FIGURE 4.4
Doubling.

doubling, if n slivers are doubled together, the C.V of the doubled material will be reduced by a factor of $1/\sqrt{n}$, or

$$C.V_{\text{after doubling}} = \frac{\overline{C.V}_{\text{before doubling}}}{\sqrt{n}}$$

where $\overline{C.V}_{\text{before doubling}}$ is the average C.V of the individual slivers before doubling.

The number of doublings lies in the range 6–8 and so is the range of draft; as a result, the input and output material is almost the same in terms of linear density. An eightfold doubling in comparison to the sixfold doubling does not cause any improvement on drafted sliver and roving unevenness% and no significant difference in the number of hooks in sliver.[2] The linear density of the output sliver is determined by the amount of total draft applied in the draft zone.

Doubling is very important for equalizing. This means well distribution of different fibers with their same or different properties all along the length of the delivery sliver. Mass variation is reduced if there are a few thick places and some amount of thin places in the same zone. The sliver irregularity can be evened by doubling slivers together at the draw frame; however, this evening action becomes ineffective as the wavelength of the irregularity increases.[9] Blending is also improved by doubling because cans from different carding machines are fed to a breaker draw frame and cans of different breaker draw frame are fed to different finisher draw frame. When doubling and drawing are combined, the input materials are doubled to reduce the long-term errors; however, new errors of shorter wavelengths are added as a result of the process of elongation. There is an exchange of relatively long-term for short-term error. The tensile strength clearly increases in ring yarn with the increase in the number of draw frame passage because of better parallelization of the fibers and the related increase in fiber–fiber friction.[2]

4.5 Influence of Draw Frame Machine Elements on Process

The quality of the draw frame sliver is highly influenced by the right selection of process parameters and quality feed material. The draw frame process can be controlled by the effective selection of process parameters and machine elements, upkeeping of machine, ambient conditions, and proper work practices. The influence of draw frame machine elements such as creel, drafting rollers (top and bottom), top arm loading system, asymmetric web condenser, and trumpet is predominant in achieving the required quality and trouble-free running. Let us discuss the role of those machine elements here.

4.5.1 Creel

The creel attached to the back of the draw frame is used to feed the sliver to the drafting section without a false draft by a pair of calender roller provided above each can, one for each sliver. The creel must be designed

- To avoid false drafts
- To stop the machine upon a sliver breakage
- To deal sliver breaks easily, comfortably, and safely

The sliver can position should be in accordance with the respective creel calender roller (as seen in Figure 4.5). The placement of feed sliver cans in the creel is

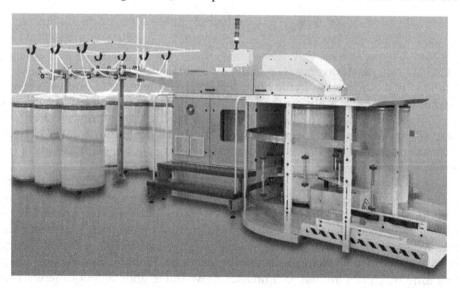

FIGURE 4.5
Draw frame creel.

important, which influences breakage and false drafts. For better draw frame efficiency, avoiding the change of cans at creel is advantageous. However, precise length of sliver at cans is the prerequisite for this. The feed sliver cans for different cards should be placed alternatively to attain homogeneity in output sliver. The sliver content in the can should be same in all feed cans to avoid soft waste or productivity loss. The can plate in the card should be leveled in order to prevent the occurrence of false draft due to tilting.

All the calender rollers fitted in the creel are equipped with electrical stop motion. It is also very important that stop motions in the creels work properly. Such stop motions stop the machine when any one of the slivers is broken or creel gets exhausted. This is achieved by the infeed roller pairs, which serve as electrical contact rollers for monitoring the sliver. If the sliver breaks, the metal rollers come into contact because the insulating sliver is no longer present between them, and the machine is stopped. Periodic checking for the correct functioning of stop motion is essential to avoid defects like singles.

4.5.2 Drafting Zone

The drafting zone is the most important element of the draw frame machine. The doubled sliver is drafted for attenuation in the drafting zone. It is responsible for producing required sliver count with good quality. The drafting zone essentially consists of drafting rollers, top arm weighing system, microdust suction system, pressure bar, etc. The draft given in the draw frame is in the range of 6–8. The drafting zone corresponding to various manufacturers and their different models have different designs such as drafting roller arrangements (3/3, 3/5, 4/4, 4/6), top arm weighing system (spring loaded, pneumatic loaded), and angle of drafting system (horizontal or curved) (as seen in Figure 4.6a and b). The drafting rollers run at different surface speed to attenuate the feed sliver to required sliver count. The pressure bar incorporated in the drafting zone applies lateral pressure, which helps to control the floating

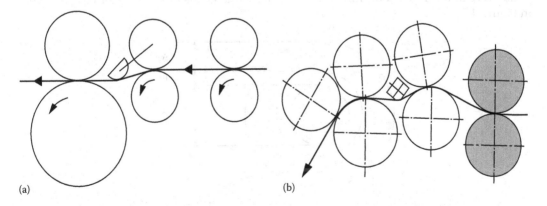

(a)   (b)

FIGURE 4.6
(a) 3 over 3 drafting system and (b) 4 over 3 drafting system.

fibers by preventing them from running fast until nipped by the front pair of rollers. Drafting components have a significant influence on yarn quality and production costs in ring spinning.

4.5.2.1 Bottom Drafting Rollers

Bottom rollers in all drafting systems are made of steel and mounted in roller, ball, or needle bearings that are positively driven. Bottom rollers have different diameters in which front bottom roller will have higher diameter in order to achieve higher productivity. These rollers have one of the following types of flutes to improve their ability to carry the fibers along:

- Axial flutes
- Spiral (inclined) flutes
- Knurled flutes

Knurled flutes are used on rollers receiving aprons to improve transfer of drive to the aprons. In draw frames, spiral fluting is used mostly because

- It offers more even clamping of the fibers and lower noise levels in running
- Its draft defects are minimized in the subsequent processes
- Top rollers can roll on spiral fluted bottom rollers more evenly with less jerking and, therefore, spiral fluted rollers are preferred for high-speed operation

4.5.2.2 Top Rollers

Top rollers are rubber cot–mounted roller that runs over the bottom roller. The purpose of cots is to provide uniform pressure on the fiber strand to facilitate efficient drafting. The top roller runs on the bottom roller by frictional contact. Ball bearings are used most exclusively in top roller mountings. The various dimensions of a top roller cot are given in Figure 4.7.

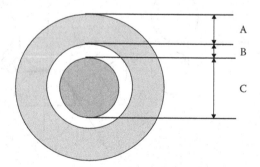

FIGURE 4.7
Top roller cot—parts. A, wall thickness of rubber portion; B, thickness of core used in rubber cot; and C, inner diameter of the cot.

TABLE 4.1

Comparison of Different Types of Top Roller Cots

Features	Plain Cots	Spring Grip Cot Thread	PVC Core Cots	Alucore Cots
Construction	Single layer	Three layers	Two layers	Two layers
Expansion after mounting (approx.) (mm)	1.5	1	0.2–0.3	0.1
Surface stress in cot after mounting	Very much	High	Negligible	Nil
Adhesive for mounting	Required	Not required	Not required	Not required
Waiting period for initial grinding after mounting	24 h (minimum)	Nil	Nil	Nil
Grinding after mounting	More	More	Minimum (because of less expansion and preground finish)	Minimum
User-friendliness	Least	Least	Excellent	Good

Hardness of the rubber cot is specified in terms of shore hardness, which is expressed in degree. Cots are available in wide shore hardness ranging from 63° to 90° shore. The selection of top roller shore hardness is based on the following parameters:

- Nature of raw material processed
- Linear density of material
- Type of application/process/mechanical conditions
- Maintenance aspects
- Ambient conditions

The top roller synthetic rubber is periodically ground (called buffing) in order to maintain the roundness and smoothness as the coatings wear out during spinning. The grinding operation has a roughening effect on the roller surface, which leads to formation of laps when processing sensitive fibers. For better smoothness after buffing, roller coatings can be treated with

- Applying a chemical film such as lacquer or another smoothing medium
- Acid treatment
- Irradiation by UV light

The comparison of different types of top roller cots used in draw frame is given in Table 4.1.

4.5.2.3 Top Roller Weighing System

Top roller weighing is essential in the attenuation process of sliver. It can be classified as spring loading, pneumatic loading, and magnetic loading. Conventionally, for many years, the weighing of drafting rollers can be accomplished with dead weight. Most of the latest draw frame machines adopt pneumatic loading of top rollers, which is easy to adjust and control precisely.

The top roller loading is necessary to provide the contact between top and bottom roller so that they turn together without slippage. The top roller loading also provides the gripping power necessary to hold the fibers at each roller bite and thus control the drafting action. The amount of top roller loading varies for different process conditions. The various factors influencing the top roller loading are listed as follows:

- Fiber bulkiness
- Draw frame speed
- Fiber fineness
- Top roller shore hardness

Bulky fiber requires a large amount of top roller loading in the drawing process. The higher delivery speed of the draw frame needs more top roller loading. A coarser and harsher variety of cotton fibers may require a large amount of top roller loading. The shore hardness of the top roller cots plays a prominent role in deciding the top roller loading. Softer cots have a larger area of contact than harder cots. In drawing process, harder cots are mostly preferred due to high speed. In modern draw frames, more pressure is often applied with reduced settings. As the settings become closer, it is necessary to increase the pressure due to increase in drafting force; otherwise, roller slippage will occur. Any fiber presented to the nip of the front pair of rollers should be immediately accelerated by that pair of rollers, and no slippage should take place. Due to this reason, front roller should have a higher pressure.

4.5.3 Web Condenser

The web condenser or trumpet helps to convert the web delivered from the drafting roller into sliver form. The trumpet is located in front of front bottom roller. The change from the web form to sliver form is the important factor at this point. It is but a rearrangement of the web where it is drawn through a small hole. The dimensions of the trumpet are about 1–1.5 in. in length and about 1.5 in. in diameter at the top and taper down to about 1/4 of an inch in diameter at the bottom. The taper in the trumpet produces lateral fiber migration and enhances sliver cohesion. The size of the trumpet varies depending on the sliver hank. The trumpet should be maintained properly without any surface damage. The selection of the trumpet influences the quality of the sliver delivered.

4.5.4 Coiler Trumpet

Coiler trumpet condenses the sliver to be deposited in the can. The dimension of coiler trumpet is smaller than the web condenser. The throat diameter (in millimeter) of the trumpet should be between $1.6\sqrt{n}$ and $1.9\sqrt{n}$ (n is the linear density in ktex) depending on the weight of the sliver. The improper selection of coiler trumpet tends to create sliver choking, which affects productivity. The coiler trumpet should be handled with utmost care so that any surface damage can be prevented. Sliver choking may occur due to improper selection and improper maintenance of the coiler trumpet.

4.6 Roller Setting

Roller setting is the important process parameter that influences the sliver uniformity and imperfections. The draw frame sliver irregularity ($U\%$) is very much dependent on the roller settings. The distance between the nips of drafting rollers is termed as "setting"

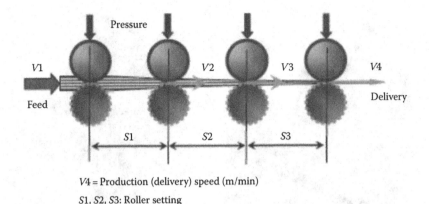

$V4$ = Production (delivery) speed (m/min)

$S1$, $S2$, $S3$: Roller setting

$S1 > S2 > S3$

FIGURE 4.8
Roller setting in the draw frame.

(as seen in Figure 4.8). Roller setting is primarily fixed on the basis of fiber length and fiber length distribution. Roller setting can be kept based on fiber's effective length or 2.5% span length or upper half mean length. The roller setting of drafting rollers is highly influenced by the following factors:

- Fiber staple length
- Feed sliver hank
- No. of drawing process
- Fiber characteristics

If the roller setting is too close, the rollers may break the longer fibers, leading to fiber rupture. If the roller setting is too far apart, the drafting will be done in an uneven manner and the result will be an uneven sliver. The fibers present in the card sliver have poor orientation and roller setting in the draw frame to be kept closer to attain the required fiber parallelization.

4.7 Top Roller Maintenance

The maintenance of top rollers plays an essential role in deciding the quality of the output sliver. The draw frame top rollers that are mounted with rubber cots should be ground once in a month or earlier depending on the process. The dirt content in the fibrous material may block the fine pores of the rubber cot, which significantly affects the cot's resiliency. The cots grinding should be done with proper leveling and without any surface damage.

4.7.1 Measurement of Shore Hardness

Hardness[10] may be defined as the resistance to indentation under conditions that do not puncture the rubber. It is called the elastic modulus of the rubber compound. These tests are based on the measurement of the penetration of the rigid ball into the rubber test piece

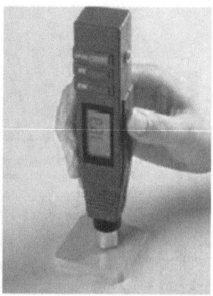

FIGURE 4.9
Measurement of top roller shore hardness.

under specific conditions. The measured penetration is converted into hardness degrees. Normally, spring-loaded pocket size durometer is commonly used for measuring the hardness of the elastomers (as seen in Figure 4.9).

Shore A durometer is used for measuring soft solid rubber compounds. Other scales such as Shore D, which is used to measure the hardness of very hard rubber compounds including ebonite, are also used. The main drawback is in the reproducibility of results by different operators. So a practical tolerance of 5° is acceptable. Better reproducibility is obtained by dead weight loading. Here, the hardness is expressed in International Rubber Hardness Degrees (IRHD). Both IRHD durometer tests require rubber specimen of definite dimensions. As per the ASTM (D 2240 [defines apparatus to be used and its sections such as diameter, length of the indentor, and force of spring], and D 1415 [defines specimen size]), DIN, BRITISH, and ISO Standards following test conditions have been laid for measuring *Shore A Hardness* of rubber products:

1. The specimen should be at least 6 mm in thickness.
2. The surface on which the measurement made should be flat.
3. The lateral dimension of the specimen should be sufficient to permit measurements at least 12 mm from the edges.

4.7.2 Berkolization of Top Roller

Laps can form on the top roller cots depending on the fiber material being processed, the climate, and the drawing machine. Reduced lapping for improved start-ups is achieved through carefully controlled cot surface treatment. Even nowadays, with "antistatic" cots widely available, a treatment of the cot surface after buffing is needed to minimize lapping and to generally optimize the running-in of freshly buffed cots. There are various methods of treatment using chemical agents such as lacquers and sulfuric acid. These treatments

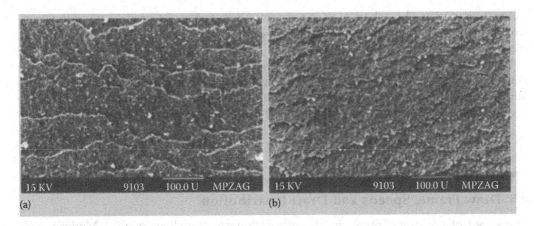

FIGURE 4.10
Top roller cot surface—(a) before and (b) after berkolization.

are very time-consuming, and they must be performed with great care, as they involve the use of chemicals and usually toxic agents. The use of these environment-polluting and hazardous chemicals can be completely eliminated by berkolizing the cots, using special UV rays.

Berkolizing is very simple, quick, reproducible, economical, and environment-friendly. Radiating the surface of the finely buffed rubber cots with specially developed UV light alters its structure. As a result, the rubber surface is less aggressive to the fibers. This considerably reduces the tendency to form laps when running in freshly buffed cots and reduces the loss of fibers during drawing to an extent. Berkolizing for too long reduces the friction coefficient of the cots, which leads to yarn breaks. The microphotographs of the surface of cots before and after berkolization operation are shown in Figure 4.10.

Each fiber and blend presents specific spinning requirements that vary depending on the characteristics of each drafting system and spinning system. In many spinning applications today, "berkolizing" top rollers have become a prerequisite for efficient and high-quality yarn production. By "berkolizing" top rollers, spinning mills achieve the following:

- Reduction of start-up problems after buffing the cots
- Reduction of work during start-up period
- Minimal lap formation
- Better running conditions
- Great reduction of fiber loss
- Positive influence on yarn quality
- General reduction of costs

4.7.3 Testing of Top Roller Concentricity and Surface Roughness

The perfect smooth running of top rollers is one of the prerequisites for producing the spun yarn of superior quality. The concentricity tester is used for checking the parallelism and smooth running of top rollers. The hardened and ground contact rollers are driven by a smooth-running electromotor. The freely moving measuring carriage runs on a precision guide, free of play. The spherical joint supports allow fast, exact positioning of the precision measuring sensors. Faults concerning smooth running, parallelism, or wear and tear can be measured with 0.01 mm accuracy.

The surface roughness has a great influence on the running behavior of the top roller cots. It can be checked with the lip, with a magnifying glass, or with a surface finish measuring device. A surface finish measuring device called perthometer has the great advantage that the coarseness can be quantified and documented with a measured value.

4.8 Draw Frame: Speeds and Draft Distribution

The speed of the draw frame significantly influences the sliver quality. Modern draw frames practically run at around 600–800 meters per minute. The increase in the delivery speed substantially increases the C.V% as well as imperfections of draw frame sliver. The delivery speed corresponding to combed process is low compared to carded process.

Sliver obtained from a carding machine normally contains 20,000–40,000 fibers in cross section. The number of fibers in the yarn cross section is approximately 100. Good drafting is difficult to achieve under high draft conditions, there generally being an optimum draft, and the total draft necessary to achieve the required sliver and yarn linear densities is normally accomplished by drafting in stages. Therefore, while converting a sliver to a yarn, fibers must be distributed over a greater length so that the cross section is gradually reduced, which is technically termed as "attenuation," and it happens by extending the fiber strand to a longer length by slippage of fibers over one another. The fibers are extended or straightened out in the drawing operation. The amount of extension of length is called draft. If there is wastage, then attenuation will be more than expected due to draft. So,

$$\text{Attenuation} = \frac{\text{Draft} \times 100}{(100 - P)} \quad \text{where } P \text{ is the percentage waste}$$

Attenuation is the "actual draft," and it can be calculated by determining the ratio of input sliver count to output sliver count. The ratio between the speeds of the delivery roller and feed roller of the drafting device is called "mechanical draft." In every drafting operation, irregularities are introduced. If the irregularities introduced are not from the same drawing zone, then total irregularity can be correlated as follows:

$$\text{C.V(total)} = \sqrt{(\text{C.V1}^2 + \text{C.V2}^2 + \cdots \text{C.V}n^2)}$$

where
   n is the number of intermediate drafts
   C.V is the coefficient of variation

Thus, the resulting irregularity will be less than the irregularity that might have been caused by drawing the strand in one drawing zone since C.V is proportional to the draft being employed in a single drafting zone. This was the reason for drafting being carried out in a number of times. Old draw frames used to have three drafting zones.

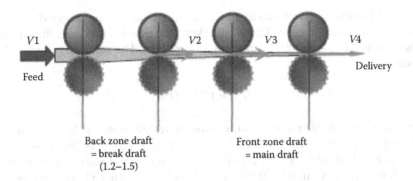

FIGURE 4.11
Draft distribution in the draw frame.

However, due to improvement in design, much better control of irregularities is achieved, and so in all the modern draw frames irrespective of the roller arrangement, basically two zones are used: the break draft zone and the main draft zone (as seen in Figure 4.11). The draft in the break draft zone is required to prepare material for drafting in the main draft zone by reducing the entanglements to some extent. The break draft should be outside the stick slip zone. For cotton, the recommended break draft is between 1.16 and 1.26 and for synthetics, between 1.42 and 1.6. The total draft is from 4 to 8. The total recommended range of draft for cotton is 7.5 and for the synthetics 8. The process of straightening is improved and accelerated when the amount of draft is increased.[11] The recommended sliver linear density for cotton is around 3.8–4.2 ktex. For synthetics, since the friction is high, sliver linear density should be below 3.8 for better drafting.

4.9 Count C.V% and Irregularity *U*%

Count C.V% is a vital factor to be kept under control in the draw frame process. C.V% has received more recognition in the modern statistics than the irregularity value *U*%. The coefficient of variation C.V% can be determined extremely accurately by electronic means, whereas the calculation of the irregularity *U*% is based on an approximation method. The sliver count C.V% affects with the following factors:

- Input sliver hank variation
- Improper no. of doubling
- Variation introduced by the machine
- Improper R.H%

In all staple spun material (yarn, rovings, and slivers), the fiber distribution along the material varies. This variation is also affected by fiber fineness, fiber fineness variations, and material type. The mass per unit length variation due to variation in fiber assembly is generally known as "irregularity" or "unevenness." A yarn with poor evenness will have thick and thin places along the yarn length, while an even yarn will have little variation in mass or thickness along the length. The irregularity *U*% is proportional to the intensity of the mass variations around the mean value. The larger deviations from the mean value are much more intensively taken into consideration in the calculation of C.V.% rather than in $U\%$ due to the squaring of the term. The relation between C.V.% and $U\%$ is given by

$$C.V.\% = 1.25 \times U\%$$

The $U\%$ is independent of the evaluating time or tested material length with homogeneously distributed mass variation. The larger deviations from the mean value are much more intensively taken into consideration in the calculation of the coefficient of variation C.V.%. The short-term irregularity ($U\%$) of the breaker drawing sliver and the long-term variation of the finisher sliver of about 0.25–0.70 m lengths depending on the count spun are the factors affecting bobbin count C.V.%.

4.9.1 Causes and Control of $U\%$ in Draw Frame

Unevenness or irregularity of the drawing sliver should be controlled as it influences the final yarn evenness and subsequently fabric appearance. The uneven sliver affects the quality and productivity of the spinning as well as fabric manufacturing processes. The various causes of the $U\%$ are summarized as follows:

- Higher $U\%$ in the input sliver
- Roller eccentricity
- Poor mechanical condition of the machine
- Improper roller setting
- Wrong selection of process parameters
- Poor work practices during piecing, can replenishing, etc.

Control measures

- Proper control of feed sliver $U\%$
- Proper maintenance of machinery
- Frequent diagnosis of roller eccentricity and rectification
- Roller setting as per recommendations
- Right selection of process parameters
- Ensuring effective functioning of autoleveler
- Right selection of fiber in mixing

4.9.2 Autoleveler

The object of an autoleveler is to measure the sliver thickness variations and then to continuously alter the draft accordingly so that more draft is applied to thick places and less to thin places with the result that the sliver delivered is less irregular than it otherwise would have been. Open-loop autoleveler can be used for the correction of fairly short-term

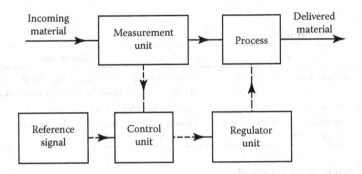

FIGURE 4.12
Autoleveler—open-loop system.

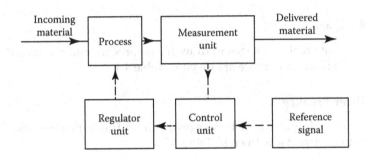

FIGURE 4.13
Autoleveler—closed-loop system.

variations. The control unit compares the measurement signal with the reference signal, which in this case represents the mean input count required (as shown in Figure 4.12).

The control unit accordingly increases, leaves unaltered, or decreases the output of the regulator, which in turn provides a variable speed to the back or front rollers of the process to give the required draft when the measured material has reached the point at which the draft is applied.

The closed-loop autoleveler is used for the correction of long-term and medium-term variations. Measurement always takes place on the material after the point where corrective action is applied (as shown in Figure 4.13). Thus, if measurement is made on the output, the correction may be applied to either the back rollers or the front rollers.

4.10 Defects Associated with Draw Frame Process

4.10.1 Roller Lapping

The various causes of roller lapping in draw frame process and the remedial measures taken to control the fault occurrence are given in Table 4.2.

TABLE 4.2

Roller Lapping: Causes and Remedies

S. No.	Causes	Remedies
1.	Poor top roller maintenance	Proper buffing and replacement of top roller cots. Also cleaning of top roller once in 2 h to free it from dirt.
2.	Improper control of R.H%	Effective control of R.H%.
3.	Damage in top roller or bottom roller due to poor work practices	Ensure knife is not used by the draw frame tenter to clear the roller lap that damages the cot.

4.10.2 Sliver Chocking in Trumpet

The various causes of sliver chocking in trumpet in draw frame process and the remedial measures taken to control the fault occurrence are given in Table 4.3.

4.10.3 Creel Breakage

The various causes of creel breakages in draw frame process and the remedial measures taken to control the fault occurrence are given in Table 4.4.

4.10.4 More Sliver Breakages

The various causes of higher sliver breakages in draw frame process and the remedial measures taken to control it are given in Table 4.5.

4.10.5 Improper Sliver Hank

The various causes of improper sliver hank in draw frame process and the remedial measures taken to control the fault occurrence are given in Table 4.6.

4.10.6 Singles

The various causes of singles in draw frame process and the remedial measures taken to control the fault occurrence are given in Table 4.7.

TABLE 4.3

Sliver Choking in Trumpet: Causes and Remedies

S. No.	Causes	Remedies
1.	Wrong selection of trumpet	Select appropriate trumpet w.r.t sliver hank.
2.	Thick and thin places in sliver	Ensure proper drafting of sliver; top arm pressure to be appropriate.

TABLE 4.4

Creel Breakage: Causes and Remedies

S. No.	Causes	Remedies
1.	Creel tension draft is high.	Optimum creel tension draft.
2.	Wrong alignment and placement of feed can.	Ensure proper positioning of feed can in the creel.
3.	More end breakages in carding process.	Control end breakages in carding process.
4.	Poor sliver piecing practices.	Proper training for tenter.

TABLE 4.5

More Sliver Breakages: Causes and Remedies

S. No.	Causes	Remedies
1.	Higher delivery speed than recommended	Maintain speed as per recommendations.
2.	Poor feed material quality	Ensure good feed material quality.

TABLE 4.6

Improper Sliver Hank: Causes and Remedies

S. No.	Causes	Remedies
1.	Feed hank sliver variation	Check the hank of input slivers and ensure they are as per plan.
2.	Wrong draft change wheels	Check the draft wheels and ensure that the wheels are put to get the required draft.
3.	Malfunction in autoleveler functioning	Check the functioning of autolevelers by sliver test method (i.e., $A\%$), and ensure that the input voltage is as per norms.
4.	Top arm pressure not sufficient	Check the pressure on top rollers and ensure them to be as per norms.

TABLE 4.7

Singles: Causes and Remedies

S. No.	Causes	Remedies
1.	Failure in stop motion functioning during can runout.	Stop motion functioning to be monitored frequently.
2.	Very high suction power of pneumafil sucks good fibers.	Keep optimum pneumafil suction pressure.
3.	Singles for a short length can also be due to partial lapping on rollers.	Proper maintenance of R.H% and frequent cleaning of drafting zone.

4.11 Process Parameters in Draw Frame

Draw frame plays a crucial role in the spinning process in determining the quality, especially on evenness. The various factors in the draw frame influencing the quality and productivity of the spinning process are as follows:

1. Total draft
2. No. of draw frame passages
3. Break draft
4. No. of doublings
5. Grams/meter of sliver fed to the draw frame
6. Fiber length
7. Fiber fineness
8. Delivery speed
9. Type of drafting
10. Type of autoleveler
11. Autoleveler settings

The total draft given in the draw frame machine is highly influenced by the factors such as fiber length, short fiber content, and feed sliver hank. The draft normally given in draw frame is in the range of 8. A higher amount of draft will affect the sliver uniformity although it improves fiber parallelization. Too much fiber parallelization is detrimental to sliver uniformity. Fiber parallelization and hook removal happens mostly at breaker draw frame than at finisher draw frame. The better yarn strength was achieved with a draft of eight and eight doublings, which gave only slight changes in yarn evenness and imperfections.[12] It was also found that on the drawing frame, the best overall results are obtained by using lightweight slivers and a draft of eight and eight doublings.[13]

The roller setting predominantly influences the output sliver quality and process efficiency. Balasubramanian[14] states that optimum roller setting depends on the length characteristics of the material, the break draft, the material bulk, and the top roller weighing. The following facts are obtained from the experimental results of some research studies:

1. Wider back roller setting will result in lower yarn strength.
2. Wider back roller setting will affect yarn evenness.
3. Wider back roller setting will increase imperfections.
4. Higher back top roller loading will reduce yarn strength.
5. Higher back top roller loading will reduce end breakage rate.
6. Wider front roller setting will improve yarn strength.

Drafting wave occurs mainly due to uncontrolled fiber movement of a periodic type resulting from the defects. With variable fiber length distribution (with more short fiber content), the drafting irregularity will be high. An increase in the number of sliver doubling reduces the irregularity caused due to random variations. Doublings do not normally eliminate periodic faults. The number of doublings depends upon the feeding hank and the total draft employed. Most of the modern draw frames are capable of drafting the material without any problem, even if the sliver fed is around 36–40 g/m.

Fiber hooks influence the effective fiber length or fiber extent, which will affect the drafting performance. For carded material, normally a draft of 8 in both breaker and finisher draw frames is recommended. For combed material, if single passage is used, it is better to employ draft of 7.5–8. The break draft setting for 3/3, or 4/3, drafting system is as follows:

1. For cotton, longest fiber + (8–12 mm)
2. For synthetic fiber, fiber length + (20%–30% of fiber length)

Break draft for cotton processing is normally 1.16–1.26. Balasubramanian and Bhatnagar[15] reported that varying the break draft had no significant influence on sliver irregularity in carded and postcombed drawings. Pressure bar depth plays a major role in the case of carded mixing and mixing meant for open end spinning process. If it is open, $U\%$ will be affected very badly. It should always be combined with front roller setting. If the pressure bar depth is high, creel height should be fixed as low as possible.

An autoleveler draw frame should be employed as a finisher draw frame to meet the present quality requirements. Most of the autoleveler draw frames are working on the principle of open-loop control system. Sliver monitor should be set properly and calibrated frequently. Intensity of leveling and timing of correction are two important parameters in autolevelers. Intensity of leveling indicates the amount of correction; that is, if 12% variation is fed to the draw frame, the draft should vary 12% so that the sliver weight is constant. Timing of correction indicates that if a thick place is sensed at scanning roller, the correction should take place exactly when this thick place reaches the correction point (leveling point). The feed variation will be higher for higher correction length; for example, if feed variation is 1%, and if the correction length is 8 mm, if feed variation is 5%, the correction length will be between 10 and 40 mm, depending upon the speed and type of the autoleveler. $U\%$ of sliver will be high, if timing of correction is set wrongly. Most of the modern autolevelers can correct 25% feed variation. It is a general practice to feed 12% variation both in plus and minus side to check $A\%$. This is called sliver test. The $A\%$ should not be more than 0.75%. $A\%$ is calculated as follows: If the number of sliver fed to the draw frame is N, check the output sliver weight with $N, N + 1, N - 1$ slivers. Then

$$A\% = \frac{g/m(N-1)-g/m(N)}{g/m(N)} \times 100$$

$$A\% = \frac{g/m(N+1)-g/m(N)}{g/m(N)} \times 100$$

The top roller shore hardness should be around 80° as the draw frame delivery speed is very high in modern draw frames. It is advisable to buff the rubber cots once in 30 days (minimum) to maintain consistent yarn quality. Coiler size should be selected depending upon the material processed. For synthetic fibers, bigger coiler tubes are used. This will help to avoid coiler choking and kinks in the slivers due to coiling in the can.

4.12 Work Practices

- Feed sliver can must be transported to drawing department by can trolley only so that sliver surface may not get damage.
- Knife should not be used to clear the roller lapping.
- Feed cans should be placed in correct position in the creel so that stretching can be avoided.
- Top rollers must be cleaned with wet cloth once in 2 h to remove the dirt adhered to the rubber cot.
- Drawing sliver cans must be wrapped properly to prevent dust accumulation.
- Drawing empty can's castor wheel should be cleaned properly.
- Drawing empty can plate should be leveled to avoid stretching.
- Sliver piecing to be done in such a way that the diameter of the mended sliver should not have much difference compared to normal sliver diameter.
- Drafting zone should be cleaned once in 3 h to reduce dust accumulations.
- Sliver from the feed can should be laid properly through guides until drafting zone to control unnecessary stretch.
- Proper functioning of stop motions in creel, drafting zone, and coiling zone should be ensured.
- Sliver from a single can should not be distributed to several cans in the creel during sliver shortage, which affects the sliver quality and also leads to more soft-waste generation.

4.13 Technological Developments in Draw Frame

The draw frame influences predominantly the quality of the final yarn in the spinning mill. The faults in the draw frame sliver result inevitably in yarn defects. Quality is highly decided in the draw frame especially at the last draw frame, and it can no longer be improved at a higher rate after the draw frame. The developments in drafting zone, coiling zone, and driving mechanisms help to achieve expected quality, productivity, and user-friendly maintenance. The autoleveler is one of the significant technological developments in draw frame machine. The modern draw frames provide consistent sliver monitoring meter by meter to meet self-set quality standards with the help of autoleveler. The finisher draw frame fitted in any spinning process must be the autoleveler draw frame. The autoleveler is a prerequisite to the control count C.V% within limits in the draw frame machine. The production rate of the draw frame machine has improved a lot in the recent decade. The delivery rate of modern draw frame machine claimed by some machinery manufacturers is 1000 m/min. The technological developments of latest draw frame machine corresponding to some machinery manufacturers are summarized in the following.

4.13.1 Rieter Draw Frame

Clean coil®: Rieter[16] has patented a coiler plate that has honeycomb-like surface and claimed to reduce cleaning frequency from 2–3 h to 1–7 days depending upon the type and quantity of finishing agents. Cleantube, a rotational plate control device, reduces the buildup of cotton trash particles and short fibers in the sliver channel. Cleantube saves up to 300 man-hours per year by reducing the need to clean trash from the sliver channel and also reduces the amount of sliver waste generated by up to 0.6%.

4.13.2 Trützschler Draw Frame

- *Auto break draft setting*—Incorrect break draft increases yarn U%, imperfections, and neps, whereas the total draft does not affect yarn quality much. The AUTO DRAFT of the Trützschler TD 03 draw frame[17] optimizes the break draft of the draw frame under the prevailing conditions. The system automatically determines a recommendation for the ideal material-related break draft in less than 1 min.

- *OPTI SET®*—Leveling quality without compromises on autoleveler draw frames, the optimum setting of the main drafting point is decisive for the leveling quality. Establishing this point usually requires extensive laboratory trials (sliver tests). With Trützschler's autoleveler draw frame TD 03, this is not necessary, since the OPTI SET self-optimizing function is a standard feature. OPTI SET determines the optimum value fully automatically by considering the current general conditions, like machine settings, material characteristics, and ambient atmosphere. The fed slivers are scanned by a sensor; this is followed by a corresponding time-lagged leveling action as soon as the material has reached the main draft zone, which is 1000 mm away. This time lag between measurement and leveling action determines the main drafting point. The exact position depends, among other things, on machine settings as well as material and ambient atmosphere. The operator starts the function at the touch screen monitor. The draw frame starts with the standard value (e.g., 1000) and successively checks slightly deviating values. During this process, the C.V values of the fed slivers and the C.V values of the delivered draw frame sliver are measured and evaluated in relation to one another.

- *Sliver focus*—Sliver focus, the output measuring funnel of the autoleveler draw frame's quality monitoring, measures every inch of the sliver prior to deposit in the can. Sliver focus transmits a warning or stops the draw frame whenever the sliver deviates in its fineness or is faulty. The limit for warning and stopping can be defined individually.

References

1. Klein, W., *The Technology of Short-Staple Spinning*, A Short Staple Spinning Series, The Textile Institute, Manchester, U.K., 1987.
2. Kumar, A., Ishtiaque, S.M., and Salhotra, K.R., Impact of different stages of spinning process on fibre orientation and properties of ring, rotor and air-jet yarns: Part 1—Measurements of fibre orientation parameters and effect of preparatory processes on fibre orientation and properties. *Indian Journal of Fibre & Textile Research* 33: 451–467, December 2008.
3. Fibre preparation for spinning, 211.67.48.5/fsx/other/SEF201-Mod3.doc, Accessed Date: November 12, 2013.
4. Draw frame—Cotton yarn market, http://www.cottonyarnmarket.net/OASMTP/THE%20 DRAWFRAME.pdf, Accessed Date: November 12, 2013.
5. Taylor, D.S., Some observations on the movement of fibres during drafting. *Journal of the Textile Institute* 45: T310, 1954.
6. Lord, P.R. and Johnson, R., Short fibres and quality control. *Journal of the Textile Institute* 76(3): 145–155, 1985.
7. Foster, G.A.R. and Martindale, J.G., Fibre motion in roller drafting. *Shirley Institute Memorial Series A* 5: 125, 1941.
8. McVittie, J. and De Barr, A.E., Fibre motion in roller and apron drafting. *Shirley Institute Memoirs* 32: 105, 1959.
9. Bowles, A.H. and Davies, I., The influence of drawing and doubling processes on the evenness of spun yarns: Part IV—Extremely long-term variations. *Textile Institute and Industry* 38–40, January 1979.
10. Sujai, B., Shore A hardness of a rubber cot and its restrictions in measuring the hardness value under mill conditions, http://www.inarco.com/pdfs/05%20technical%20information/ publications/Shore%20A%20hardness%20of%20a%20rubber%20cot.pdf, Accessed Date: November 12, 2013.
11. Balasubramanian, N. and Bhatnagar, V.K., The effect of disorientation of hooks in the input sliver on drafting irregularities and yarn quality. *Textile Research Journal* 41: 750–759, 1971.
12. Sands, J.E., Little, H.W., and Fiori, L.A., Yarn production and properties. *Textile Progress* 3: 34, 1971.
13. Sands, J.E. and Fiori, L.A., Yarn production and properties. *Textile Progress* 3: 35, 1971.
14. Balasubramanian, N., Improving regularity of material at drawframe and speedframe, *Indian Textile Journal*, 34, May 1994.
15. Balasubramanian, N. and Bhatnagar, V.K., Factors affecting irregularity of post combed drawing sliver, *Proceedings of the 14th Joint Technological Conference* (ATIRA/BTRA/SITRA), Coimbatore, India, 1973, p. 22.
16. Draw frame—Technical Brochure, Rieter, Winterthur, Switzerland, 2013.
17. Draw frame—Technical Brochure, Trützschler, Mönchengladbach, Germany, 2013.

5

Process Control in Comber and Its Preparatory

5.1 Significance of Combing Process

Cotton fibers have a distribution of fiber lengths ranging from the longest fiber group to the shortest fiber group. Short fibers of length less than 12.5 mm do not contribute to the mechanical properties of the yarn but increase yarn hairiness, which adversely affects the yarn and fabric appearance. The combing process aims to remove short fibers from the group of fibers. The sliver produced in the carding process has poor fiber orientation and higher short-fiber content. The main objective of combing process is to remove short fibers and to improve fiber orientation. Combing is a key process that makes the difference between an ordinary yarn and a quality yarn. The various parts of a comber machine are given in Figure 5.1.

Combing is an intermittent operation carried out between carding and draw frame. The combing process is carried out in order to improve the quality of the sliver coming out of the card. Combed sliver has a better luster compared with carded sliver because of the improved fiber alignment. The materials extracted from the feed lap in the combing process are called noil. Noil contains short fibers, neps, and impurities. The amount of noil produced may be expressed as percentage noil. Depending upon the yarn quality requirements, the noil% in the combing process is set, ranging from 12% to 25%. The feed material for the combing machine is in the form of lap, so lap preparation process is needed for the combing process. Card slivers are normally doubled to produce a sliver lap, and six such sliver laps are doubled to produce a ribbon lap. Sometimes, a super lap machine is used for lap preparation. One or two drawing passages are followed either before or after the combing process. During combing, a series of fine and closely spaced needles (Unicomb) are passed through the fibers projecting from gripped nippers. By means of this operation, shorter fibers, neps, and dirt are removed and the fibers are oriented.

The combing process is especially of value with longer cottons where the extreme variation in fiber length is too high for the best spinning conditions. The combing process is normally used to produce smoother, finer, stronger, and more uniform yarns (Figure 5.2). Combing has been utilized for upgrading the quality of medium staple fibers. The removal of short fibers in combing facilitates the better binding of long fibers in the yarn, which ensures greater strength.

The uniformity of the combed yarn is superior, which contains fewer weak places. Hattenschwiler et al.[1] claimed that combed cotton yarns contain only approximately 1/10th the number of neps found in carded yarns. The yarn produced from the combed cotton sliver needs less twist than a carded yarn. However, these quality improvements

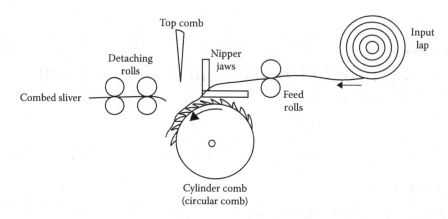

FIGURE 5.1
Combing machine—parts.

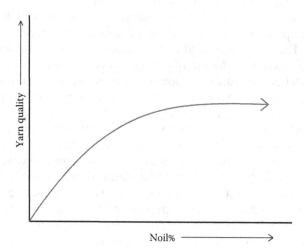

FIGURE 5.2
Effect of noil removal on yarn quality.

are obtained at the cost of additional expenditure on machines, floor space, and personnel, together with a loss of raw material. Yarn production cost is increased depending on the intensity of combing.

5.2 Lap Preparation

5.2.1 Lap Preparation Methods

Lap preparation is the important step in the combing process, which influences the quality of the combed sliver. The requirements of the lap for the good performance of combing are good fiber orientation, leading hooks presentation, and uniform thickness. The combers are fed with fiber lap produced by doubling several slivers and webs.

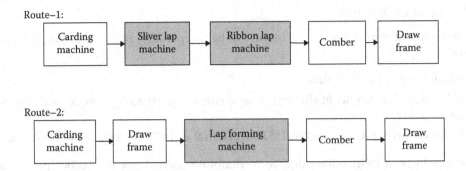

FIGURE 5.3
Comber preparatory process—lap preparation methods.

The lap prepared with proper doubling and draft may enhance the uniformity of the output sliver. The lap sheet with higher degree of evenness produces the highly uniform combed sliver. Good fiber orientation in the feed lap along the length of the feed lap is a prerequisite for obtaining better sliver quality. Lap can be produced by different routes as pointed in Figure 5.3.

Sliver lap and ribbon lap machines are used to produce the lap by doubling several numbers of slivers and webs, respectively. Another route has sliver doubling machine, also called super lap, used for lap preparation. Route-2 is adopted in the modern lap preparation process. Route-1 is also still in use in many spinning mills. In the route-1, the card slivers in the range of 16–32 numbers are doubled and drafted in the sliver lap machine and the delivered lap sheet is wound onto the cylindrical spool, which is termed as "sliver lap." The dimension details of sliver lap are given as follows:

- Lap grams per square meter (GSM) of 50–70 g/m
- Width of 230–300 mm
- Diameter of 500 mm
- Weight of up to 27 kg

5.2.2 Precomber Draft

The draft given in the sliver lap machine is commonly in the range of 1.5–2.5. As there is no waste removed in the sliver lap machine, the actual and mechanical drafts are theoretically the same. Six sliver laps are then doubled and drafted in the ribbon lap machine and the resulting lap sheet is wound onto the cylindrical spool, which is termed as "ribbon lap." Laps from the sliver lap machine are taken to the ribbon lap machine and thin sheets from the heads are led down over a curved plate, which turns at right angles, inverts them, and superimposes one upon the other. The draft given in the ribbon lap machine is commonly in the range of 2–4. The ribbon lap produced is fed to the comber consisting normally of eight heads. The web from eight ribbon laps are fed, combed, detached, doubled, and drafted in the comber machine and delivered as a single combed sliver. Variation between the slivers can cause drafting errors, so it is important to use quality slivers. It is also important that the fiber web is wrapped on the lap at the correct tension.

5.2.3 Degree of Doubling

The total number of doublings done in the comber preparatory machine is summarized as follows:

- Sliver lap has 24 no. of slivers.
- Ribbon lap has 144 no. of slivers, that is, 6 sliver laps for each 24 no. of card slivers, $24 \times 6 = 144$.
- Combed sliver produced from 1152 slivers, that is, $144 \times 8 = 1152$ no. of doublings.

There should be a tremendous degree of doubling carried out in comber preparatory machines to attain the uniformity of the sliver as well as for hook reversal purpose. Within the limits imposed by the capacity of the cans in the creel, the short-term evenness should be improved. The coefficient of variation should be $1/\sqrt{n}$ of the average value in the input slivers, where n is the number of slivers in the lap ribbon.

5.3 Factors Influencing the Combing Process

5.3.1 Fiber Properties

Fiber length and its uniformity significantly influence the combing performance. If the given cotton fiber is having higher short-fiber content, then the yarn quality is enhanced by removing a significant amount of short fibers from it, which increases the cost of the final yarn. The higher moisture content in the fiber tends to complicate the combing operation as fibers stick with each other and also with the machine parts, which may obstruct the fiber passage and lead to fiber breakage and poor combing. If the moisture content is less, then the fibers become dry and lack cohesion, which makes the fibers difficult to move from one machine part to another, and tend to fly. The bending properties of fiber are important in the combing operation, which requires fiber to undergo a lot of bending. Stiffer fiber will be more prone to fiber breakage than flexible fiber. Fiber fineness influences the selection of process parameters such as speed and setting in combing operation. The lap made of finer fibers produces fine-quality combed sliver compared with the lap made of coarser fibers. Trash particles are mostly removed in the blow room and carding process. The higher amount of trash in the feed lap may damage the half lap and top comb needles.

5.3.2 Lap Preparation

Menzi and Gahweiler[2] stated that the performance of the combing machine depends not only on the combing machine itself but also on the preparation of the material prior to combing. Poor lap preparation leads to excessive parallelization, poor lap runoff, and improper lap thickness. The fiber orientation and parallelization in the ribbon lap or feed lap is of primary importance in deciding the combing efficiency. If the fibers are parallel and well oriented, the load on the half lap and top comb needles gets reduced. The degree of parallelization of the lap fed to the combers should be optimum. If fibers are overparallelized, lap licking will be a major problem. Because of fiber-to-fiber adhesion, mutual separation of layers within the sheet is very poor. The lap thickness plays a crucial role in deciding the load on the combing needles and also on combing efficiency. Thicker laps tend to impart more stress on combing needles, resulting in inefficient combing. Yarn imperfections and hairiness index increase with thicker lap. A thick sheet always exerts a greater retaining power than a thin one. To certain extent, the bite of the nipper is more effective with a higher sheet volume. The lap with optimum thickness is necessary to minimize the stress on the needles as well as to maintain production rate. The uniformity of lap across the width is highly important for smooth combing operation. If there are too many thick and thin places in the lap, then it will affect the productivity as well as quality of output sliver. Combing operation removes the leading hooks present in the feed lap preferentially.

The size of the feed lap (ribbon lap) influences the productivity and web piecing irregularities. The higher the speed of the comber machine, the longer the time for the machinist to change lap. Increasing the size of the lap on a high-speed comber machine, the time of changing the lap, piecing frequency of the lap, and labor intensity of the worker can be reduced, to improve the production efficiency. The time required for changing the lap decreases in the range of 25% if the lap length is changed from 300 to 400 meters. When the lap thickness increases, the elasticity of the lap will enlarge during nipper opening, which helps in the rise of the fiber bundle as well as in detaching and piecing.

Lack of proper web tension generates a soft lap that is prone to damage during handling and also occupies more space. Too high a tension makes it difficult to unwind the lap at combing, especially the last few layers. In addition, the lap preparation should ensure the following:

- Lap with less number of piecing points
- Lap without tendency of licking
- Lap with longer length (bigger laps)

5.3.3 Machine Factors

The combing cylinder is the heart of the combing process, which removes short fibers and neps. Due to this operation, the cotton fibers are very much parallelized. The term "cylinder" is used to designate the combing assembly. Combing cylinder usually consists of half lap and segment. Half lap is the needle portion of the cylinder. The segment is a longitudinal section of another hollow cylinder, which is attached to the cylinder opposite the half lap. The purpose of the segment is to assist the detaching rolls in drawing away the fibers that have been combed. In addition, the segment serves to give sufficient weight opposite the half lap to balance its weight so that the cylinder will run without vibration. The needles of the combing cylinder play a primary role in attaining the proper combing. The factors such as needle sharpness, arrangement of needles, and needle density in the combing cylinder heavily influence the combing performance. The needles of the bars become finer, closer, and shorter from the first to the last bar. The first rows of needles are inclined at a steep angle with a radial line so that they will definitely catch the lap projecting from the nippers and lead the fibers well down to their base. Successive bars have less of this inclination. Once the fibers have been led to the base of the needles of the first row, less inclination is needed to keep them at the base, and so less inclination is used on the succeeding rows of needles. The more cylindrical the tooth, the more short fibers, neps, and impurities are eliminated, and the degree of separation, straightness, parallel degree of fiber, and yarn quality are better.

The purpose of the nippers is to grip the ribbon lap securely so that the end may be combed without removing long fibers as waste, and at the same time to hold the end of the ribbon lap as close to the combing cylinder so that combing may be as complete as possible. The top and bottom nipper should be accurately made and adjusted; otherwise, the irregularities of the nippers will not grip the lap perfectly all along their width. The setting of each set of nippers should be carried out with utmost care. The bottom nipper should be made just as near parallel to the top nipper as possible.

The function of the top comb is to comb the portion of the fibers held in the nippers during the cylinder combing. As it is impossible for the comb cylinder to comb the entire length of the fiber, the top comb is located so that it will comb the tail end of the fibers as they are being drawn away by the detaching rolls. The top comb parameters such as needles/cm and top comb position influence the quality of the delivery sliver. The purpose of the detaching mechanism is to grasp the combed fibers that project farthest from the nippers and draw them away, overlapping the previously detached cotton to produce a continuous sheet of fibers. The detaching is accomplished by three rolls and the combing cylinder segment. In the detaching zone, the factors such as detaching roller's rubber cot shore hardness, detaching roller pressure, cot uniformity, and detaching setting influence output sliver quality. The condensation of the combed web emerging out of the detaching rollers is carried out with the help of an asymmetric web condenser, which reduces the short-term irregularities in the slivers. The combing process introduces an additional irregularity known as piecing irregularity. The draw box found at the delivery side of the machine consists of drafting rollers that attenuate the condensed web obtained from each comber head and convert it into a single combed sliver. The factors such as drafting zone setting, top roller shore hardness, and delivery speed influence the performance of the comber.

5.4 Setting Points in Comber Machine

The setting between the machine elements in the comber machine plays a critical role in deciding the quality as well as processability of the material. The setting of machine elements influences the amount of waste removal and life of the machine elements. Good combing performance combined with higher production is possible with optimum and accurate setting between machine elements. The following settings usually have a high impact on the quality of combing.

5.4.1 Feed Setting

5.4.1.1 Type of Feed

Feeding of fiber sheet can be done either by forward feed or backward feed. Feeding of lap carried out during the forward movement of nipper is termed as forward feed or concurrent feed. Feeding of lap carried out during the backward movement of nipper is termed as backward feed or counterfeed. Counterfeed is used when the better quality of product is required at the loss of more noil (12%–25%). Concurrent feed is preferred for higher production rate with less noil desired (5%–12%). Concurrent feed is mostly used over all staple ranges for achieving noil levels from 8% to 18%. According to Charles Gegauff's noil theory,[3] the percentage noil ($N\%$) is related to the detachment setting (D), feed distance (F), and the longest fiber length (L), according to the following formulas:

$$N\% = \left(\frac{D - F/2}{L}\right)^2 \times 100 \quad \text{(for concurrent feed)}$$

$$N\% = \left(\frac{D + F/2}{L}\right)^2 \times 100 \quad \text{(for counter feed)}$$

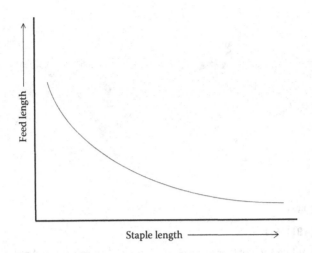

FIGURE 5.4
Effect of fiber staple length on feed length.

Gupte and Patel[4] reported that the percentage of short-fiber removal and the percentage of mean length improvement are higher with forward feed than backward feed at any noil level.

5.4.1.2 Amount of Feed per Nip

It indicates the amount of lap sheet fed per nipping cycle. The amount of lap feed per nip has a noticeable effect on noil%, combing quality, and production rate. A higher amount of feed per nip can increase the output web thickness and reduce the web hole, improving the production of the combing machine. At the same time, longer feeding reduces the comber noil and impairs the quality. However, noil% and feed rate relationship are different for forward and backward feed systems. Feed distance also depends very much on the staple length (Figure 5.4).

A mill study reported that the small change in the amount of feed per nip does not significantly affect fiber characteristics, nep removal efficiency, and noil extraction%. However, the study suggested that better combing took place at lower feed amount. Yarn imperfections have increasing trend with an increase in the feed amount. Subramanian and Gobi[5] reported that the increase in the amount of feed increases the amount of fiber handled by the combing mechanism which reduces the combing efficiency, resulting in higher short fiber%.

5.4.2 Detachment Setting

This is one of the main settings responsible for the change of noil percentage. This setting refers to the distance between the bite of the nippers and the nip of the detaching rollers when nipper assembly is at the most advanced position. A closer setting is associated with lower noil level. Detachment setting normally lies in the range of 15–25 mm. A large setting (more than optimum) does not produce further improvement in quality but results in more loss of noil. As such, optimum setting has to be found out depending on quality and production level desired for a particular machine and the material processed.

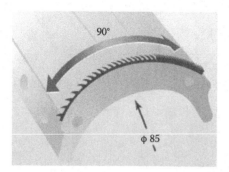 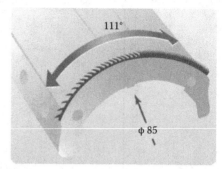

FIGURE 5.5
Circular comb—wire angle.

TABLE 5.1

Applications of Combing Surface with Different Wire Angles

Combing Surface	No. of Sections	Applications
90°	4	Short/medium staple
	5	Medium/long staple
	6	Long staple
111°	5	Medium/long staple

5.4.3 Point Density and Wire Angle of Comb

Efficient combing is realized by improving the density of the cylinder tooth and adopting larger angle cylinder needle so as to increase the combing area of cylinder. Sawtooth clothing is used in circular comb and needles are used in the top comb.

Fineness and point density depend on the raw material processed. Usually, top comb point density falls in the range of 23–32 needles per centimeter. Fewer needles are used when higher production is needed together with lower waste elimination. Wire points of combing cylinder are usually inclined at 75°, 90°, and 111° in different machines (Figure 5.5). The applications of combing surface with different wire angle are given in Table 5.1.

5.4.4 Top Comb Parameters (Depth of Penetration and Needle Density)

The depth of penetration of the top comb as shown in Figure 5.6 inside the lap has a major influence on the amount of noil extraction. During the forward swing, the top comb penetrates through the fiber fringe and on its reverse movement comes out. The position of the top comb with respect to the nipper plates will influence noil% extracted. Setting the top comb close to the nippers will reduce the noil% extracted. Lowering of the top comb by about 0.5 mm is followed by an increase in the noil of about 2%.

The main improvement with the increase in top comb depth is seen in the elimination of neps. Deep penetration of the top comb disturbs fiber movement during piecing and so optimum penetration is very important. A mill study found that the increase in top comb penetration depth from –0.5 to +1.0 mm increases comber noil% and nep removal efficiency. The study also concludes that fiber length and short-fiber content of the combed sliver are not much affected with the change in top comb penetration depth. Apart from the depth of penetration, top comb spacing from the detaching rollers is also important and can be adjusted.

FIGURE 5.6
Top comb.

TABLE 5.2

Top Comb Needle Density for Different Applications

Top Comb Needles (cm⁻¹)	Application
26	Short/medium staple
30	Medium/long staple

For a micronaire of less than 3.6, needles per centimeter are usually 30. For a micronaire of more than 3.8, it can be less, usually 26 (Table 5.2). A study by Jayaram[6] reported that more number of needles per inch in the top comb produce slivers with improved mean length, less short-fiber content, and better C.V%. Top combs with higher needles per inch have better cleaning capacity and give a lesser number of neps per gram in the output sliver. Subramanian and Gobi[5] reported that the increase in top comb penetration with a larger number of fibers reduces the imperfections but increases the classimat long faults.

5.4.5 Timing

Timing refers to regulating the various individual actions of the comb so that they occur in the proper sequence and at the correct moment in the combing cycle. The settings on a comb may be perfect, but if the timing is not correct, poor combing will result. Correct timing is obtained by the use of an index gear.

5.4.6 Nips per Minute (Comber Speed)

The speed of the comber is normally expressed as "nips per minute," which means the number of times the nippers close per minute. As the cylinder makes one turn for each combing cycle, the nips per minute equals the turns of the cylinder shaft per minute. For every rotation of the main combing cylinder, one cycle of combing comprising feeding, combing, detaching, and top comb operations is completed. The comber has undergone a lot of changes in the recent decade because of technological developments. Due to the improved machine design and technology, the speed of the comber has now attained up to 500 nips/min. The reason behind this higher speed, apart from others things, is the better manipulation of the comber technology, kinematics, and air control.

5.4.7 Piecing

Detaching rollers perform a back-and-forth movement in order to piece up the newly combed web with the web combed and detached in the previous cycle. The forward component (V) is larger than the backward component (R). The constant basic rotation of the detaching rollers (B) is given from the comb shaft. An intermittent rotation (A) is superimposed on this basic rotation. Even at high nipping rates, the top detaching rolls must guarantee perfect piecing of the web. The detaching rolls perform a steady back-and-forth movement matching the nipping rate. After the operation of the circular combs is finished, they feed back part of the previously formed web. The nippers lay the newly combed fiber fringe onto the portion of the web that has been drawn back by the detaching rolls. During piecing, the individual strips of web must be laid on top of one another by the detaching rolls so that a continuous sliver is formed.

5.5 Draft

The draft given in the comber is similar to the draft given in the carding machine. The waste removal in the comber process is significantly higher than any other spinning process. Because of this, there is a great difference between the actual draft and the mechanical draft. The actual draft of the comber machine normally ranges between 60 and 80 and the corresponding mechanical draft ranging from 50 to 70. The latest high production combers offer an actual draft up to 100.

5.6 Noil Removal

The waste from a comber is termed as "noil." Noil consists of shorter fibers and neps. The amount of noil removed in the combing process may be varied to suit the circumstances and is usually expressed in percentage based on the original weight of laps fed into the machine. A higher noil% always improves the imperfections in the final yarn. Mangialardi[7] stated that a noil level of 10% gives rise to a nep reduction of 65% in yarn neps. But the strength and other quality parameters improve up to a certain noil%; further increase in noil results in quality deterioration. In combing, if circular comb is not cleaned properly, then it gets loaded and combing suffers. In all modern combers, the combing cycle is slowed down (to one-fifth of normal speed) at preset intervals for better cleaning purpose. The brush below the circular comb continues to rotate at full speed and thereby cleans the comb effectively when the cycle speed is reduced. The noil percentage from a comber depends upon the following:

- Short-fiber content
- Detaching distance
- Feed length
- Top comb penetration
- Fiber extent and hook type

5.6.1 Combing Efficiency

Combing efficiency is calculated based on the improvement in 50% span length, expressed as a percentage over 50% span length of the lap fed to the comber multiplied with waste percentage:

$$\text{Combing efficiency} = \left(\frac{S-L}{L \times W}\right) \times 100$$

where
  S is the 50% span length of comber sliver
  L is the 50% span length of comber lap
  W is the waste percentage

The head-to-head and comber-to-comber variations in waste should not exceed ±1.5% and ±0.5% from the average, respectively. A study conducted by Jacobsen et al.[8] concluded that neps found in combed sliver samples appeared to be slightly smaller than card neps and uniform in size. Another study conducted by Sriramulu and Shankaranarayanan[9] concluded that the amount of noil extracted is affected by fiber disorder and the number of trailing hooks in the feed lap.

5.6.2 Degrees of Combing

The percentage noil removal in the combing process depends on the short fiber% present in the cotton, the respective end use of the yarn produced, and the process economics. With respect to the amount of noil extraction, combing can be classified as follows:

- *Scratch combing*: Up to 5% noil is extracted, and this does not improve yarn properties significantly but lowers the end breakage rate in spinning and weaving.
- *Half-combing*: Up to 9% noil is extracted, which enhances the yarn uniformity and spinning performance.
- *Ordinary combing*: Noil between 10% and 18% is extracted, which is beneficial in finer count ranges.
- *Full combing*: Noil greater than 18% is extracted to attain highest-quality yarns.

5.7 Nep Removal in Combing

The main objective of the combing process is the removal of short fibers from the cotton. Further and several other researchers claimed that combing represents the final possibility of significantly reducing the nep levels, which depends on noil level and cotton type. Neps are removed to a great extent (about 75%) by combing. Assessment and control of neps in combers is similar to that of the carding process either by monitoring the remnant neps in the combed sliver or by monitoring the nep removal efficiency. Control of remnant neps in the combed sliver is very important since comber is the final process stage for nep removal. Bogdan[10] reported that the nep removal is influenced by the setting of combers such as the needling of cylinder and top comb, setting of half lap, unicomb and top comb, and maintenance.

Variability in performance among different heads within the same comber is an often-noted aspect in many combers leading to an overall drop in performance. The following are the remedial measures for improving the nep removal efficiency:

- Replacement of poor top combs
- Correction of improper top comb depth setting
- Correction of top comb to back top detaching roller setting

The nep value of the individual heads in a comber can be a good indicator of the condition of the machinery components in a comber. The following design features of modern combers help to improve the nep removal efficiency:

- Higher cylinder diameter
- Half laps covering 120° of the cylinder
- Circumferential nipper that helps to maintain constant setting between nipper and half lap

When nep value deviates beyond control limits, besides checking the settings involved, it is important to inspect the condition of machinery components such as top comb, and comb cylinder and replace them promptly if it is worn out.

5.8 Hook Straightening in Comber

If the fibers are parallel to the yarn axis, yarn properties become better. When fibers exist in intermediate products in a hooked form, under certain condition, it may impair the comber performance. Again, the working length of a hooked fiber into the yarn axis becomes less and acts as a short fiber. The number of hook fibers in the yarn determines the yarn quality to a large extent. Thus, hooks not only determine the yarn quality and comber performance but also reduce the price realization.

Fibers must be presented to the comber so that leading hooks predominate in the feedstock. The way of feeding hooks toward the comber machine depends upon the selection of machineries between carding and combing. After passing through a machine, the direction of the strand is reversed. As stated earlier, a comber machine can straighten out

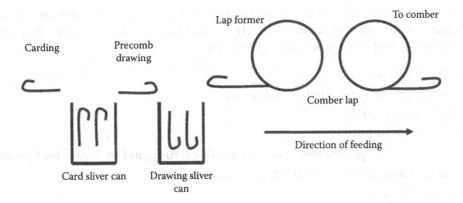

FIGURE 5.7
Hook straightening in the combing process.

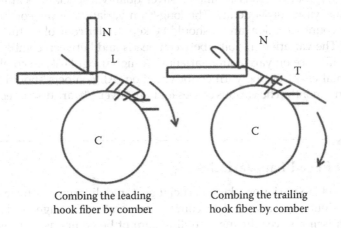

FIGURE 5.8
Effect of the combing process with trailing hooks and leading hooks. N, nippers; C, cylinder; L, lending hook; T, trailing hook.

the leading hooks only so that, to present the majority of hooks in the leading form, there must be an even number of passages between carding and combing (Figure 5.7).

The nippers grip the fibers at the tip, and the circular comb straightens out the hooks at the leading end as it sweeps the fiber fringe. But if the fiber hooks are present as trailing hooks, then the hooked end may be gripped or may not be gripped at all by the nippers (Figure 5.8). The fiber then will go to the delivery sliver in a hooked form, or the fiber is treated as a short fiber and will be wasted.

5.9 Sliver Uniformity

In the drafting system of the comber, the individual slivers of the comber heads are condensed to a single sliver and drawn to the sliver count required. The drafting zone is a highly significant part, determining the running behavior of the fibrous material and accounting for a high machine output with even sliver. High aging stability, good abrasion resistance, rapid recovery, and minimum tendency to lap formation are properties that are absolutely essential for optimal drafting. The various factors influencing the sliver uniformity in the combing process are given as follows:

- Fiber properties
- Precomber draft and number of doublings
- Feed lap uniformity
- Card sliver uniformity
- Comber process parameters—speed, detaching setting, and amount of feed per nip
- Proper maintenance of comb cylinder and top comb
- Proper R.H%
- Proper work practices

The short-term irregularity tends to impair sliver quality, especially within-bobbin variation in roving and yarn subsequently. The long-term variation of the comber sliver affects between-bobbin count variation, and it should be kept under control within a C.V of 3% for 1 m wrappings. The variation in noil% between heads and between combers will not have any significant influence on yarn count variation. A high irregularity in comber sliver could have a detrimental effect on the yarn count variation. All combers should be checked for sliver U% once in a month. The various causes for high sliver U% are discussed in Section 5.11.

5.10 Control of Feed Lap Variation

The control of feed lap variation plays a crucial part in delivering uniform combed sliver. The number of doublings involved in comber preparatory is huge, and any variations arising due to missing sliver during doubling cannot be compensated and lead to high variability in sliver. The proper functioning of stop motions in the creel of sliver lap and ribbon lap should be ensured to prevent the occurrence of singles. The top arm loading arrangement of drafting systems in sliver lap and ribbon lap machines should be checked frequently, especially with the deadweight and hook system, to avoid the generation of thick and thin places in the lap due to the lack of top arm pressure. The conversion of deadweight and hook system to direct spring loading system provides an even distribution of pressure over drafting top rollers tend to ensure better control of lap C.V%. The path of sliver and sliver can position in the creel of the sliver lap machine should be kept in a proper manner to avoid unnecessary stretch, which leads to count variation. The lap winding mechanism should be checked frequently to control the stretch generation. Lap weight should be monitored frequently to control hank variations.

5.11 Defects and Remedies

5.11.1 Inadequate Removal of Short Fibers and Neps

The various causes of inadequate removal of short fibers and neps in the combing process and the remedial measures taken to control the fault occurrence are given in Table 5.3.

TABLE 5.3

Inadequate Removal of Short Fibers and Neps

S. No.	Causes	Remedies
1	Head-to-head variation in comber noil%	Check noil% in individual heads and rectify the head that deviates.
2.	Uncombed portions due to slippage under feed roller	Check feed roller setting in individual heads and rectify.
3.	Slippage of fibers under detaching rollers	Check the condition of detaching roller cots and condition of the gears driving bottom detaching rollers.
4.	Plucking of fibers by half lap from nipper grip	Check the machines thoroughly for bent and hooked needles on half lap and top comb.
5.	Web disturbance due to air currents due to defects in brush or/in aspirator	Check for damaged air seals in the aspirator box.

TABLE 5.4

Short-Term Unevenness

S. No.	Causes	Remedies
1.	Piecing waves	Check for eccentric detaching rollers and correct it.
2.	Drafting waves	Check U% and make use of spectrogram diagram to identify the source of the problem. Check for eccentric rollers in drafting zone and correct them. Check for the drafting rollers setting.
3.	Poor fiber control due to worn-out top roller cots in draw box	Proper maintenance of top roller cots.
4.	High or low tension draft and improper settings	Correct the setting of tension draft and other settings as per norms.

5.11.2 Short-Term Unevenness

The various causes of short-term unevenness in sliver in the combing process and the remedial measures taken to control the fault occurrence are given in Table 5.4.

5.11.3 Hank Variations

The various causes of hank variations in the combing process and the remedial measures taken to control the fault occurrence are given in Table 5.5.

5.11.4 Higher Sliver Breaks at Coiler

The various causes of higher sliver breaks at coiler in the combing process and the remedial measures taken to control the fault occurrence are given in Table 5.6.

5.11.5 Coiler Choking

The various causes of coiler choking in the combing process and the remedial measures taken to control the fault occurrence are given in Table 5.7.

TABLE 5.5

Hank Variations

S. No.	Causes	Remedies
1.	Singles due to improper functioning of stop motions in the creel.	Check the functioning of stop motions frequently.
2.	Improper selection of tension draft.	Proper selection of tension draft.
3.	Rough surface of the sliver table that creates stretch in sliver.	Proper polishing of the sliver table.
4.	Variation in the feeding lap.	Correct the variations in feed lap.
5.	Lap licking while unwinding.	Set correct precomb draft. Maintain correct R.H%.
6.	Hank variation between combers is high.	Check the combers for variations in lap roller feed per nip, draft wheels on draw box, tension drafts at tables, draw box and coiler, and noil level variations.

TABLE 5.6

Higher Sliver Breaks at Coiler

S. No.	Causes	Remedies
1.	Rough surface of sliver guides	Sliver guides to be maintained properly.
2.	Eccentric coiler calender rollers	Check for eccentricity in coiler calender roller and rectify it.
3.	Tension draft too high	Set appropriate tension draft.
4.	Higher precomber draft, i.e., excess parallelization of fibers in the sliver	Optimum precomber draft.

TABLE 5.7

Coiler Choking

S. No.	Causes	Remedies
1.	Wax and trash deposited in coiler obstruct sliver passage and chocks.	Proper cleaning of coiler parts where sliver passes with a rope.
2.	Too many thick places, undrafted places in the sliver.	Proper maintenance of drafting zone, top rollers, correct setting of drafting rollers.
3.	Wrong sliver mending procedure after drafting while attending breakage.	Proper training of tenter for correct sliver mending procedures.
4.	Higher R.H%.	Maintain recommended R.H%.

5.11.6 Web Breakages at Drafting Zone

The various causes of web breakages at drafting zone in the combing process and the remedial measures taken to control the fault occurrence are given in Table 5.8.

5.11.7 Breakages in Comber Heads

The various causes of breakages in comber heads in the combing process and the remedial measures taken to control the fault occurrence are given in Table 5.9.

TABLE 5.8

Web Breakages at Drafting Zone

S. No.	Causes	Remedies
1.	Spreading of web too much	Correct positioning of web width guides.
2.	Defects in gears, burrs in bottom rollers	Check gears for any fault and rectify it. Check for burrs and rectify.
3.	Improper tension drafts	Set tension draft as per norms.
4.	Wax deposition and trash accumulations at trumpet	Clean the trumpet regularly with petrol.
5.	Poor buffing of top roller cots	Ensure proper buffing procedures.

TABLE 5.9

Breakages in Comber Heads

S. No.	Causes	Remedies
1.	Tight or slack web	Maintain lap tension properly.
2.	Improper positioning of web trays	Proper positioning of web trays.
3.	Unclean web trays	Clean the web trays frequently.
4.	Improper functioning of clearer rollers in detaching section	Clean the clearer rollers frequently.
5.	Trumpets set too far away from the nip of calender rollers	Correct the setting of trumpet from the nip of calender rollers.

TABLE 5.10

Excessive Lap Licking and Splitting

S. No.	Causes	Remedies
1.	Improper tension drafts and roller setting	Set optimum tension draft and roller setting.
2.	Higher precomber draft	Set optimum precomber draft.
3.	Uneven lap and tight winding	Ensure good lap quality and set proper winding tension during lap winding.

5.11.8 Excessive Lap Licking and Splitting

The various causes of excessive lap licking and splitting in the combing process and the remedial measures taken to control the fault occurrence are given in Table 5.10.

5.12 Technological Developments in Comber and Its Preparatory

5.12.1 Rieter® Comber and Lap Former

Rieter developed the E80 comber machine which claims that with the help of computer-aided process development (C•A•P•DQ), outstanding fiber selection, and optimal machine running behavior, the comber achieves superior-quality values with maximum economy. Gentle, controlled fiber treatment is achieved by the optimal coordination of comb movements with the help of C•A•P•DQ and the largest combing area developed by Rieter.

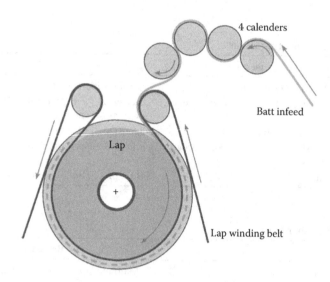

FIGURE 5.9

Rieter E35 OMEGAlap. (From Rieter, Rieter comber manuals, Rieter, Winterthur, Switzerland, 2013. With permission.)

Rieter's ROBOlap fully automated lap changing and piecing system is a unique feature and still sets the standard for modern combing operations. The design of the E80 comber with a 45% higher combing area enables customers to focus on productivity, raw material utilization, or quality, as required. The 3-over-3 cylinder drafting system with pressure bar and variable break and main draft distance ensures precise fiber guidance for all staple lengths. Rieter provides "Ri-Q-Top" top comb with high self-cleaning effect. The perfect interplay of the new Ri-Q-Comb circular comb developed by Rieter and other technology components enables cleaning to be performed at the highest level. The distinguishing feature of the new Ri-Q-Comb high-quality circular comb is a combing area some 45% larger. The increase of up to 60% in the number of points on the new generation of combs makes uniquely intensive extraction of noil from the fiber tuft possible.

Rieter's OMEGAlap E35, as shown in Figure 5.9, offers gentle web guidance from the beginning to end of the lap and uniform pressure distribution over up to 75% of the lap circumference.[12] The path of the belt around the lap (wrapping) resembles, in cross section, the Greek letter "Ω" standing on its head. The name OMEGAlap has been derived from this. The batt being fed is passed over 180° of the lap circumference at the start of the lap and over 270° of the lap circumference by the end of the lap.[13] The contact pressure necessary for lap buildup is thus distributed ideally over the outside diameter of the lap. The OMEGAlap produces at a constant speed of 180 m/min, regardless of raw material and lap diameter. Rieter's fully automatic lap transport system E26 SERVOlap can carry eight laps at a time from the UNIlap/OMEGAlap to the comber.[14]

5.12.2 Trutzschler® Comber

Trutzschler[15] has redesigned the comber frame for the dynamic load, which results in (Comber TCO1) 500 nips/min. A special software simulates the dynamic loads during the running process, and then individual frame components are exactly designed for these alternating loads. This gives sound frame structure without vibrations, impact loads, or uncontrolled distortions even at 500 nips/min. Optimized kinematics also leads to lower mechanical work and lower losses through friction and vibrations that consequently lower down the power and energy consumption. Magnesium–aluminum alloys are used to make nippers lighter, which reduces vibrations and noise level substantially. The top combs, during operation, get loaded with short fibers and impurities. Thus, the machine is stopped intermittently for cleaning, which results in production loss. Trutzschler comber is equipped with self-cleaning top combs. An extremely short compressed air blast of a few milliseconds purges the needles from top to bottom and detaches the adhering fibers. The cleaning frequency can be adapted to the respective degree of soiling.

5.12.3 Marzoli® Comber

Marzoli also claims that comber CM600N can run at 480 nips/min. This becomes possible due to new kinematics of the nipper unit that reduces the closing speed of the nipper jaws, thus giving the slowest force of impact during the closing of jaws. The comber frame has been redesigned using 3D-CAD software. This leads to considerable reduction in the mechanical stress and machine noise; thus, the operation becomes smooth without vibrations and heavy shocks.

5.12.4 Toyota® Comber

Toyota® also claims that lightweight aluminum differential arm along with improved "Cam-less Mechanism Link Motion" for detaching roller drive reduces the amount of inertia in detaching motion and enhances operational speed. The combing will be better if the bottom nipper's lower surface remains at a fixed set distance from the bottom comb during the entire circular combing. This is achieved by using the standing pendulum principle.

5.13 Work Practices

- Check the functioning of stop motion in the sliver lap, ribbon lap, and combers to control soft-waste generation and quality problems.
- Materials such as sliver can and laps should be transported with utmost care without any stretch and surface damages.
- Tenter should have knowledge about the process such as no. of doublings, hank feed, hank delivery, lap weight, and color codes.
- Comber can's castor wheels should be maintained in a clean condition to prevent the falling of can due to fiber accumulation in castor wheels, which generates a huge amount of soft waste.
- Tenter should check the top comb frequently for cleanliness.
- The drafting top rollers should be cleaned with water once in a shift to free them of dirt deposition. The dirt blocks the fine pores of rubber cots and reduces its resiliency.
- Using a picker gun, the dust accumulations in the individual comber heads, drafting zone, and coiler zone should be cleaned during stoppage time.
- Mending of web and sliver in the comber process should be done in a proper manner so that thick-place occurrence is avoided.

- Sticking between layers of lap (lap licking) is a serious problem. It should be reported immediately to the shift in-charge for remedial action.
- R.H% and dry bulb temperature should be maintained properly for the good working of combers.
- The rubber cots of detaching rollers should be cleaned frequently.
- Usage of knife and hooks for clearing roller lapping and trumpet chocking should be strictly avoided.
- Tenter should check the noil delivery from each comber head frequently so that blockage in any comber head due to brush roller jamming gets noticed.
- Lap unwinding tension should be maintained properly without any slackness.

References

1. Hattenschwiler, P., Muller, H., and Wetter, B., Quality and production control in the spinning mill, SE432-1, Zellweger Uster, Uster, Switzerland, 1986.
2. Menzi, E. and Gahweiler, E., No high-performance without hi-performance technology, Rieter Spinning Systems, Winterthur, Switzerland, October 1992.
3. Klien, W., *A Practical Guide to Combing and Drawing*, Manual of Textile Technology—Short-Staple Spinning Series, Vol. 3, The Textile Institute, Manchester, U.K., 1987, Chapter 2.
4. Gupte, A.A. and Patel, B.A., Comber performance with forward feed and backward feed, *Proceedings of the 27th Joint Technological Conference*, NITRA, Ghaziabad, India, 1986, p. 08.
5. Subramanian, S. and Gobi, N., Effect of process parameters at comber on yarn and fabric properties. *Indian Journal of Fibre & Textile Research* 29: 196–199, June 2004.
6. Jayaram, V.S., *Proceedings of the Fifth Joint Technological Conference*, SITRA, Coimbatore, India, 1963, p. 24.
7. Mangialardi, G.J., Saw-cylinder lint cleaning: Effects of saw speed and combing ratio on lint quality. *USDA-ARS Technical Bulletin* 1418, 1970.
8. Jacobsen, K.R., Grossman, Y.L., Hsieh, Y.-L., Plant, R.E., Lalor, W.F., and Jernstedt, J.A., Neps, seed-coat fragments, and non-seed impurities in processed cotton. *The Journal of Cotton Science* 5: 53–67, 2001.
9. Sriramulu, V. and Shankaranarayanan, K.S., *Proceedings of the 18th Joint Technological Conference*, BTRA, Mumbai, India, 1977, p. 31.
10. Bogdan, J.F., A review of literature on neps. *Textile India* 114(1): 98, 1950.
11. Quality meets flexibility, E80 Comber, http://www.rieter.com/en/spun-yarn-systems/products/spinning-preparation/e-80-comber/, Accessed Date: May 1, 2013.
12. Rieter OMEGA lap, http://www.docstoc.com/docs/74485322/Rieter-OMEGA-lap, Accessed Date: May 1, 2013.
13. OMEGAlap E 35: Revolutionary Lap Winding Technology, Rieter Textile System, Information brochure, Accessed Date: May 1, 2013.
14. UNIlap E 32: The Preparation for High Performance Combers, Rieter Textile System, Information brochure, Accessed Date: May 1, 2013.
15. Comber—Technical brochure, Trutzschler, Germany.
16. Rieter, Rieter comber manuals, Rieter, Winterthur, Switzerland, 2013.

6

Process Control in Speed Frame

6.1 Significance of Speed Frame

Speed frame process normally comes after comber in the combing process and draw frame in the carding process. Speed frame is also called simplex or roving frame. The speed frame process minimizes the sliver weight to a suitable size for spinning into yarn and inserting twist, which maintains the integrity of the draft strands. It is impossible to feed the sliver to ring frame for yarn production due to limitation in draft in ring frame. The irregularity of the output strand increases as the draft increases due to the noticeable effect of the drafting wave. Due to this reason, the draw frame sliver hank must be reduced in two steps so that an acceptable yarn quality is achieved. Cans of slivers from finisher drawing or combing are placed in the creel, and individual slivers are fed through two sets of rollers, the second of which rotates faster, thus reducing the size of the sliver. Twist is imparted to the fibers by passing the bundle of fibers through a roving "flyer." The product is now called "roving," which is packaged on a bobbin (Figure 6.1). A roving is a long and narrow bundle of fibrous strand. Roving is an intermediate product produced from sliver, and it is normally used as a precursor for yarn.

Roving is also distinguished as the first process in which material is wound on a bobbin. Faulty roving preparation has a drastic effect on the spinning performance. The process parameters adopted in the roving process have a significant influence on spinning quality and production. The speed frame machine essentially comprises between 60 and 132 spindles, each containing a drafting system and flyer twister. The rotation of flyer imparts twist to the fibrous strands. Flyer rotational speed is limited because of mechanical design difficulties. The defect arising in the drafting of roving introduces the short-term irregularity in the yarn produced from it. The wrong selection of twist in the roving affects the spinning performance by either higher creel breakages or higher undrafted ends. The improperly built bobbin in roving leads to end breakage in ring frame and higher slough-off during material handling. The preparation of roving bobbin for the yarn spinning is of paramount importance for a spinning mill.

FIGURE 6.1
Speed frame.

6.2 Tasks of Speed Frame

The tasks of a speed frame machine may be divided into a number of individual operations, each of which is almost independent of the others. The major tasks of the speed frame process are listed as follows:

- *Drafting*: to reduce the size of the strand
- *Twisting*: to impart necessary strength
- *Laying*: to put the coils on the bobbins
- *Winding*: to wind successive layers on the bobbin at the proper rate of speed
- *Building*: to shorten successive layers to make conical ends on the package of roving

In a sliver of 3 ktex, approximately 20,000 fibers are present in its cross section. The draft of 10 would reduce the number of fibers to 2000 in the cross section, and a small amount of twist would be required to provide sufficient cohesion for suitable handling. Drafting is normally carried out by a draft system with double apron capable of working with entering sliver counts of 0.12–0.24 Ne and counts of the delivered roving of 0.27–3 Ne. The draft given in the roving process is normally calculated from the hank sizes involved. The draft given in the roving process will be in the range between 4 and 20 and can work fibers of a length of up to 60 mm.

In addition to accomplishing drafting, the operation inserts a slight amount of twist to give the roving the required strength and puts the strand in a special type of package to facilitate handling. The insertion of twist in the roving must not create any difficulty in the ring frame drafting operation by developing a high drafting force. The roving with higher twist requires higher draft in the ring frame. The increase of draft beyond the recommendations deteriorates the yarn quality.

Lay refers to the arrangement of the roving coils wound around the bobbin in any given layer. The closeness of the lay is measured in "coils per inch," which means the number of roving coils wound around the bobbin per inch parallel to the axis of the bobbin. The purpose of the laying operation is to put the successive coils of roving side by side in a uniformly spaced arrangement. This regular, uniform arrangement is achieved by making the bobbins move up and down at a uniform rate of speed for each layer.

Winding is the process by means of which the roving is drawn from the front roll through the flyer and onto the bobbin. The rate of winding compared to the rate of delivery at the front roll controls the winding tension. The building motion is controlled by the steady upward and downward movements of the bobbin rail containing the bobbins and spindles.

6.3 Importance of Machine Components in Speed Frame

6.3.1 Creel Zone

Creel is the place situated at the back of the machine where the raw material is placed to be fed to the drafting zone. Cans from comber or drawing machine are kept at the creel in an orderly fashion to utilize the floor space effectively. The position of the can in the creel corresponding to its drafting head is crucial in controlling the false drafts. The method of feeding slivers into speed frames and the condition of cans are important. Slivers crossing each other, damaged edges of drawing sliver cans, etc., would disturb the free withdrawal of slivers from cans. Sliver stretch in the creel in speed frames due to too high a creel draft has to be avoided. Optimum creel tension draft should be selected to control sagging or stretch in drawing sliver. Creel normally consists of four to six rows of guide rollers fitted with smooth plastic sliver guides running along the length of the speed frame. The draw frame sliver cans are arranged in four or six rows in the creel zone. The position of the plastic sliver guides in the guide rod corresponding to the drafting head is important in the control of stretch. The sliver guides should be checked during machine maintenance to ensure the production of uniform roving. The speed of the creel guide roller is crucial for combed sliver as it lacks cohesion. The wrong selection of creel tension draft for combed sliver may lead to more creel breakages. In modern fly frames, the creel transport rollers are arranged without vertical supporting rods, and these types of creels are called telescopic creels. This type of arrangement enables placement of cans without any hindrance, and also the movement of the machine operators is easy.

If a speed frame has 120 spindles, the cans fed at the creel may be in batches of 30 with different sliver content in the cans. This makes the sliver can replenishment time as minimum as possible, which does not affect productivity much. Sliver distribution from a single can during severe back material shortage is to be avoided to ensure better quality roving. Topping cans with the last few layers of the previous cans lead to production loss. When the slivers are spliced, the mass is usually not acceptable, leading to a quality stop. The malfunction of any one creel guide roller may affect the roving quality in terms of false drafts. The correct functioning of all creel guide rollers should be ensured periodically. The drive systems of creel rollers should be maintained properly during cleaning.

6.3.2 Drafting System

The purpose of drafting is to attenuate or reduce the weight per unit length of the feed material to the required fineness or count. The drafting system in the speed frame, as shown in Figure 6.2, usually consists of top rollers, bottom rollers, top arm, and the associated parts. The important parts of drafting system of speed frame are given as follows:

- Bottom rollers
- Rubber-covered top rollers
- Aprons (top and bottom)
- Spacer
- Condenser (inlet and floating)
- Top arm

Shankaranarayana[1] reported that eccentric drafting rollers, hard cots, lower pressure on top rollers, and high roving irregularity increase the thick and thin places in the yarn.

6.3.2.1 Bottom Rollers

Bottom rollers have opposite helix flutes for zero axial thrust. The exact circularity of bottom rollers and top rollers prevents roving breakages, and it can be duly interchanged. The rollers are completely hard chrome plated that helps to reduce the lapping. The bottom

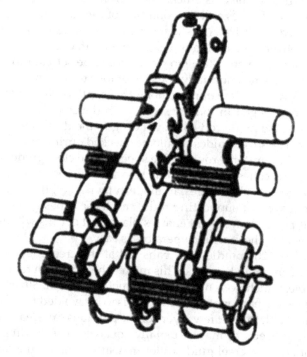

FIGURE 6.2
Speed frame—drafting systems.

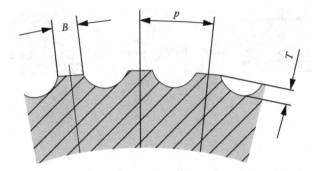

FIGURE 6.3
Bottom roller—flute parameters.

rollers of speed frame exercise immense influence on the quality of roving. Eccentric or damaged bottom rollers, especially front bottom rollers, are the most common cause of unnecessary roving faults and excessive roller laps. Precise concentricity of a bottom roller is a function of improved roving quality (evenness and strength). By means of electronic straightening, bottom rollers reach maximum concentricity. The running conditions of top and bottom rollers are equally influential to roving breakages, etc. Bottom rollers having narrow tolerances of dimensions B, p, and T (Figure 6.3) produce fault-free roving.

The spectrogram of yarn evenness produced from a faulty bottom roller and a fault-free bottom roller is shown in Figure 6.4a and b, respectively.

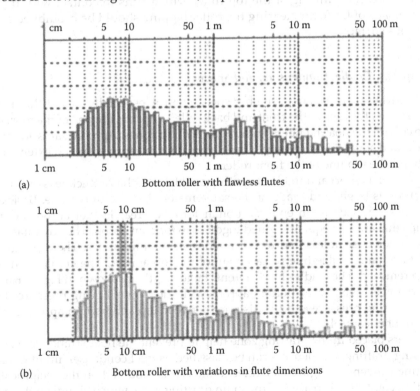

FIGURE 6.4
Spectrogram of bottom roller: (a) flawless flute and (b) flawed flute.

FIGURE 6.5
Top roller.

6.3.2.2 Top Rollers

The smooth running of the top roller with its direct contact to the roving influences the drafting result and therewith the yarn quality achieved. Top rollers are held strictly parallel to and in perfect alignment with bottom rollers. Top rollers (Figure 6.5) covered with rubber cots play a significant role in the control of drafting irregularities. The shore hardness of the rubber cots should be as per recommendations to control the fibers effectively. The shore hardness of top rollers depends upon the type of material and the process. Grinding of top roller cots should be performed with utmost care. Maintenance of top rollers such as top roller greasing and cots grinding should be done as per norms. The diameter should not be reduced by more than 3 mm to ensure sufficient loading pressure. Top roller setting can be adjusted with the top arms in their loaded position. The fiber or dust accumulation in the top roller neck should be cleaned frequently using picker gun. The dust accumulation in the neck of the top roller may resist the running of the top roller which runs in contact with the bottom roller. The usage of knife for clearing the roller lapping should be prohibited as it damages the cot's surface.

6.3.2.3 Aprons, Cradle, Condensers, and Spacer

Efficient drafting requires effective fiber speed control, especially that of the short fibers floating between the nips of the front and back rollers. Aprons are one of the most effective means to control the floating fibers within the drafting zones. Apron wear is accelerated by high drafts and sliver linear density. It is essential that the aprons should extend as closely as possible to the nip line of the front rollers.

The top apron is short and made of synthetic rubber that has a thickness of about 1 mm. Bottom apron is larger and made of the same material as the upper one. Basically, synthetic aprons are made in an endless tubular form whereas leather aprons are made in open strips that are subsequently glued together to form an apron. The advantage of tubular construction is seamless and uniform along its circumference. The top apron cradle ensures quick and trouble-free replacement of the top aprons. The cradles can be easily fitted and removed. The cradles for different staple length are shown in Figure 6.6. The top aprons are forced against the bottom aprons with the help of spring pressure. The combination of spring pressure and the distance between top and bottom aprons decides the intensity of fiber control.

Condensers placed in the drafting zone help to prevent the fiber strand from spreading apart during drafting. Condensers can be classified as inlet condenser, middle condenser, and floating condenser (Figure 6.7). Inlet condenser is mounted on the reciprocating bar, and floating condenser is placed in the main drafting zone, which significantly influences

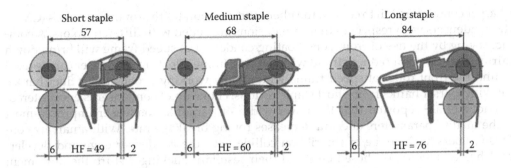

FIGURE 6.6
Cradles for different staple length.

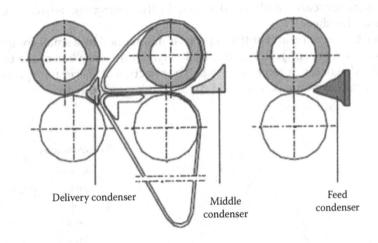

FIGURE 6.7
Types of condensers.

the quality of the roving produced as well as the running performance of the speed frame machine. The condenser size is selected based on the fibers being processed and the thickness of the material passing through the draft zone (Table 6.1). Reciprocating movement of the bar helps to spread the wear over the whole width of the top roller rubber cots. The selection of smaller size condenser for a coarser hank material leads to uncontrolled stretching and fiber accumulations.

TABLE 6.1
Condenser Size and Color for Different Sliver Hanks

Condenser Type	Sliver Hank: 0.16–0.12		Sliver Hank: 0.12–0.10	
	Size (mm)	Color Code	Size (mm)	Color Code
Inlet	12	Black	14	Red
Middle	10	Grey	12	White
Floating	10	Grey	12	White

Larger condensers fail to control the fibers and deteriorates the roving evenness (C.V%). The condensers with respect to their dimensions are coded with different colors. A compact roving by the use of front zone floating condenser at speed frame will bring down hairiness, as this will reduce strand width at ring frame. Floating condenser can be used behind the front roller at speed frame without any working problems for finer hanks, but with coarser hanks from short staple cottons, choke up of condenser is encountered. Ishtiaque et al.[2] reported that the lower width of a middle condenser improves most of the quality parameters and but decreases roving breakage rate. With a narrow condenser, fibers at the edge of the ribbon collide with the condenser edge and decelerate. The deceleration of these fibers and their resultant bucking disturb the movement of other fibers during drafting, which deteriorates the roving evenness. With a very wide condenser, the width of the fiber ribbon becomes too large, and, as a consequence, interfiber frictional contacts decrease, which leads to drafting irregularities. Yarn imperfections decrease with the decrease in condenser size. Ishtiaque et al.[2] also reported that the yarn tenacity increases with the decrease in the roving condenser size due to the compactness of the fibers.

The distance between top and bottom aprons is maintained by a small component called "cradle spacer" or "spacer," which is inserted between the nose bar of the bottom apron and the cradle edge of top apron (Figure 6.8). The selection of spacer for a process depends on the hank of the sliver, break draft, and roving hank (Table 6.2).

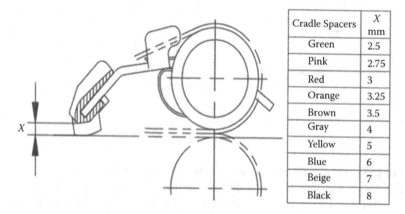

Cradle Spacers	X mm
Green	2.5
Pink	2.75
Red	3
Orange	3.25
Brown	3.5
Gray	4
Yellow	5
Blue	6
Beige	7
Black	8

FIGURE 6.8
Cradle spacer and its dimensions.

TABLE 6.2

Cradle Opening or Spacer Size for Different Roving Hanks

Roving Hank	Spacer Size X (mm)
Up to 1.0	7–8
1.1–1.8	6–7
1.8–2.5	4–6
2.5 and above	3–4

TABLE 6.3

Advantages and Disadvantages of Reducing the Spacer Size

Advantages	Disadvantages
Improves the uniformity of roving	Affects the running behavior of speed frame
Reduces the imperfection level	Drafting problems
	Generation of slubs due to overcontrol of fibers

The reduction in the size of the spacer may have several advantages and disadvantages as mentioned in Table 6.3.

6.3.2.4 Top Arm Loading

Spring loaded top arms are normally adopted in speed frame to get optimum pressure on top rollers to achieve required quality and performance. The top arm pressure and roller setting influence the roving quality and subsequently yarn quality.

Yarn unevenness and imperfections show an initial decrease to an extent with an increase in the two aforementioned parameters. In general, the moderate level of top arm pressure and roller setting gives better results. With the PK5000 (TEXparts®),[3] the weighting pressures on the top rollers are adjusted infinitely and centrally using a compressed air treatment system, which provides constant loading at all spinning positions of the roving frame (Figure 6.9). The individual weighting arms are linked by connecting hoses to each other and to the air supply system. End pieces at both ends of the roving frame (first and last weighting arms) close off the air supply system. The pressure setting and system monitoring are performed centrally at the pneumatic unit installed in the machine control. The closed-circuit compressed air system of the PK5000 ensures the same loading conditions in all arms and elements. The centralized pressure setting permits infinite and rapid adaptation to the technological requirements of the material to be spun (Figure 6.10).

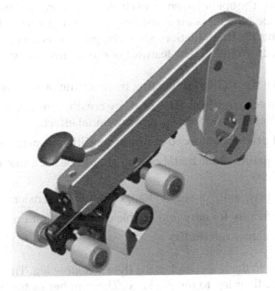

FIGURE 6.9
PK5025 top arm for speed frame drafting system.

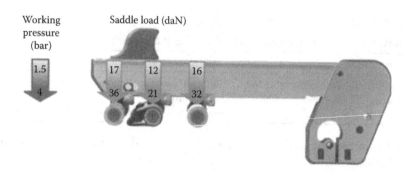

FIGURE 6.10
Top arm loading—working pressure.

A study on the influence of top roller loading conducted by Ishtiaque et al.[2] reported that the increase in top roller loading initially decreases the roving U% and then increases it. The reason attributed to this trend is that the initial increase in top roller loading narrows down the gap between the pressure fields of back and front beard of fibers and exerts better control over the fibers, which leads to reduction in U%. At higher top roller loading, there may be overlapping between the front and back pressure fields in the main drafting zone, which hinders the smooth fiber motion, which results in high roving U%. This study also reported that the increase in top roller loading first reduces and then increases the roving breakage rate and imperfections.[2] Top arm loading in speed frame does not have any significant influence on yarn tenacity as reported in this study.[2]

6.3.3 Flyer and Spindle

The rotation of flyer facilitates false twister fitted on its top to impart twist to the roving. A perfect balance during the operation ensures consistent roving and prevents wear on parts. The flyers' special shape offers less air resistance, preventing roving breakages. Antistatic coating on flyer prevents fly accumulation, and light, precision-cast construction withstands the rigors of high-speed revolution. The features of a good flyer are summarized as follows:

- It should improve the quality of twisting by inserting false twist.
- No chance of the occurrence of false draft by creating minimum resistance to the flow of roving, that is, minimum surface frictional effect.
- The flyer should produce balanced running condition, especially at higher speed.
- The design and quality of the metal of flyer should be such that there is no chance of spreading of flyer.
- There should be provision for slight changes in roving tension.
- There should be facility for easy and simple doffing operation.
- It should be maintenance friendly.

Presser arm is attached to the lower end of the flyer's hollow leg. The arm has to guide the roving from the exit of flyer leg to the package. The number of turns around the presser arm determines the roving tension and package hardness (Figure 6.11). If it is high, then a compact package is obtained. The number of turns depends upon the fiber type and twist in the roving.

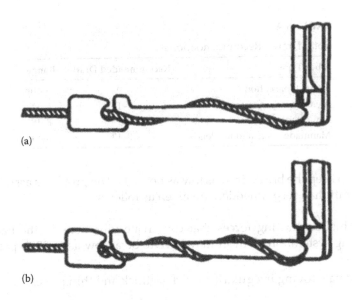

FIGURE 6.11
Presser arm: (a) two wraps and (b) three wraps.

The spindle is a long vertical cylindrical shaft on which the flyer is mounted at the top. It is a long steel shaft mounted at its lower end in a bearing and supported in the middle by a vertically reciprocating shaft of the package tube acting as the neck bearing. Flyer through spindle is driven by a set of gears housed in the spindle rail, which is stationary. Spindle speed and flyer speed are same. The bobbin or the package over which the roving material is wound is loosely mounted over the spindle. The bobbin gets its drive from another set of gears housed in the bobbin rail, which moves vertically up and down. The laying of roving coils onto the surface of the bobbin is achieved by the movement of bobbin rail. The flyer speeds employed normally range from about 1000 to 1400 rpm. According to the studies conducted by Shulz[4] and Bohmer,[5] roving irregularity increases with the increase in flyer speed due to higher flyer vibration. The reason attributed to the aforementioned trend is the fact that an increase in flyer vibration increases the centrifugal force and air resistance in the flyer leg, and there will be a shearing action of roving with the flyer eye. Ishtiaque et al.[2] reported that the increase in flyer speed increases the roving force in conjunction with the friction condition in the flyer leg, which is responsible for more breakages.

6.4 Draft Distribution

Drafting takes place by fiber straightening, fiber elongation, and fiber sliding (relative movement). The draft given in the speed frame is usually calculated based on the roving hank produced. In a 3-over-3 drafting system, the first drafting zone, referred to as "break-draft," is in the range of 1.03–2.03, while the main draft is much higher, the total draft being the product of the two, generally ranging from about 5 to 18 (Table 6.4).

The break draft facilitates the reduction of interfiber cohesion and frictional forces, thereby facilitating the fibers sliding past each other during the subsequent drafting. It is

TABLE 6.4

Total Draft—Recommendations

Fiber Type	Recommended Draft	Range
Short staple cotton	6–9	5–10
Medium staple cotton	7–12	6–14
Long staple cotton	9–18	8–18
Manmade fibers and its blends	8–13	8–14

recommended to keep the break draft as low as possible. The problems associated with the higher break draft than recommended are given as follows:

- Requires higher drafting forces that can create vibrations in the back zone of the drafting system. Break draft may have to be as low as 1.022 to prevent roller vibrations.
- Tends to create roving irregularities such as thick and thin places.

Drafting generally introduces its own unevenness, increasing sliver unevenness to varying extents. Good drafting requires effective fiber speed control, particularly that of the short fibers floating between the nips of the front and back rollers. Aprons represent one of the most effective and popular means of controlling the movement of the floating fibers within the drafting zones. The process of the attenuation of linear fiber assemblies by roller drafting causes a tension to be generated in the fibers in the drafting zone. Dutta et al.[6] reported that the drafting force and its variability are important characteristics that determine the irregularity added during drafting, the number of faults generated, and the drafting failures. The force necessary to give rise to the average tension in the moving fiber mass in the drafting zone is referred to as the drafting force. Drafting force of roving has been found to affect spinning efficiency. The various fiber parameters influencing the drafting force are given in Table 6.5.

According to the investigation by Das et al.,[7] the drafting force initially increases with the draft and then declines sharply as the draft increases further. This is due to the fact that at the lower level of draft, very little fiber slippage occurs due to elastic behavior of fiber strand and the fibers are simply straightening out (removal of crimp and hooks). The maximum drafting force is observed at different drafts for different roller settings. With further increase in draft, the principal mode of roving deformation is the sliding of fibers relative to one another, because the static friction is fully overcome, and hence, after the peak region, the drafting force declines quickly.

TABLE 6.5

Fiber Parameters Influencing Drafting Force

Fiber length
Fiber fineness
Fiber-to-fiber friction
Fiber parallelization
Packing factor
Twist
Fiber irregularity

TABLE 6.6

Machine Parameters Influencing Drafting Force

Draft ratio
Drafting speed
Roller setting

The various machine factors influencing the drafting force are given in Table 6.6. At the higher level of draft, the drafting force generated is due to the dynamic friction of fiber that is lesser than the static friction. At higher draft and drafting speed, the control over the fibers goes down, and chances of fiber shuffling become less, thereby reducing the drafting force. Das et al.[7] reported that the drafting force always reduces with the increase in roller setting. The aforementioned trend is due to the fact that at lower roller setting, the control over the movement of floating fibers becomes more because of the high interfiber cohesion. But as the roller setting increases, the interfiber cohesion goes down and, hence, the drafting force reduces.

6.5 Twist

Twisting in the speed frame is the process of rotating the fibrous strand about its own axis so that the fibers are arranged in a spiral form and thus bind each other together. The purpose of providing twist in roving is to give the strand sufficient strength to withstand the strain during unwinding in the creel of the ring frame. The insertion of twist is achieved by the rotation of the flyer. Twist level depends on flyer speed and delivery speed of the speed frame. The increase in twist reduces the productive capacity of the machine, so it is generally used in a range as limited as possible. The relationship between the twist and the aforementioned factors is given as follows:

$$\text{Twist} = \frac{\text{Flyer speed or spindle speed (rpm)}}{\text{Delivery speed (m/min)}}$$

False twisting devices as shown in Figure 6.12 are used on the flyers to add false twist when the roving is twisted between the front roller and the flyer. Because of this supplementary twist, the roving is strongly twisted, and this reduces the breakage rate. False twisting device is also called "twist crown."

The level of twist imparted in the speed frame process varies with the staple of the cotton and the hank of the roving. Longer cotton requires less twist because individual fibers extend further in the strand and thus help to bind them together more securely than do short fibers. Finer roving requires more twist compared with coarser rovings. In mill practice, the level of twist is normally judged by the way the roving acts. Roving must have sufficient twist to give strength to turn the bobbin in the creel of the ring frame without having it break. The lower twist level reduces the production and efficiency of the machine due to higher roving breakages. Higher twist in speed frame reduces the production rate and thus increases the cost of production. Ishtiaque and Vijay[8] reported that the increase in roving twist increases the interfiber friction due to more contact areas, which creates a problem during ring frame drafting and ultimately

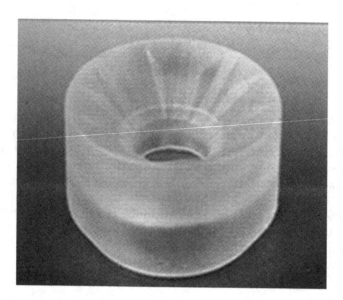

FIGURE 6.12
False twisting device.

deteriorates the yarn quality. Ishtiaque et al.[9] have concluded that the yarn irregularity and total imperfections increase with an increase in fiber-to-fiber friction. In another study, Ishtiaque et al.[9] have shown that the increase in bottom apron slippage is a result of an increase in roving twist multiplier (TM). Balasubramanian[10] states that one of the reasons for the stretch of strand in the creel is low roving twist. Bay and Baier[11] reported that high twist roving can be withdrawn without creel stretch in ring frame. The high twist roving has the advantage of large bobbin size and fewer roving brakes in speed frame so that roving uniformity can be improved. Su and Lo[12] reported that the high twist roving could not be sufficiently drafted in the back roller zone, resulting in thick places in the spun yarn. Basu and Gotipamul[13] reported that with the increase in break draft, the yarn imperfection also increases, and they further state that the yarn imperfection increases due to the increasing roving twist multiplier for the same yarn count. The problems associated with the high roving twist w.r.t process are as follows:

- Roving may not get properly drafted in the ring frame break draft zone.
- Undrafted roving will be pulled through the system and called "hard ends."
- More ends down due to hard ends.
- Ends down at the roving frame.

The speed frames are generally fitted with two rows of the flyer. The amount of false twist inserted by the flyer inlet is dependent on the contact angle of the roving length between the flyer and the front drafting rollers; the smaller the angle, the higher the false twist. The back row flyers are nearer to the drafting rollers and have the greater contact angle (Figure 6.13). The false twist given by the back row flyer does not provide sufficient cohesion to the roving, which reduces the roving hank slightly. There is a significant count variation between the rovings produced from the front row and back row flyers. In order to overcome the above problem, modern speed frames have the back row flyer fitted with a raised false twister attachment providing the contact angles of the two row to almost similar.

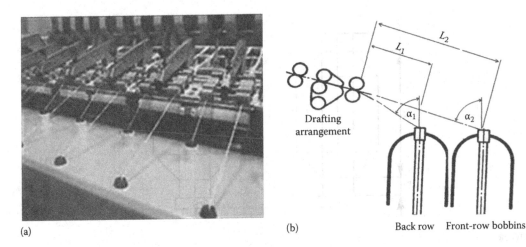

FIGURE 6.13
Contact angle—front and back row flyers (a) and (b).

6.6 Bobbin Formation

Bobbin formation is one of the important tasks performed in the speed frame machine. The roving is drawn from the front roller through the flyer and onto the bobbin. The bobbin has a higher surface speed than the flyer, which winds the twisted roving onto the bobbin (Figure 6.14). The centrifugal forces increase as the bobbin diameter increases. The rate of winding, compared with the front roller delivery rate, plays a major role in controlling the roving tension. There is no precise method for measuring the roving tension.

It is customary to say in spinning mills that tension is "tight" if there is considerable pull between the front roller and the bobbin. On the other hand, the tension is "slack" or "loose" when the bobbin hardly winds the roving delivered by front roller. The setting of proper roving tension is crucial in the bobbin formation, which depends mainly on judgment and experience.

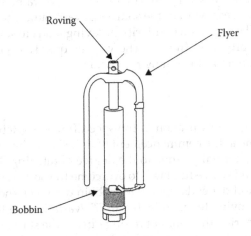

FIGURE 6.14
Flyer and bobbin.

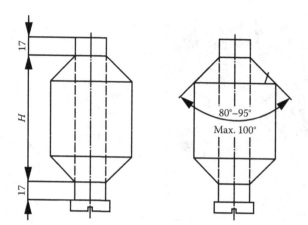

FIGURE 6.15
Roving bobbin—taper and dimensions.

6.6.1 Taper Formation

The bobbin obtained from speed frame is a cylindrical package with conical ends. The drafted fibrous strand is wound onto an empty package, which is usually made of plastic. Each roving layer is made of helical coils arranged close to each other, which are attained by the up-and-down movement of the bobbin. After the completion of one layer, the bobbin's direction of movement is reversed to start a next layer. The up-and-down movement of the bobbin for layering a layer is called "traverse." The taper is achieved by reducing the amount of traverse after the completion of each layer (Figure 6.15).

6.7 Quality Control of Roving

The control of roving quality is very essential since speed frame is the final stage of the spinning preparatory process. Most of the sources of yarn faults are mainly due to the bad roving bobbin quality. The higher irregularity of roving tends to severely affect the tensile properties of the yarn produced from it. The various quality aspects of roving are discussed in a detailed manner in the following sections.

6.7.1 Ratching

The main source of long-term variation in the speed frame is ratching. The tension draft given in the speed frame at the commencement of doff should be less than 1%. The initial layers of the bobbin are normally prone to the degree of ratching. The bobbin rail should be set up or down to wind with a full layer to overcome the occurrence of ratching. Roving tension at the start of doff depends upon bare bobbin diameter and cone drum belt position. A 5% ratching in roving tends to increase 15% variations in the yarn. Ratching can be determined by comparing the count of roving during the start and end of the doff. The position of the cone-drum belt at the time of full doff also gives an indication of ratching. If the belt is near its extreme position, the ratchet wheel is almost correct.

6.7.1.1 Procedure to Determine Ratching% in Roving

- Collect the full bobbins from the spindles during doff in the speed frame.
- Run the machine after the doff for 2–5 min.
- Collect the bobbins with two initial layers from the same spindles where full bobbins were collected previously.
- Sample length for evaluating roving hank is 15–30 yards.
- Find the roving hank from the bobbins with two initial layers ($H1$).
- Find the roving hank from the full bobbin ($H2$).
- Ratching% = $((H1 - H2)/((H1 + H2)/2)) \times 100$.
- Ratching should not exceed 1.5%.

6.7.2 Roving Strength

Roving strength is an important quality parameter that is required for trouble-free unwinding in ring frame. It plays a significant role in the optimization of roving twist multiplier. Some of the poor work practices attributed for poor roving strength are as follows:

- Improper sliver piecing at creel
- Improper roving piecing
- Stretch at creel
- Poor condition of spindle

6.7.3 Count C.V%

Roving hank or count C.V% can be categorized as within-bobbin C.V% and between-bobbin C.V%. The sample length taken for testing roving count C.V% is normally 15 yards. The count C.V% of roving falls in the range of 1.5%–2.0% under good working conditions. Ratching within lengths of about 100–300 m of rove would produce a higher effect on within-bobbin C.V%, while ratching over higher lengths would affect between-bobbin C.V% to a great extent. Tension variations from start to end of the doff causes roving hank variation. The various reasons for the occurrence of the count variation (within- and between-bobbin) are summarized in Table 6.7.

6.7.4 Unevenness

Unevenness or irregularity of roving has a predominant influence on the yarn quality. The rovings with high irregularity cause more erratic movement of fibers, resulting in more

TABLE 6.7

Reasons for Higher Count C.V% (Within-Bobbin and Between-Bobbin)

Various Reasons for Count C.V%	
Within-Bobbin	**Between-Bobbin**
Uneven tension during bobbin build-up	Variation in drafting pressure
Sliver splitting and stretching at creel zone	Quality of sliver (1 m C.V%)
Roller lapping	
Defective spindle	
High C.V% in sliver	

drafting force variability. The control of short-term irregularity and end breaks is very important in speed frame. The various factors influencing the unevenness of the roving are as follows:

- Quality of cots
- Aprons
- Spacer size
- Draft
- Condenser size
- Break draft
- Top arm pressure

6.8 Defects in Roving

6.8.1 Higher *U*% of Rove

The various causes of higher roving unevenness in speed frame and the remedial measures taken to control the fault occurrence are given in Table 6.8.

6.8.2 Higher Roving Breakages

The various causes of higher roving breakages in the speed frame and the remedial measures taken to control it are given in Table 6.9.

6.8.3 Soft Bobbins

The various causes of soft bobbins in the roving process and the remedial measures taken to control the fault occurrence are given in Table 6.10.

TABLE 6.8

Higher Roving *U*%

Causes	Remedies
Inadequate top arm pressures	Refer to the spectrogram and check the
Improper settings	spindle and the feed material before
Worn-out gears or bearings	taking any action. If any machine part is
Grooved top rollers, tilted top rollers	found faulty, correct it.
Wrong selection of condensers	
Worn-out aprons	
Poor cleaning of draft zone	
Higher stretch	
Uneven feed material	
Sliver splitting in creel, jerks in creel movement	
Vibrations in the machines	

TABLE 6.9

Higher Roving Breakages

Causes	Remedies
Uneven sliver	Supply of high-quality sliver with lower C.V%
Worn-out parts, damaged machine parts, and vibrations	Replacement of worn-out parts; proper maintenance of machine parts
Insufficient twist	Imparting twist as per norms
Improper draft distribution	Proper draft distribution
Fluctuations in R.H%	Proper follow-up of R.H%
Rough surface in the flyer tube	Polishing the surface of flyer tube
Improper build of bobbins	Proper maintenance of builder motion
Improper piecing of sliver	Proper work practices for correct piecing
Uncontrolled air current, etc.	Arresting uncontrolled air currents
Higher spindle speed than recommended	Speeds and settings as per recommendations

TABLE 6.10

Soft Bobbins

Causes	Remedies
Too finer hank	Roving hank should be optimized.
Singles or a finer drawing hank	Avoid singles; optimize drawing hank
Less number of turns on presser arm	Number of turns on presser arm as per recommendations
Belt shift on cone drum faster than required	Optimize the rate of belt shift on cone drum
Lower twist	Optimum twist
Lower relative humidity	Maintain proper R.H%

TABLE 6.11

Lashing-In

Causes	Remedies
Broken end joins adjacent end and creates lashing-in.	Provision of separators and roving end catcher solve this problem.
	Set the suction tube near the front roller nip.
	Reduce the end breakage rate.

6.8.4 Lashing-In

The various causes of lashing-in in the roving process and the remedial measures taken to control the fault occurrence are given in Table 6.11.

6.8.5 Hard Bobbins

The various causes of hard bobbins in the roving process and the remedial measures taken to control the fault occurrence are given in Table 6.12.

6.8.6 Oozed-Out Bobbins

The various causes of oozed-out bobbins in the roving process and the remedial measures taken to control the fault occurrence are given in Table 6.13.

TABLE 6.12

Hard Bobbins

Causes	Remedies
Coarser hank	Optimize the roving hank.
Doubles or coarser draw frame hank	Correct doubles and maintain proper draw frame hank.
Lower top arm loading	Optimize twist.
Higher twist	Proper rate of movement of belt on cone drums.
Lesser movement of belt on cone drums	Recommended number of turns on flyer presser arm.
More number of turns on flyer presser arm	Maintain proper R.H%.
Higher R.H%	

TABLE 6.13

Oozed-Out Bobbins

Causes	Remedies
Malfunction of reversing bevels in the builder motion	Correct the malfunction in builder motion.
Stopping the machine when the bobbin rails are in extreme positions	Stop the machine when bobbin rails are in middle position.
Jumping bobbin	Proper quality of empty bobbin to be maintained.

TABLE 6.14

Higher Roving Count C.V%

Causes	Remedies
Insufficient top arm pressure.	Set top arm pressure, break draft and condenser guides as per norms.
Too high a break draft.	
Improper selection of condenser guides.	Check the proper seating of bobbin on pin.
Vibration of roving bobbin while running.	Control stretch at creel.
Sliver stretch at creel.	Proper selection of wheels.
High variation in empty bobbin diameter.	Check the belt position on cone drum and correct.
Wrong selection of winding-on wheel and ratchet wheel.	Proper cleaning of drafting zone.
Improper shifting of cone drum belt.	Proper maintenance of flyer and bobbin rail.
Disturbance in the movement of aprons.	
Rough spots on flyer.	
Bobbin rail movement is not uniform.	

6.8.7 High Roving Count C.V%

The various causes of higher roving count C.V% in the roving process and the remedial measures taken to control the fault occurrence are given in Table 6.14.

6.8.8 Roller Lapping

The various causes of roller lapping in the roving process and the remedial measures taken to control the fault occurrence are given in Table 6.15.

TABLE 6.15

Roller Lapping

Causes	Remedies
Too high spindle speed and draft	Set speeds, settings, and drafts as per norms.
Cuts or damages in the top roller cots	Avoid usage of knife while clearing roller lapping.
Damages in apron, condenser, bottom roller	Proper maintenance of top roller cots, apron, condenser,
Wrong choice of spacer	bottom roller.
Wider setting at the back zone	Select spacer size as per norms.
Improper R.H%	Maintain R.H% as per norms.

TABLE 6.16

Slubs

Causes	Remedies
Too high end breakages	Minimize end breakages.
Waste accumulation at creel, drafting zone, and flyer	Proper cleaning of machine parts during shifts.
Wrong selection of spacer	Proper selection of spacer, break draft, and back
Too low break draft	zone setting.
Closer setting at the back zone	Ensure proper running of clearer roller.
Top and bottom clearer not functioning properly	

6.8.9 Slubs

The various causes of slubs in the roving process and the remedial measures taken to control the fault occurrence are given in Table 6.16.

6.9 Technological Developments in Speed Frame

The machinery developments in the speed frame are significantly low compared with other machines of the spinning process. This is proved by the fact that the spindle speed of the speed frame has attained only 1500 rpm as on date compared with 600 rpm in the 1950s.[14] In terms of production and quality, increase in the roving bobbin diameter from 4" to 7" and lift from 8" to 16", the use of straight cone drum instead of hyperbolic cone drum for better control over the roving tension, etc., are the significant developments that occurred in the last two decades. Almost all the latest speed frames are fitted with closed (AC type) flyers that are used to overcome the problem of air drag on roving. These flyers are aerodynamically balanced and are lightweight. The roving frames are equipped with autodoffing system that, apart from avoiding man handling, reduces doffing time. Even the Toyota claims to have autodoffing on FL100 roving frame in the record time of 3½ min.[15] Roving bobbins autodoffing and transportation, to the ring spinning through overhead rails, becomes a standard feature of the roving frame.

FIGURE 6.16
Roving tension sensor—Rieter F35 Roving frame. (From Rieter, Rieter speed frame manual, Rieter, Winterthur, Switzerland, 2013. With permission.)

In multimotor drive system, drafting rollers, flyers, bobbins, and bobbin rail are driven directly by individual servomotors and are synchronized throughout package build by the control system. The advantages of this system include no need of heavy counter weight for bobbin rail balancing and differential gear, reduced maintenance, and lower energy consumption.

Roving tension sensors, as shown in Figure 6.16, measures and controls the roving tension (constant) throughout the bobbin build. These tension sensors do not actually contact the roving while measuring the tension. The tension is measured at periodic intervals, and the required change in tension is actuated by changing the bobbin speed through servomotor. Rieter F15/F35 roving frame, Zinser 668 roving frame, Marzoli FTN roving frame, Lakshmi LFS 1660 speed frame, and Toyota FL100 roving frame have incorporated roving tension sensor on their machines.[15-19]

Automated bobbin transport, as shown in Figure 6.17, offers the advantages of labor savings and a substantial increase in bobbin quality. The roving bobbin is one of the most delicate intermediate products to handle for two reasons: the roving wound around the bobbin is completely unprotected and is therefore highly susceptible to damage; and all roving defects are transferred to the yarn and cannot be corrected. Automated bobbin transport eliminates the need to handle the bobbin or touch the textile product and to maintain intermediate storage areas, where bobbins can accidentally age, get dirty, and deteriorate.[20] The train of bobbins is automatically transported from the speed frame to the storage area and the respective ring frame by selecting the appropriate program in the PLC. Empty bobbins in the ring frame are manually interchanged with full bobbins from bobbin transport system (BTS) by the operator, and empty bobbins are transported back to the speed frame automatically.

FIGURE 6.17
Roving transport system.

References

1. Shankaranarayana, K.S., *Proceedings of the 14th Joint Technological Conference* (ATIRA/BTRA/ SITRA), Coimbatore, India, 1973, p. 55.
2. Ishtiaque, S.M., Rengasamy, R.S., and Ghosh, A., Optimization of speed frame parameters for better yarn quality and production. *Indian Journal of Fibre & Textile Research* 29(1): 39–43, January 2004.
3. PK5000 Innovative drafting system technology, Texparts Information brochure, http://texparts. saurer.com/fileadmin/Texparts/Dokumente/Texparts_PK_5000_Series_en.pdf, Accessed Date: May 1, 2013.
4. Shulz, G., Automation of the flyer frame in conjunction with the ring spinning machine, *Melliand Textilberichte* 68(8): E238, 1987.
5. Bohmer, I., Modernization of flyers, cotton and combed yarn ring spinning machines, *Melliand Textilberichte* 73(6): E242, 1992.
6. Dutta, B., Salhotra, K.R., and Qureshi, A.W., Blended textiles, Paper presented at the *38th All India Textile Conference*, Mumbai, India, November 1981.
7. Das, A., Ishtiaque, S.M., and Kumar, R., Study on drafting force of roving: Part I—Effect of process variables. *Indian Journal of Fibre & Textile Research* 29: 173–178, June 2004.
8. Ishtiaque, S.M. and Vijay, A., Optimization of ring frame parameters for coarser preparatory. *Indian Journal of Fibre & Textile Research* 19(4): 239–246, 1994.
9. Ishtiaque, S.M., Das, A., and Niyogi, R., Optimization of fiber friction, top arm pressure and roller setting at various drafting stages. *Textile Research Journal* 76(12): 913–921, December 2006.
10. Balasubramanian, N., Controlling count variation in yarn, http://Business.vsnl.com/ balasubramanian/index.html, 2006, Accessed Date: May 1, 2013.
11. Bay, E. and Baier, F., Modern drafting system for improving flexibility in ring spinning. *International Textile Bulletin* 42: 64–70, 1996.

12. Su, C.I. and Lo, K.J., Optimum drafting conditions of fine—Denier polyester spun yarn. *Textile Research Journal* 70(2): 93–97, 2000.

13. Basu, A. and Gotipamul, R., Effect of some ring spinning and winding parameters on extra-sensitive yarn imperfections. *Indian Journal of Fibre & Textile Research* 30(2): 211–214, 2005.

14. Singh, R.P. and Kothari, V.K., Developments in comber, speed frame & ring frame. *The Indian Textile Journal*, May 2009, http://www.indiantextilejournal.com/articles/fadetails.asp?id=2069, Accessed Date: May 1, 2013.

15. Toyota Roving Frame FL100, Toyota Textile Machinery Division, Information brochure, Accessed Date: May 1, 2013.

16. Zinser 668: The roving frame solution for simplified setting and more flexibility, Zinser Textilmaschinen GmbH, Information brochure, Accessed Date: May 1, 2013.

17. Roving Frame F 15/F 35: The solution for your spinning preparation, Rieter Textile System, Information brochure, Accessed Date: May 1, 2013.

18. Spinning Section: Roving Frame FTN, Marzoli Spa, Italy, Information brochure, Accessed Date: May 1, 2013.

19. Spinning Value: Speed Frame LFS 1660, Lakshmi Machine Works Ltd, Information brochure, Accessed Date: May 1, 2013.

20. Bullio, P. and Rodie, J.B., A question of change. *Textile World*, November–December 2006, http://www.textileworld.com/Issues/2006/November-December/Features/A_Question_Of_Change, Accessed Date: November 12, 2013.

21. Rieter, Rieter speed frame manual, Rieter, Winterthur, Switzerland, 2013, Accessed Date: May 1, 2013.

7

Process Control in Ring Spinning

7.1 Significance of Ring Spinning Process

Spinning is the process of producing continuous twisted strands (yarn) of a desired size from fibrous materials. Spinning can be categorized based on staple length as long staple spinning and short staple spinning. Short staple spinning machines process fibers such as cotton, polyester, viscose, and its blends thereof. The yarn formation is usually accomplished by any one of the spinning systems. Spinning systems presently employed include ring, rotor, self-twist, friction, air jet, and twistless and wrap spinning, with the first two systems by far the most important for cotton spinning, together accounting for over 90% of the cotton yarn produced globally.

The ring spinning machine (Figure 7.1) invented by John Thorpe in 1830 has been very successful in producing yarns from staple fibers. Since its inception, ring spinning had remained unchallenged for almost 150 years. Ring spinning accounts for some 75% of global long and short staple yarn production. The main reason attributed for the success of ring spinning over other spinning systems is the superior quality, notably strength and evenness, of ring-spun yarns over those produced by other systems.[1] Other spinning technologies being developed are higher in productivity but are lacking in many aspects of the yarn's desirable characteristics. Ring spinning remains a popular spinning system due to its versatility in terms of yarn count, fiber type, superior quality, and yarn characteristics as a result of good fiber control and orientation. The major reason that limits the twisting rate is the heat generation due to traveler friction with the stationary ring.

7.1.1 Ring Spinning Machine

The ring spinning machine consists of a roller drafting unit, a yarn guide (lappet), a ring and traveler assembly, and a bobbin mounted on a spindle (driven by a tape). The ring spinning machine performs basic operations such as

1. Attenuation (drafting) of the roving to the required yarn count
2. Imparting cohesion to the fibrous strand, usually by twist insertion
3. Winding the yarn onto an appropriate package

Drafting system helps to achieve the attenuation process by means of drafting rollers, top arm, spacer, apron, and other guiding elements. Spindle along with ring and traveler system facilitates the twist insertion process. The winding of yarn on the package is accomplished with the aid of spindle, ring rail, and builder motion mechanisms. A modern spinning machine consists of a huge number of spindles up to a maximum of 1200.

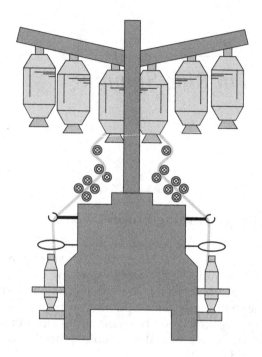

FIGURE 7.1
Ring frame—cross section.

The attenuation of roving is accomplished through a 3-over-3 double-apron drafting system with pneumatic top arm loading. The roller stand is inclined at 45°–60° to the horizontal in order that twisting starts the moment the fiber strand leaves the first draft cylinder. The drafting system is capable of processing fibers of up to 60 mm (cotton, synthetics, and blends) with draft values between 10 and 80. The roving is first subjected to a break draft with values between 1.1 and 1.5 and successively a main draft in the apron zone until the desired count is achieved. The counts spun can be normally in the range of 5–150 Ne. The spindle speed on a modern spinning machine reaches 25,000 rpm, with the possibility of twist insertion ranging 4–80 TPI.

7.2 Influence of Ring Spinning Machine Components on Spinning Process

The machine components of the ring spinning machine have a predominant influence on the productivity and quality of the ring spinning process. Wrong selection of the component and its setting in the ring spinning machine may significantly deteriorate the yarn quality, which in turn affects the productivity also. The influence of the machine components on the ring spinning process is discussed later in a detailed manner.

7.2.1 Creel

The creel has significant effects on ring spinning machine performance in several ways. The creel is the zone where roving bobbins are conveniently suspended over the drafting system. The creel zone of the ring spinning machine essentially consists of bobbin holder

FIGURE 7.2
Ring frame creel—bobbin holders.

(Figure 7.2) and several guide rods. Klein[2] stated that the creel can influence the number of faults in yarn, in particular if the roving bobbin does not unwind perfectly, then false drafts can arise or even ends can break. The main function of a bobbin holder is to control the roving draw-out tension and maintain uniform roving delivery.

The maintenance of the bobbin holder is crucial in controlling yarn faults, especially thin places. The inbuilt braking system of the bobbin holder ensures the aforementioned function. An opposition force that is required to control the overrotation of a bobbin holder for a uniform delivery of roving is called the brake force of the bobbin holder, and it is provided internally under the dust cap.[3] The brake force depends on the material processed, count spun, roving hank fed, lift of the roving bobbin, weight of the roving bobbin, and the spindle speed. The smooth surface finish of the roving guide eyelets, roving hooks, tension rods, position of the tension rods, and smooth movements of the traverse bars are the factors that enhance the performance of the bobbin. In order to maximize the weight of the roving in the creel, the bobbin holders and roving guides have to be precisely located to prevent abrasion of the roving as it moves to the drafting system. The important process control measures to be considered in the creel zone for effective functioning are summarized in Table 7.1.

The fluff protector enhances the performance of the bobbin holder by preventing microdust entry. The rotation of a bobbin holder should be smooth and uniform. The main yarn fault contributed by the bobbin holders is long thin fault caused by creel stretch. Generally in CLASSIMAT results, the reading of H1 and I1 faults denotes the long thin faults. Long thin fault as per CLASSIMAT system is defined as faults that have 30%–40% cross-sectional area of normal cross-sectional area and extends to 32 cm.

TABLE 7.1

Process Control Measures in the Creel Zone

Roving must not rub other roving or roving bobbin.
Bobbins should be loaded evenly with half full bobbins especially while hanging 165 mm diameter bobbins in five rows in a creel.
Roving guide rods must be positioned at correct height to control the roving tension. Higher roving tension leads to count variation in the yarn.
Creel vibration arises due to imbalance of some components.
Creel vibrations accentuate the tendency of the roving to stretch in its path to the drafting zone. This can lead to yarn irregularities and count variation.
The bobbin holders and bobbin brakes, if used, should be clean and function correctly.
Roving tension in the creel should be as uniform as possible.
Overhead cleaner should be maintained properly to prevent the creel from fly accumulation.

TABLE 7.2

Process Control Measures in Roving Guide

The surface of the roving guides must be smooth with low friction.
Roving guides should not scuff the roving.
The width of the roving guides should be sufficient enough to allow free passage of the roving.
The dimension of roving guide should be selected in such a way that it should not create resistance while passing. The resistance of roving while passing through roving guide creates increased yarn hairiness, roving stretch, and subsequently end breakage in spinning.

7.2.2 Roving Guide

Roving guide or roving inlet condenser is mounted on the traverse bar located behind the back pair of drafting rollers. The traverse bar moves slowly to feed the roving over an extended area to spread the wear of the components such as cots and aprons so that their lifetime increases. The traverse motion should be set in such a way that it should ensure that the fibers do not get too close to edges. The important process control measures to be considered in roving guide for effective functioning are summarized in Table 7.2.

7.2.3 Drafting Elements

Drafting or attenuation is the action of progressively increasing the rate at which fibers pass through an operation so that the bulk of the strand is uniformly reduced without breaking its continuity. The attenuation process is carried out in the apron drafting system as in the roving process. Drafting system (Figure 7.3) is the key element in ring spinning, which influences the yarn quality to a greater extent. The roving that enters the back roller nip of the drafting system should be attenuated slightly in the back draft zone so that the mild twist imparted to the roving at speed frame is released and the fibrous strand is prepared for the main drafting in the front drafting zone. In the main drafting zone where the fibrous strand is attenuated to a greater extent, the fibers are guided up to the front roller nip by means of top and bottom aprons. This facilitates better evenness of yarn, controlling the floating fibers. The quality and efficiency of drafting is highly influenced by optimum selection of top and bottom rollers, apron, spacer, top roller rubber cots, top arm pressure, speeds, and settings of drafting rollers. Balasubramanian[4] reported that the irregularities in drafting rose mainly from uncontrolled movement of fibers in the drafting zone when mechanical faults were kept down to a minimum.

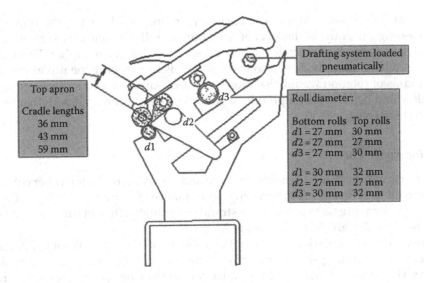

FIGURE 7.3
Ring frame—drafting system.

7.2.3.1 Bottom Rollers

The top and bottom rollers among the drafting elements have a significant influence on yarn quality. Precise concentricity is a function for yarn quality (evenness and strength). The periodic irregularities that are found in the spun yarns may be the result of machinery defects such as eccentric drafting rollers, variability in the covering of drafting rollers, inaccurately cut or worn-out drafting rollers, and the vibration of drafting rollers. An increased level of roller eccentricity results in a higher nip movement. Movement of the roller nip is identified as the primary mechanical cause of irregular drafting with defective fluted rollers of ring frames. In consequence, any forward movement of the nip makes the drafted roving thinner and any backward movement makes it to a large extent thicker, which increases the intensity of the resulting periodic fault and results in a higher index of irregularity (Figure 7.4). The eccentric fluted roller produces uneven and weaker yarn with

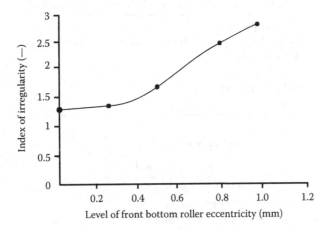

FIGURE 7.4
Effect of bottom roller eccentricity on yarn irregularity.

more thick and thin places and also increases yarn breaks. Moreover, the imperfections also increase significantly, as the level of front bottom roller eccentricity increases.

The eccentricity of a bottom roller is more severe than that of a top roller. This is because the bottom roller is directly driven so that, as well as introducing the nip movement, the varying radius of rotation would also cause the roller surface speed to fluctuate regularly. The result would be of larger amplitude of periodic fault as compared with that caused by an eccentric top roller.

7.2.3.2 Top Roller Cots

Among the various drafting elements, the type and condition of the top roller cot is crucial in deciding the yarn quality and spinning performance. The purpose of top roller cots is to provide uniform pressure on the fiber strand to facilitate efficient drafting. The dimensions of the top roller are shown in Figure 7.5.

The essential characteristics of a top roller rubber cot are given in Table 7.3. Front top roller cot in ring spinning should also offer sufficient pulling force to overcome drafting resistance. The various demands of top roller cots in the ring spinning process are given in Table 7.4.

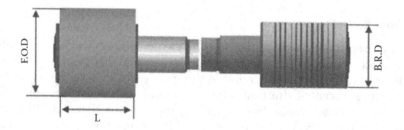

FIGURE 7.5
Top roller cots.

TABLE 7.3

Essential Characteristics of a Top Roller Rubber Cot

Resilience properties
Surface characteristics such as grip offered on fiber strands
Abrasion resistance
Tensile strength
Swelling resistance
Color

TABLE 7.4

Demands of Top Roller Cots

Good fiber guiding
No lap formation
Long working life
Good ageing stability
Minimal film formation

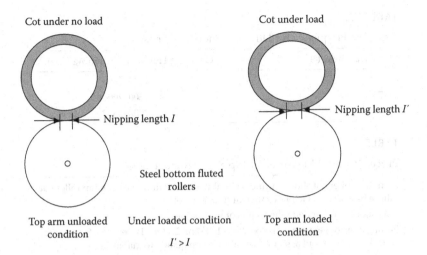

FIGURE 7.6
Top roller cots—load and no-load conditions.

Arc of contact or the nipping length as shown in Figure 7.6 made by top roller cot with fluted roller (*I*) is inversely proportional to the shore hardness of the rubber cot.[5] In general, the lower the shore hardness, the higher will be the contact area with steel bottom roller better so that there will be positive control on fiber's strand producing the yarn with better mass uniformity, lesser imperfection levels. Under identical condition, a cot measuring 56° shore hardness will make larger arc of contact with steel bottom than a cot measuring 90° shore hardness. The recommended shore hardness level to front and back top roller cots for processing cotton and synthetics is given in Table 7.5.

After a period of time, the surface of the cots is damaged and cracks in its circumference. These cracks and uneven surface of the cots will cause the slipping of the fibers during the drafting process. The common way to remove the cracks and roughness of the cots surface is to grind the surface of the cots by a grinding machine. The unevenness of the yarn decreases with the decrease in top rollers' diameter up to an optimum diameter and after that the unevenness of yarns increases rapidly as the top rollers' diameter decreases. The optimum cots, diameter is very important in the period of the cots grinding operation. With the new top roll cots, it is essential to remove at least 0.3 mm in diameter with each buffing cycle. It is advisable to maintain a surface roughness of 0.8–1.0 μm as per standard grinding instructions. The grinding frequency for different shore hardness levels of the top roller cots is given in Table 7.6.

The various process control measures to be considered with respect to top roller cots are summarized in Table 7.7.

TABLE 7.5

Recommended Top Roller Cot Shore Hardness

Fiber Type	Front Roller Cots Shore Hardness (°)	Back Roller Cots Shore Hardness (°)
Cotton	65–75	75–80
Synthetics	75–85	65–75

TABLE 7.6

Grinding Frequency for Different Shore Hardness Levels

Shore Hardness (°)	Grinding Frequency (Running Hours)
60–70	1000–1500
71–75	1500–2000
75–90	2000–2500

TABLE 7.7

Process Control Measures in Top Roller Rubber Cots

Front top roller cot shore hardness should be lower than the back top roller cot shore hardness to facilitate efficient fiber control.

Cots should be grinded when signs of wear appear.

Buffing or grinding time varies between 1000 and 2000 h of operation. Sometimes, buffing is carried out at short interval based on quality requirements.

Care should be taken while greasing top roller as the greasy cots, surface deteriorates the yarn quality.

Softer cots on delivery roller improve fiber control but increase lapping tendency while processing synthetics and its blends.

The back top roll buffing interval can be up to four times longer than that of the front roll cots.

Cradle roller (middle top roller) cots must never be grinded.

7.2.3.3 Top Arm Loading

Drafting quality can also be improved by increasing the pressure on top rollers, as this would help to reduce the incidence of slippage of the material (under the rollers). The weighting arm is required to allow different loads to be set on the top rollers. Normally, this is achieved by helical springs that can be adjusted mechanically in steps. The top rollers are mostly loaded by pneumatic means in the modern spinning machine. The underlying concept of the new weighting arm PK 6000 (SKF®) is to use pneumatic pressure instead of metal springs to generate the load on the top rollers, which can be adjusted infinitely.[6] The pneumatic principle guarantees constant load on all top rollers at every spindle position. The load remains constant even when the diameter of the top rollers decreases as a consequence of grinding the rubber cots of the top rollers. The air pressure is set centrally for all arms at just one point on the ring frame. Thus, load setting and the partial load reduction, necessary during longer standstills, can be done in a second instead of having to operate hundreds of weighting arms. This ensures that the top rollers are not deformed during standstill.

The top arm pressure for fine roving or low-twist roving can be lower compared with coarser roving or high-twist roving or fibers offering higher drafting resistance. Yarn unevenness decreases with the increase in top arm pressure and traveler mass. The increase in top arm pressure consolidates fibrous strand in the drafting zone and fibers move in a more controlled manner so that the erratic movement of floating fibers is restricted, which reduces the yarn unevenness. Higher top roller pressure increases the normal force over the fibers, and this may reduce the fiber slippage and cause some fiber straightening. Also, with the increase in top roller pressure, there will be better control over the movement of fibers during their sliding. These factors are responsible for greater yarn strength.

7.2.3.4 Spacer–Apron Spacing

Fiber control in the drafting zone is critically influenced by the apron spacing. Reduction of apron spacing up to certain level enhances the fiber control. The reduction of apron spacing beyond a certain level leads to frequent drafting faults. Optimum apron spacing would highly depend on top arm loading. The apron spacing in combination with top arm loading has greater influence on imperfections. Apron spacing lies at lower level at higher top arm loading conditions, which tend to improve the evenness of the yarn. Apron spacing, that is, distance between top and bottom apron in the drafting zone, is decided by a small component called "spacer" or "distance clip" or "cradle spacer" (Figure 7.7).

Optimum control of fibers in the drafting field is decided by the appropriate selection of spacers. The yarn quality requirements and spinning performance are highly influenced by the spacer. Thin spacers usually produce better yarn quality in terms of imperfections and evenness. However, the usage of too thin spacer leads to strong restraining forces in between the aprons, which overcontrol the fibers that may severely affect the yarn evenness (Table 7.8).

Caveny and Foster[7] stated that the widened apron spacing deteriorates the regularity and strength of yarn, but the effect was more noticeable only with higher break drafts. Bannot and Balasubramaniam[8] reported that the evenness and total imperfection could be improved by closing down the apron spacing. Basu and Gotipamul[9] reported that the increase in spacer size from 3.0 to 3.5 mm, the imperfections, both at normal level and at extra sensitivity level, decrease, but there is a marginal increase in CLASSIMAT short thick faults.

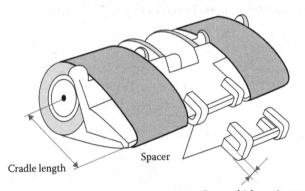

Cradle length · · · Spacer · · · Spacer thickness in mm

FIGURE 7.7
Cradle and spacer.

TABLE 7.8

Problems Associated with Too Thin Spacer

End breakage due to hard ends
High thick and thin places
Variation in drafting cohesion

7.2.4 Lappet

Lappet or pigtail guide acts as guide through which twisted fibrous strand flows from traveler to front roller nip. The adjustment of lappet height has a direct effect on hairiness. The reduction of lappet height reduces the balloon length, which in turn reduces the hairiness tendency in the yarn. If lappet to bobbin tip distance is high, balloon will be longer. This will reduce twist flow due to increase in the area of contact between yarn and lappet. As a result, hairiness will be higher. While setting lappet height, care should be taken to ensure that the yarn does not abrade the empties. Abrasion of yarn against worn-out or grooved lappet aggravates hairiness. Some manufacturers have come out with glass finish lappet, which minimizes friction and thereby reduces hairiness:

$$\text{Lappet height} = 2D + 5 \text{ mm} \quad \text{where } D - \text{ring dia. in mm}$$

Using lesser lift and lesser ring diameter will lead to direct and significant reduction in hairiness. Height of lappet above the ring bobbin has to be optimized to reduce not only end breaks but also hairiness.

7.2.5 Balloon Control Ring

With the developments of ring frame in terms of long bobbins and spindles arranged in narrow gauge, it is necessary to use balloon control rings to control the size of the balloon. The usage of heavy travelers will contain the balloon size, but this leads to high yarn tension and also high ring worn-out. The stability of the yarn balloon formed between the yarn guide and the traveler ring is crucial to the success and economics of the ring spinning process. Balloon control ring as shown in Figure 7.8 is used to contain the balloon by

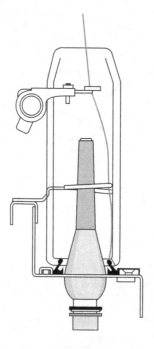

FIGURE 7.8
Balloon control ring.

reducing the yarn tension and decreasing the balloon flutter instability. Flutter instability refers to the uncontrolled changes in a ballooning yarn under dynamic forces, including the air drag.[10] Due to the significant variation in the length and radius of the balloon during the bobbin formation process, the optimal location of the balloon control ring is crucial in influencing the end breakage rate. The balloon control ring moves together with the lappet from the beginning of winding until the yarn is wound to 40% of full bobbin. From that point, it starts moving with the ring. This means that the balloon control ring always works effectively to stabilize the balloon.

The balloon control ring divides the balloon into two smaller balloons and facilitates spinning with relatively low tensions. The yarn abrades against the surface of the balloon control ring at speeds similar to the traveler speed (35–40 m/s). The balloon control ring should not disturb the movement of the yarn. It should be maintained cleanly without any surface irregularities. Balloon control rings can cause roughening of the yarn and fusing of heat-sensitive fibers and fiber shedding. In ring spinning, the maximum yarn tension can be reduced by up to two-third with the use of a single balloon control of the right size and position. The research study conducted by Tang et al.[10] on the effects of balloon control ring on ring spinning suggests that to achieve the optimal results from balloon control rings in ring spinning, it is advisable that

1. A single balloon control ring should be used
2. The balloon control ring radius should be the same as the traveler ring radius
3. The balloon control ring should remain approximately half way between the yarn guide and the ring rail during the entire spinning process

7.2.6 Ring and Traveler

Ring and traveler are the dominant elements in the ring spinning process. Ring and traveler play a crucial role in the twisting process.[11] The correct selection of ring and traveler has a predominant influence on the spinning performance. Traveler selection influences the end breakage rate. It is very important for the technologist to understand this and act on them to optimize the yarn production and quality. The characteristics of a good ring are listed in Table 7.9.

In ring spinning, the energy to drive the twisting mechanism is derived from the bobbin, but the level of twist is controlled by the traveler. Each revolution of the traveler

TABLE 7.9

Characteristics of a Good Ring

A good ring in operation should have
Best quality raw material
Good, but not too high, surface smoothness
An even surface
Exact roundness
Good, even surface hardness, higher than that of the traveler
Should have been run in as per ring manufacturers, requirement
Long operating life
Correct relationship between ring and bobbin tube diameters
Perfectly horizontal position
Precise center position relative to the spindle

TABLE 7.10

Factors Influencing Ring Life

Type of fiber processed
Yarn count (traveler weight)
Spindle speed
Traveler running time
General conditions (centering of rings, etc.)

inserts one turn of twist into the yarn. The mass of the traveler has to be balanced against the yarn linear density, and the so-called "traveler weight" is an important factor in determining the yarn tension. The yarn tension, in turn, is an important factor in determining balloon size as well as the end breakage rate. The bobbin rotates faster than the traveler and the trailing yarn drags the traveler behind it. The difference in speed causes the yarn to wind onto the constant speed bobbin. The positional stability of the traveler during the cop running time influences yarn quality, ends-down rate and traveler wear. This was measured indirectly by means of the peaks of the tensile strength.

With high traveler wear (burnt travelers) the ring lifetime is reduced (Table 7.10). On conventional rings, microwelding damages the running track and reduces the ring life dramatically. A worn-out ring surface influences the yarn quality, specially the yarn hairiness.

7.2.6.1 Load on Ring and Traveler

During the spinning process, there is always a high load on the ring running track. The traveler's centrifugal force (F_c) depends on the traveler weight (m), the ring radius (r), and the traveler linear speed (v). The centrifugal force is calculated with the following formula:

$$F_c = \frac{\left(m \times v^2\right)}{r_{Ring}}$$

This leads to very high values compared with the relatively small weight of a traveler. The centrifugal force can reach a load that is up to 8000 times the traveler weight. These high loads create heat and lead stress to the ring surface. In order to prevent premature wear on the running track when working under extremely high loads or heavy conditions, it is recommended to use a ring treatment with very high wear resistance.[11]

The lifetime of rings and travelers depends on two main parameters:

1. Raw material processed (lubrication potential)
2. The mechanical and thermal load on ring and traveler (speed, ring diameter, and traveler weight)

The centrifugal force increases in square in proportion to the traveler speed. The traveler temperature in the contact area of traveler ring increases in cube in proportion to traveler speed (Figure 7.9). The characteristics of a good traveler are listed in Table 7.11.

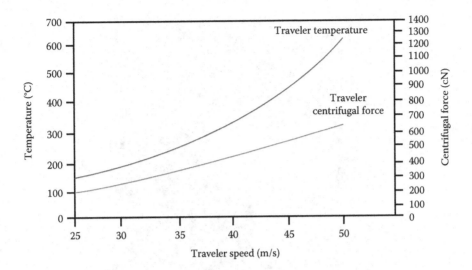

FIGURE 7.9
Influence of traveler speed on its temperature and centrifugal force.

TABLE 7.11

Characteristics of a Good Traveler

Generates as little heat as possible
Quickly distributes the generated heat from the area where it develops over the whole volume of the traveler
Transfers this heat rapidly to the ring and the air
Is elastic, so that the traveler will not break as it is pushed on to the ring
Exhibits high wear resistance
Is less hard than the ring because the traveler must wear out in use in preference to the ring

7.2.6.2 Shape of the Traveler

The traveler must be shaped to match exactly with the ring in the contact zone so that a single contact surface with the maximum surface area is created between the ring and the traveler. The cross section of the traveler is shown in Figure 7.10. The bow of the traveler should be as flat as possible in order to keep the center of gravity low and thereby improve smoothness of running.[13] However, the flat bow must still leave adequate space for passage of the yarn. If the yarn clearance opening is too small, rubbing of the yarn on the ring leads to roughening of the yarn, a high level of fiber loss as fly, deterioration of yarn quality, and formation of melt spots in spinning of synthetic fiber yarns.[12]

7.2.6.3 Traveler Friction

The traveler has, among other duties, the function to regulate the spinning tension.[13] This spinning tension must be high enough to keep the thread balloon stable and, on the other hand, not too high in order to avoid yarn breaks due to tension. At high speeds, normal friction systems only work with additional lubricants.

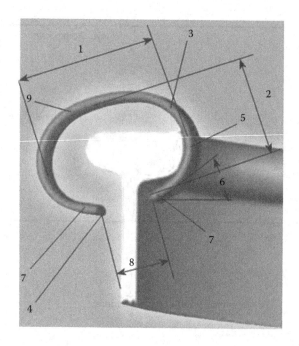

FIGURE 7.10
Traveler cross section. 1, inner traveler width; 2, height of bow; 3, yarn passage; 4, wire section; 5, traveler—ring contact surface; 6, angle of toe; 7, Toe; 8, opening; 9, upper part of traveler bow.

The ring/traveler system (Figure 7.11) does only function with this so-called fiber lubrication. As a result, the fibers protruding from the yarn body between the ring and the traveler are crushed and form a steady regenerating lubrication film. Depending on the fiber (dry or strong, wax-containing cotton, or softening agents on synthetics), the resulting coefficient of friction varies.[12] The coefficient of friction with fiber lubrication can vary from 0.03 and 0.15:

$$R = \text{Coefficient of friction} \times N$$

where
  R is the traveler friction in mN
  N is the normal force $>= (F_c \times ML \times V \times V)/(R)$
  F_c is the centrifugal force
  ML is the mass of the traveler in mg
  V is the traveler speed in m/s
  R is the radius of the ring (inside)

7.2.6.4 Traveler Mass

Traveler mass[11] determines the magnitude of frictional forces between the traveler and the ring, and these in turn determine the winding and balloon tension.[13] Mass of the traveler depends on

- Yarn count
- Yarn strength
- Spindle speed
- Material being spun

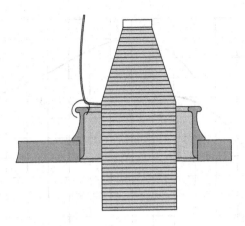

FIGURE 7.11
Traveler friction.

TABLE 7.12

Factors Considered for Traveler Selection

Yarn count
Ring flange
Type of ring
Life of ring
Material
Spindle speed

If traveler weight is too low, the bobbin becomes too soft and the cop content will be low. If it is unduly high, yarn tension will go up and will result in end breaks. If a choice is available between two traveler weights, then the heavier is normally selected, since it will give greater cop weight, smoother running of the traveler and better transfer of heat out of traveler. High contact pressure (up to 35 N/mm²) is generated between the ring and the traveler during winding, mainly due to centrifugal force.[2] Ishtiaque et al.[14] reported that the yarn imperfection level reduces with the increase in traveler mass. Barella et al.[15] found that using a too heavy or too light traveler increased the hairiness of yarns. The various factors considered for the selection of traveler for a particular process are listed in Table 7.12.

7.2.6.5 Traveler Speed and Yarn Count

When the spindle speed is increased, the friction work between the ring and the traveler (hence the buildup) increases as the third power of the spindle rpm. The traveler speed increases with the increase in yarn count up to 40ˢNe in carded process and decreases for yarn count above 50ˢNe in combed process[16] (Figure 7.12).

If the traveler speed is raised beyond normal levels, the thermal stress limit of the traveler is exceeded; a drastic change in the wear behavior of the ring and traveler ensues. Owing to the strongly increased adhesion forces between the ring and the traveler, welding takes place between the two. These seizures inflict massive damage not only to the traveler but to the ring as well.[16]

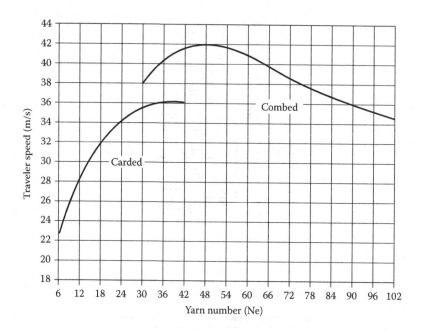

FIGURE 7.12
Traveler speed and yarn count. (From Riener—360° Performance, Rieners traveller—Product information brochure, Reiners + Fürst GmbH u. Co. KG, Mönchengladbach, Germany, 2011, http://www.reinersfuerst.com.)

7.2.6.6 Traveler and Spinning Tension

Lunenschloss[17] reported that the hairiness of yarns was considerably influenced by the spinning tension. Stalder[18] has shown that by increasing spinning tension from 2 to 4 cN/tex for 60ˢNe cotton yarn, the hairiness decreases by 50%, whereas thin places and elongation of the yarn deteriorate. The spinning tension is proportional

- To the friction coefficient between the ring and the traveler
- To the traveler mass
- To the square of the traveler speed

and inversely proportional

- To the ring diameter
- To the angle between the connecting line from the traveler–spindle axis to the piece of yarn between the traveler and the cop

7.2.6.7 Traveler Clearer

Traveler clearer (Figure 7.13) is an excellent method for removing all fiber fly that accumulates on the outer part of C or El traveler. Failure to use a clearer that is not set up tight enough may result in traveler blockage due to fiber clogs, or its performance may be severely negatively affected. This leads to elevated end breakage rates and a decline in yarn quality. The traveler clearer should have the right distance to the outside ring flange. A distance of about 0.5 mm between clearer and the traveler (in operating position) is recommended.[19] When adjusting the distance between outside ring flange and clearer, the size of the traveler should be taken into consideration.

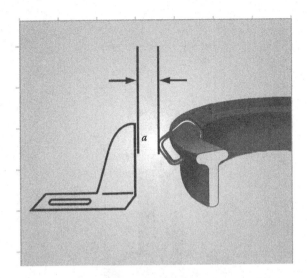

FIGURE 7.13
Traveler clearer.

7.2.6.8 Prerequisites for Smooth and Stable Running of Traveler on Ring

- Faultless condition of the support and guide of the ring rail as well as a steady and smooth traverse motion. Concentric position of the ring and spindle as well as antiballooning ring and yarn guide as shown in Figure 7.14.[19]
- Spindle rotation without vibration and correct concentricity of bobbin tube. Ring with exact roundness and firm seating in horizontal position.
- Correct setting of the traveler clearer as shown in Figure 7.15. Space *a* should be around 0.3 mm.
- Favorable ratio of ring diameter to tube diameter as shown in Figure 7.16.
- The diameter of the empty tube onto which the yarn will be wound has a minimum size of at least 45% of the ring size; otherwise, excessive yarn tensions would be generated. Recommended ratio is $D{:}d = 2{:}1$ (ring diameter: D; tube diameter: d).
- Faultless condition of ring race way.

7.2.6.9 Traveler and Spinning Geometry

The ring traveler, together with the yarn as a pull element, is set into motion on the ring by the rotation of the spindle.[19] Yarn tension will be too high if the direction of pull deviates too much from the running direction of the traveler ($\alpha < 30$). The pulling tension can be reduced by adapting the ring or tube diameter ($\alpha > 30$) during the winding up on the tube (Figure 7.17).

The tube length determines (with the yarn guide) the maximum balloon length. This is an important factor for the performance of a ring spinning machine. The shorter the balloon, the higher traveler speeds can be achieved.[13] In practical use, the ideal ratio of tube length to ring diameter has been shown to be between 4.5:1 and 5:1.

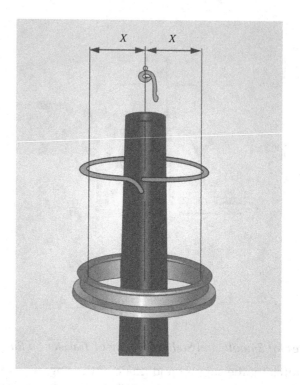

FIGURE 7.14
Ring and spindle position.

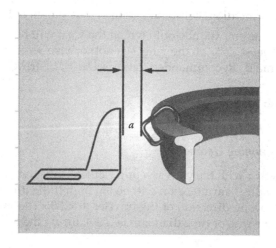

FIGURE 7.15
Traveler clearer setting.

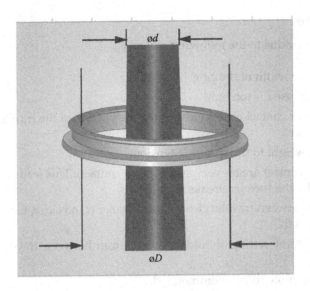

FIGURE 7.16
Ratio of ring diameter to tube diameter.

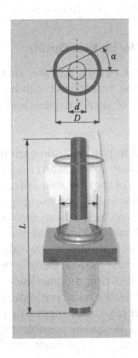

FIGURE 7.17
Traveler and spinning geometry.

7.2.6.10 Traveler Fly

Traveler fly can occur due to the following reasons[19]:

1. Reduced flange width of the ring
2. If ring traveler used is too light
   a. Ring traveler contact area is close to the toe portion of the traveler. Hence, traveler fly occurs.
   b. Improper weight to the spinning tension.
3. Ring traveler contact area is very small (point contact). This leads to extreme wear out and finally the traveler breaks and flies.
4. If the setting between traveler clearer and traveler is too close, the traveler will hit the clearer and fly.
   a. Traveler clearer setting should be 0.2–0.3 mm between traveler and traveler clearer.
5. Cop content is more than recommended.
6. Lesser winding length results in faster movement of ring rails. Hence, there is a chance of traveler fly.
   a. Increase the winding length with respect to count and spindle speed.
7. If the center of gravity is higher for the required speed, there is a chance that traveler has an unstable running. It leads to traveler fly.

7.2.6.11 Impact of Ring and Traveler on Yarn Quality

Yarn quality is affected by several factors, and the ring, as well as the traveler, has an impact. The correct selection of ring and traveler can have a positive impact on the spinning results, in particular in terms of yarn hairiness and evenness.

7.2.6.11.1 Hairiness

Ring and traveler do have an effect in particular on yarn hairiness. The wear condition and the centering of the ring play a primary role. A worn ring surface always results in increased yarn hairiness results. The centering of the ring is also of critical importance for minimal yarn hairiness, and its importance increases with increasing spindle speed.[12] Also, the smaller the ring diameter, the more important is centering. Even at an eccentricity of 0.3 mm, the theoretical traveler speed fluctuates considerably. This makes the traveler buzz, and thus, yarn hairiness increases. The correct selection of the traveler weight is also important for good yarn hairiness results. If the traveler weight is too low, the balloon may bulge a lot, which results in increased friction on the balloon control ring and thus leads to increased yarn hairiness.[12] If the spinning rings are worn, the ring–traveler friction is reduced as a result of the damaged ring surface. In this case, heavier traveler rates can temporarily remedy the problem. Nonetheless, the rings should always be replaced as soon as possible if this is the case. Choosing the right shape of traveler and wire profile will yield optimum yarn hairiness results.[13]

7.2.6.11.2 Neps

Neps are extremely short mass fluctuations that usually arise from the spinning preparation process. In some cases, push-up neps, which may develop on the traveler, can also lead to an increased number of neps. The cause may be an unsuitable traveler or a heavily

TABLE 7.13

Problems Associated with Wrong Selection of Ring/Traveler in Ring Spinning Process

Problems	Causes	Remedies
Poor ring traveler life	Improper matching Poor ring condition	Correct traveler selection Good ring condition
Less yarn elongation	Heavier ring travelers	Lighter ring travelers
More yarn hairiness	Low bow height traveler Improper temperature and air humidity	High bow height travelers Correct R.H% and temperature Heavier travelers
Unable to increase the spindle speed	Improper spinning geometry Poor ring/lappet centering Spindle tube vibration Improper selection of travelers	Proper lift w.r.t. count The ratio between ring diameter 1:5 Ring diameter to tube length ($2d$ + 5) of tube diameter for lappet setting Correct centering of ring/lappet Vibration free spindle/tube Proper combination of ring travelers
Fluff accumulation	Improper traveler clearer setting Higher room temperature Poor R.H% Poor housekeeping	Setting should be 0.2–0.3 mm Traveler (operating position) Optimum room temperature Better R.H% Housekeeping should be proper
Pushed-up neps	Less yarn clearance Higher wear-out Traveler clearer setting	High clearance traveler finish/profile Use best traveler combination/finish/profile

worn traveler.[11] In this case, a suitable traveler shape or shorter traveler replacement intervals can improve the situation. If the number of neps is increased significantly and is caused by push-up neps, the C.V value may be increased as well.

Some of the quality problems occurred due to the wrong selection of ring–traveler particulars are summarized in Table 7.13.

7.2.7 Roller Setting

The principle of roller setting is that one pair of rollers should release its grip on one end of the cotton fiber as the next pair of rollers takes hold of the other end of it. The fiber control during drafting operation and the production of perfect fibrous strand are highly influenced by proper drafting roller settings. The roller setting is influenced by several factors as given in Table 7.14.

The distance between the back bottom roller and the middle bottom roller is referred to as the break draft zone (Figure 7.18). The ratio of surface speed of middle bottom roller to the surface speed of back bottom roller is termed as "break draft." The bottom fluted

TABLE 7.14

Various Factors Influencing Setting of Drafting Rollers in a Ring Frame

Staple length of fiber
Hank of roving
Fiber characteristics
Type of drafting system

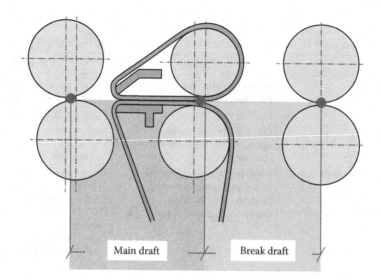

FIGURE 7.18
Roller setting.

TABLE 7.15

Drafting Zone Roller Setting for Different Staple Lengths of Cotton

Fiber Staple Length	Break Draft Zone Setting (mm)	Main Draft Zone Setting (mm)
Carded cotton (<30 mm)	60–70	43
Combed cotton (up to 30 mm)	65–75	43
Long stapled cotton (>30 mm)	65–75	48

rollers are usually of 27 mm and lie with their centers in the same plane. The back and front top rollers mounted with rubber cots have a diameter usually of 30 mm. The distance between the middle bottom roller and the delivery bottom roller is referred to as the main draft zone. The ratio of surface speed of delivery bottom roller to the surface speed of middle bottom roller is termed as "main draft." The normal clearance between the nose of the apron and the front top roller is 0.5–1.0 mm. The recommended drafting zone settings for processing different staple length cotton are given in Table 7.15.

7.2.7.1 Top Roller Overhang

The spinning triangle is a critical region in the spinning process of yarn and its geometry influences the distribution of fiber tension in the spinning triangle, thus affecting the properties of spun yarns, especially the yarn strength, torque, and hairiness. The distance between the apron nose and the front roller nip should be as small as practically possible to ensure the best fiber control during drafting. Liang[20] stated that the quality of ring spinning yarn is improved by changing the spinning triangle shape actively, especially the horizontal offset of the twisting point. The overhang of the front top roller as shown in Figure 7.19 has a significant influence on yarn hairiness.[21] Haghighat et al.[22] reported that with the increase in roller overhang, yarn hairiness initially decreases and then increases. The top front roller almost never lies vertically above the associated bottom roller. Usually, the top roller is shifted 2 mm forward. This gives a rather smoother running, because the

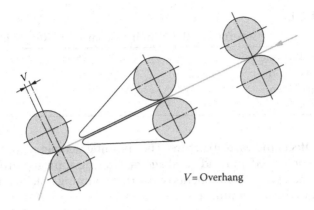

V = Overhang

FIGURE 7.19
Roller overhang.

weighting force acts as a stabilizing component in the running direction so that swinging of the top roller is avoided. Furthermore, the angle of warp is reduced, and the spinning triangle is made shorter; consequently, yarn hairiness is decreased. On the other hand, the overhang must not be made too large, otherwise the distance from the exit opening of the aprons to the roller nip line becomes too long, resulting in poorer fiber guidance and increased yarn hairiness.

7.2.7.2 Spinning Geometry

The fibrous strand coming out from the drafting unit passes at different inclination angles through the lappet, balloon control ring, and ring traveler that are placed far from each other and are not aligned as shown in Figure 7.20. This path decides the spinning geometry

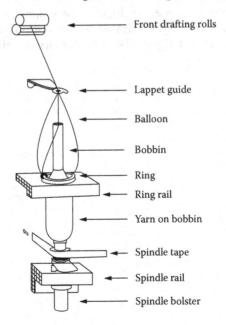

Front drafting rolls

Lappet guide

Balloon

Bobbin

Ring

Ring rail

Yarn on bobbin

Spindle tape

Spindle rail

Spindle bolster

FIGURE 7.20
Spinning geometry.

TABLE 7.16

Various Factors Governed by the Spinning Geometry in Ring Spinning

End breakage rate
Evenness of yarn
Binding of fibers
Hairiness

that significantly affects the yarn structure. The spinning geometry highly influences the quality of the yarn produced and end breakage rate in ring spinning process (Table 7.16).

The positions of the various components of the ring spinning machine must be carefully set in order to obtain the maximum spinning yield.

7.2.8 Spindle and Its Drive

Spindle as shown in Figure 7.21 is a short light shaft that may be rotated on its own lengthwise axis to insert twist and to wind yarn into the desired form of package. Spindles are generally made from aluminum alloys, whereas the bolsters are made from cast iron or steel. They hold the bobbin somewhat loosely but are tight enough to prevent slippage. Spindles of modern ring frame are driven by spindle tape through ABS pulley. The direction of twist in the yarn spun is decided by the direction of spindle rotation. Spindles have a predominant influence on the ring spinning machine's energy consumption and noise level. The chamber between the blade and the bolster is filled with lubricating oil, which is generally replaced after 10,000 h. In order to prevent vibrations of the spindle, damping spiral is provided below the spindle tip. Spindle oil topping and spindle oil replenishing are the important maintenance activities done to keep the spindle in perfect working condition.

New generation spindles are designed with reduced lift in order to get better speeds due to reduced yarn tension, package weight, etc. The major requirements of spindle are given in Table 7.17. The top parts are designed to suit auto-doffing, and spindle brakes are provided to remove operator fatigue. The concentricity of spindle with respect to ring

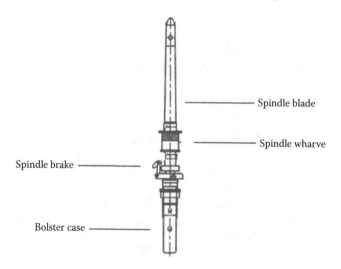

Spindle blade

Spindle wharve

Spindle brake

Bolster case

FIGURE 7.21
Spindle—parts.

TABLE 7.17

Major Requirements of Spindle

True running even at high speed
Less power consumption
Less noise level at higher speeds

plays a critical role in deciding the yarn quality and end breakage rate. Eccentric spindles in the ring frame significantly increase the hairiness in the yarn and affect strength and elongation also. An eccentric spindle can increase the end breakage markedly because of the once-per revolution cycle of tensions produced.[23] The mechanical balance limits with respect to eccentricity, dimensional accuracy, etc. Bigger wharve diameter requires higher driving speed and hence higher power. Therefore, the diameter of the wharve underwent a gradual reduction from 25.4 to 19 mm.

The drive from ABS or tin roller pulley is transferred to the spindle by means of spindle tape as shown in Figure 7.22. The drive to the spindle is not positive and may have some slippage. The spindle tape tension is maintained with the help of tension or jockey pulley setting. The particulars of spindle tape used in ring spinning machine are given in Table 7.18.

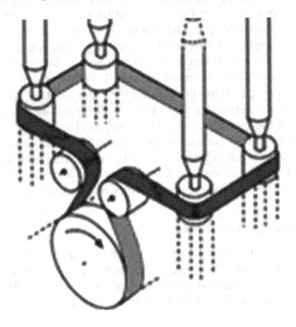

FIGURE 7.22
Spindle tape.

TABLE 7.18

Particulars of Spindle Tape

Tape thickness (mm)	0.55
Tape weight (kg/m²)	0.56
Construction	
Wharve side	Mixed fabric
Pulley side	High-friction elastomer

7.2.8.1 Spindle Speed

The spindle speed is an important process parameter that characterizes the technological process on ring spinning machine and determines the technical and economic indices of operation of the machines for processing of fibers. The increase in spindle speed results increase in yarn tension, which can lead to yarn breakage over a level recognized as standard. Sokolov[24] proposed an empirical equation for calculating the admissible spindle rotation rate (n_{sp}, min^{-1})

$$n_{sp} = \frac{26,000\sqrt{1,000/T}}{\sqrt{D_v}}$$

where
n_{sp} is the spindle speed (rpm)
T is the linear density of the fiber (tex)
D_v is the ring diameter (mm)

Yarn quality gradually decreases with the increase in spindle speed. The increase in spindle speed in turn increases the noise pollution as well as energy consumption in the ring frame. The spinning tension varies from the bottom to the top of the ring bobbin and it is proportional to the spindle speed for the constant ring diameter and traveler mass. Optimization of spindle speed at different ring bobbin positions is essential for achieving uniform spinning tension, which would reduce the end breakage rate and also result in uniform quality of yarn. With the frequency-controlled inverter drives and variator speed control systems, different speeds are possible for different lengths of time. Ishtiaque et al.[14] reported that the increase in traveler mass and spindle speed increases the end breakage rate due to the fact that the increase in spindle speed and traveler mass increases the spinning tension, which may often exceed the safe spinning tension limit, resulting in more number of end breakages. Walton[25] found that hairiness increases with spindle speed. At higher spindle speeds, the hairiness increase may be attributed to the increase in the centrifugal and air drag forces and the rubbing velocity of yarn over lappet, ring, and traveler.

7.2.8.2 Ring Spinning Empties or Tubes

The selection of the ideal ring spinning tube has predominant influence on performance of the ring spinning machine. The correct selection of ring spinning empties facilitates increase in spindle speed, preventing empties-related quality issues. The following parameters[26] usually govern the selection of ideal spinning tubes:

1. *Overall length*: The total length of the ring tube is known as the overall length. The overall length of the tube depends on the spindle length and ideally the difference between the overall length of the tube and the spindle length should not be more than 20 mm.

2. *Lift of the tube*: The distance for which yarn could be wound on the tube is called the lift of the tube.

3. *Tube overhang (tube clearance)*: The clearance between the top of the ring tube and the top of the spindle is called as tube overhang or tube clearance. Ideally, it should be kept as low as possible, and it should not be more than 20 mm, so that there are no breakages at higher speeds.

4. *DUI/DOI*: The bore diameter of the tube required to fit on the bottom of the spindle is known as diameter under insertion (DUI), and the bore diameter required to fit on the top of the spindle is known as diameter over insertion (DOI).

5. *Taper ratio*: Taper ratio is a measure of the rate at which the diameter of the tube decreases in terms of the length of the tube. For example, if the taper ratio is 1:40, there will be a decrease of 1 mm in the tube diameter for every 40 mm length of the tube. The various taper ratios available are 1:38, 1:40, and 1:64.

6. *Wall thickness*: Wall thickness is nothing but the thickness of the tube. It is always better to stick to the machinery manufacturer's specification on wall thickness so that the machine could be run at high speeds and the ring tubes will also have good lifetime. The weight of the tube is directly proportional to its wall thickness.

7. *Eccentricity*: The eccentricity is the measure of variation in the circularity of the ring tube and is measured by the variation in the radius of the tube when in rotation. The lower the eccentricity of the tube, the better will be the results.

8. *Pressing force*: Ideally, the ring tube should press on the spindle with a pressure of 0.75–1.50 kg, which would ensure better life for both the spindle and the tube. Precise dimensions, proper taper, and smooth internal surface are the key factors for achieving this.

7.3 End Breakage Rate

In a modern spinning mill, achieving the productivity and quality is a quite challenging task.[27] Productivity of the spinning mill is decided by several factors such as spindle speed, work allotment, energy consumed, and end breakage rate. End breakage rate is a deciding factor that influences the productivity as well as the quality of the yarn produced in the ring spinning process. The end breakage is a critical spinning parameter that also affects the maximum spindle speed, the mechanical condition of the machines, and the quality of raw material. The ring frame performance and higher spindle speed are achieved by controlling the end breakage rate. A high end breakage rate points to a combination of machine, material, and human faults. The end breakage rate is usually higher at start and end of the cop-build process compared with the remaining period of the cop buildup (Figure 7.23).

7.3.1 Occurrence of End Breakage

In ring spinning, the end breakage occurs due to the imbalance in the tension imposed on the yarn and the yarn strength at the weakest portion.[27] It is an observed fact that almost all end breaks in the ring frame take place just after the delivery from the front nip in the spinning zone, that is, between the front rollers' nip and the thread guide. Therefore, an end will break when the spinning tension exceeds the strength of the weakest portion of the yarn in the spinning zone.[28] The end breakage phenomenon in ring spinning is absolutely slippage dominated, that is, there is no evidence of fiber breakage. A yarn break during the spinning process occurs when the actual spinning tension is higher than the actual yarn strength.

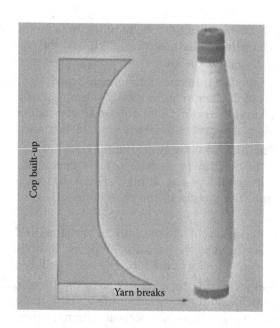

FIGURE 7.23
Yarn breaks during cop buildup. (From Riener—360° Performance, Rieners traveller—Product information brochure, Reiners + Fürst GmbH u. Co. KG, Mönchengladbach, Germany, 2011, http://www.reinersfuerst.com.)

Twist reduction affects the stability of the spinning process. This negative effect is even increased by the following factors:

- High yarn tension
- Smaller deviation radius of the yarn at the yarn guide
- Smaller diameter of the yarn guide material
- Reduced twist multiplier
- Smaller elasticity module of the fiber

End breakage will always occur just at the weakest point of the yarn within the described section. This may be either in the spinning triangle itself at the front roller pair of the drafting system or in the subsequent yarn section between spinning triangle and yarn guide.

7.3.2 Conditions in the Spinning Triangle

In a spinning triangle, fibers are always subjected to uneven load due to the spinning tension, while maximum load is exerted on marginal fibers.[29] The wider the spinning triangle, the more different is the pretension of the marginal fibers at the moment of twist impartation (Figure 7.24). As a result of this pretension, especially the marginal fibers are prevented from migrating between the different layers of the yarn cross section.[30]

7.3.2.1 Forces in the Yarn during the Spinning Process

The spinning tension is only 10%–15% of the yarn strength. On the other hand, the yarn strength during spinning is also reduced (only about 85% of the twist reaches the spinning

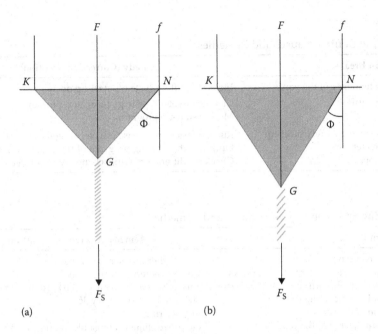

FIGURE 7.24
Spinning triangle. (a) Short triangle and (b) long triangle.

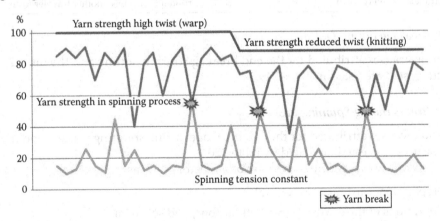

FIGURE 7.25
Yarn strength–spinning tension–yarn breaks.

triangle).[27] Disturbances such as thin places, slubs (fiber fly), foreign fibers, peaks in the spinning tension, or most likely a combination of two or more of the previous mentioned parameters lead to yarn breaks (Figure 7.25). Consider also the twist multiplier (warp or knitting).

7.3.3 Causes of Yarn Breaks

In general, we distinguish between two kinds of yarn breaks:

1. Breakage during doffing
2. Breakage during the spinning process

TABLE 7.19

Breaks during Doffing: Causes and Remedies

Cause of Yarn Breaks	Remedy (Correction/Solution)
High curling tendency of the yarn	Increase drafting system's start-up delay
Unthreading yarn	Change traveler type (e.g., form, and profile) Modify start-up program
Balloon stability too slowly built-up	Run-up faster or increase traveler weight
Start-up characteristics	Optimize the starting program of the ring spinning machine
Traveler jammed	Check condition of the ring, change type of traveler

TABLE 7.20

Breaks during Spinning Process: Causes and Remedies

Cause of Yarn Break	Remedy (Correction/Solution)
Unsuitable traveler type	Try another traveler form, change traveler profile
Uneven spinning tension, yarn tension peaks	Correct centering of the rings
Spindle speed too high, yarn strength insufficient	Yarn twist too low, increase the twist or reduce the speed
Too long running time of the travelers	Shorten the changing cycle
Poor condition of the rings	Replace rings
Climatic conditions not optimal (fiber fly)	Optimize climate, adjust blower nozzles
Yarn evenness (roving, impurities)	Optimize spinning preparation
Unfavorable raw material	Check raw material composition
Ring traveler severely worn	Reduce running time, test another traveler form

7.3.3.1 Breakage during Doffing

The various causes attributed for the end breakage during doffing and the remedial measures suggested are listed in Table 7.19.

7.3.3.2 Breaks during Spinning Process

The various causes attributed for the end breakage during spinning process and the remedial measures suggested are listed in Table 7.20.

Yarn breaks are caused through many other factors such as

- Preparation (carding, and roving)
- Conditions of the drafting unit (cots, aprons, and settings)
- Centering of yarn guiding parts (snarl wire, balloon control rings, spindles, and rings)

If the end breakage rate goes beyond manageable levels, idle spindles will increase. Repeated occurrence of end breaks in a few spindles is often cited as the reason for poor ring frame performance.

7.3.4 Effects of End Breakage

An increase in end break rate severely affects the productivity of the mill in terms of

- Increase in pneumafil waste
- Decrease in the number of spindles per spinner hence increasing labor costs
- Increase in costs in subsequent processes, that is, higher the end break rate in spinning higher the end break rate in weaving

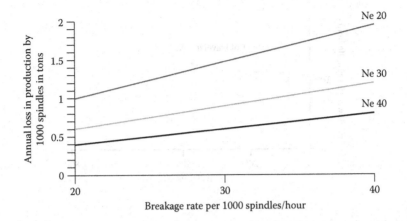

FIGURE 7.26
Effect of end breakage rate on productivity. (From Riener—360° Performance, Rieners traveller—Product information brochure, Reiners + Fürst GmbH u. Co. KG, Mönchengladbach, Germany, 2011, http://www.reinersfuerst.com.)

- Not being able to increase machine speed due to work load, waste, and yarn quality especially with yarn strength and variation
- Probability distributions of applied stress and tenacity of the strand

Therefore, reducing the number of end break rate eliminates many problems in spinning. The causes of end break differ from mill to mill and type of raw material used even fiber-to-fiber friction may be the main cause. The annual loss of production[12] in ring spinning due to higher end breakage rate is shown in Figure 7.26.

7.3.5 Influence of Various Parameters on End Breakage Rate in Spinning

7.3.5.1 Yarn Count

The spindle speed increases with the increase in yarn count. The influence of yarn count on the end breakage rate is shown in Figure 7.27.

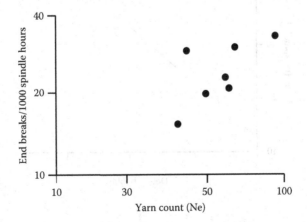

FIGURE 7.27
Effect of yarn count on end breakage. (From Lord, P.R., *Handbook of Yarn Technology: Technology, Science and Economics*, Woodhead Publishing Ltd., Cambridge, U.K., p. 308, 2003. With permission.)

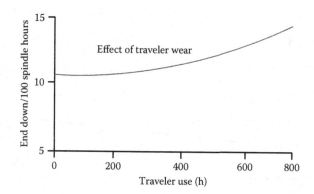

FIGURE 7.28

Effect of traveler use on end breakage rate. (From Lord, P.R., *Handbook of Yarn Technology: Technology, Science and Economics*, Woodhead Publishing Ltd., Cambridge, U.K., p. 308, 2003. With permission.)

7.3.5.2 Traveler Wear

The increase in running period of traveler beyond recommendations increases the traveler wear rate, which in turn increases the end breakage rate (Figure 7.28).

7.3.5.3 Defective Feed Bobbin

The end breakage in ring spinning is highly influenced by the feed bobbin quality.[23] There is a linear relationship between end breakage rate in ring spinning and defective roving bobbin (Figure 7.29).

7.3.5.4 Operator Assignment

Operator assignment is strongly affected by the end breakage rate in processing.[23] As the number of "thins" increases, so does the number of weak spots, and an operator can only

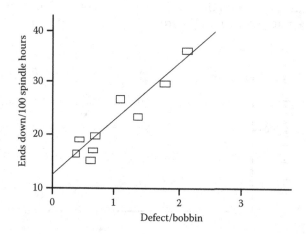

FIGURE 7.29

Effect of defective bobbin on end breakage rate. (From Lord, P.R., *Handbook of Yarn Technology: Technology, Science and Economics*, Woodhead Publishing Ltd., Cambridge, U.K., p. 308, 2003. With permission.)

serve a smaller set of spindles. In other words, the assignment has to be lowered. The product of the end breakage rate and the duration of the downtime enable an estimate to be made of the lost production. Ends down denotes those spindles where end has broken and is waiting for piecer to mend it. Ends down loss is given by

$$d = e \times 0.75t$$

where
    d is the ends down %
    e is the end breakage rate (breaks/100 spindle hours)
    t is the patrol time of piecer in hours

Further patrol time of piecer also increases with end breakage rate. As a result, ends down loss increases exponentially with the increase in end breakage rate.

7.3.6 Control of End Breakage in Ring Frame

Control of end breakage rate at the ring frame is the first step for improving the ring frame productivity. It not only leads to ends down loss but also restricts spindle speed. The following are the various methods to control the end breakage rate[27,29]:

1. The width of the drafted ribbon at the front roller nip should be reduced.

2. Measures to be taken to reduce mass irregularity of yarn straight after carding.

3. A reduction in friction between the ring and the traveler could reduce the peak tension during the rotation of the traveler.

4. Since the end breakage in ring spinning is related to slippage of fibers at the spinning triangle as a result of peaks occurring in the spinning tension fiber, the grip at the front drafting rollers should be increased by having a higher top roller pressure. The use of softer cots also enhances the grip at the front roller. If the total pressure on the rollers cannot be increased, the grip at the front rollers nip can be improved by reducing the width of cots.[31]

5. The permutation and combination of speeds at different stages of cop build in ring frames can help a lot in reduction of breaks.

6. In case of spring loading top arms in ring frames, when you restart the ring frame after a mill holiday, the breakages are very high. The practice of inching at interval of 8–10 h on holiday without releasing the top arm pressure reduces the end breaks to a great extent.[31]

7. In case of pneumatic loading top arms in ring frames, when you restart the ring frame after a mill holiday, the breakages are very high. The practice of tightening the material in drafting zone by reversing the back and middle rollers with the help of back roller wheel before starting the machine helps in reduction of end breaks during starting of the ring frames.

8. The lowering of spindle speed by 5%–6% just after putting new travelers for 2 h for running-in of travelers helps in reduction of end breaks to a large extent.

9. The use of higher twist multipliers will reduce breaks even at higher spindle speeds than at lower spindle speeds with low twist multipliers.

10. The higher twist than required in roving reduces the creel breaks in ring frames.

11. The delay in change of travelers will increase end breaks in spinning. Hence, timely change of travelers is a must.

12. The use of common tube suction system improves the suction of pneumafil system further without increasing power consumption compared to individual suction tube system in ring frame and reduces end breakages to a large extent.

13. Any compromise in right selection of fiber properties may lead to higher end breakages in spinning ring frames.

14. The frequent stoppage of overhead traveling cleaners accumulates fly and results in higher breaks due to loading of traveler with fly.

7.3.7 End Breakage and Economics

In ring spinning, much of that cost is caused by end breakage. One factor influencing the end breakage rate is the CV of strand strength.[23] The word strand is taken to include the weak point in the twist triangle in ring spinning. An analysis of cotton yarns from a variety of mills showed a relationship between the ends down rate and the CV of yarn strength (Figure 7.30). As the yarn tension increases, the stress in the strand increases and so does the end breakage rate. The probability of an end break depends on the probability distributions of applied stress and tenacity of the strand.

Quality control from a purely economic standpoint becomes ever more important as the count rises. The number of bobbins with an end break leaving the ring frame is one important factor in determining the work load for the winder. If the end breakage rate in spinning is high, there is probably a high rate of intolerable yarn faults. The chance of a piecing in a bobbin is low at fairly low yarn counts. On the other hand, the chance of an end break within a bobbin of high count yarn is much higher.

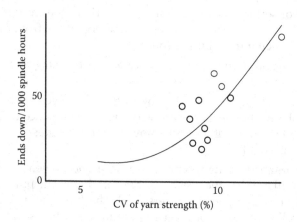

FIGURE 7.30
Effect of yarn strength variation on end breakage rate. (From Lord, P.R., *Handbook of Yarn Technology: Technology, Science and Economics*, Woodhead Publishing Ltd., Cambridge, U.K., p. 308, 2003. With permission.)

7.4 Draft Distribution

Drafting at ring frame significantly influences yarn evenness, appearance of fabric, and rejections due to yarn faults. The amount of draft given at the ring frame has a considerable influence on yarn quality and ring performance. The major problems faced due to improper selection of draft parameters and drafting zone setting are listed in Table 7.21.

The draft given at the ring frame is divided into break draft and main draft. The recommended amounts of draft with respect to yarn count and roving hank for different staple length cotton are given in Table 7.22. The break draft given in the back zone is expected to break the binding action of the twist so that a consistent form of fibrous strand is presented to the main draft zone. Break draft is adjusted according to the twist imparted to roving and the total draft required as follows:

- Normal twisted roving, break draft = 1.14–1.25
- Strong twisted roving, break draft = 1.3–1.5
- If total draft exceeds 40, break draft = 1.4–2.0

In roller drafting, the significance of drafting force is necessary to overcome interfiber frictional forces, which generate due to interfiber contacts, and also to counteract fiber bending. Das et al.[32] stated in a study that the drafting force is determined by interfiber friction, crimps, fiber parallelization, incidence and direction of hooks, and twist of roving. Das et al.[33] reported that the drafting force increases sharply with the increase in roving twist multiplier due to the increased fiber-to-fiber cohesion as the roving twist multiplier is increased. The drafting force reduces as the roving hank is increased due to the reduction in the number of fibers in the cross section, but it again increases with the increase in roving hank for a certain roving TM due to the better binding of fibers. Das et al.[33] also reported that the drafting force increases with the increase in fiber-to-fiber friction.

TABLE 7.21

Various Causes of Irregularity due to Drafting-Related Problems

Inadequate control over the movement of short and floating fibers
Slippage of strand and fibers under the drafting rollers
Variations in speed of drafting rollers
Mechanical faults

TABLE 7.22

Recommended Draft Range for Different Staple Length Cottons

Cotton Type	Roving Hank	Yarn Count in Ne	Draft Range
Short staple cotton	0.5–1.5	Up to 30ˢ	15–35
Medium staple cotton	0.6–1.8	30ˢ–70ˢ	25–80
Long staple cotton	1.0–3.0	50ˢ–150ˢ	35–80

7.5 Twist

Twist is inserted to the staple yarn to hold the constituent fibers together, thus giving enough strength to the yarn and also producing a continuous length of yarn. Twist is imparted to staple yarns to induce lateral forces.[34] Friction created by these forces acts to control fiber slippage in a strand under tension. The degree of twist given to fibrous strand in the ring spinning process is dependent on the rate of revolving the strand and the rate at which the strand is delivered from the drafting rollers:

$$\text{Twist Per Inch (TPI)} = \frac{\text{Spindle speed}}{\text{Front roller delivery (in./min)}}$$

$$\text{TPI} = \text{TM} \times \sqrt{\text{Count in Ne}}$$

When the twist increases, the yarn strength increases up to a certain level, beyond which the increase in twist actually decreases the strength of staple yarn[35] (Figure 7.31). Fine yarns require long staple fibers and high twist (Table 7.23). Coarse yarns can be produced with short fibers and low twist. Twist multiplier for cotton usually ranges between 3 and 5. The twist of weft yarns is approximately 4%–5% below the twist of warp yarns. The twist of hosiery yarns is approximately 12%–15% below the twist of warp yarns.

Knitted yarns normally have low twist multiplier (Table 7.24) as it is intended for application that needs better and softer handle. SITRA[36] developed an expression for optimum twist multiplier in ring spinning for maximum yarn strength (T_{max}) for the present day cottons is found to be linearly related to 50% span length (L) and micronaire value (F):

$$T_{max} = \frac{(54 - L + F)}{9}$$

$$T_{max} \text{ (compact yarns)} = \frac{(46 - 0.85(L - F))}{9}$$

Yarn twist also determines the productivity of a mill because more twist means less productivity. There is also a relationship between twist and diameter, density, hairiness,

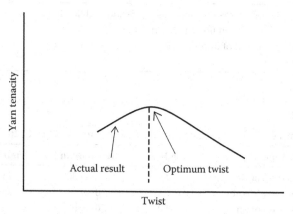

FIGURE 7.31
Effect of twist level on yarn tenacity.

TABLE 7.23

Factors Influencing the Amount of Twist in Ring Spinning

Count of the yarn spun
End-use application of the yarn
Staple length of the fiber being used

TABLE 7.24

Twist Multiplier for Different End-Use Applications

End-Use Application	Characteristics	Twist Multiplier
Knitting	Soft twist	2.5–4.0
Weaving (weft)	Normal twist	3–4.5
Weaving (warp)	Hard twist	3.5–5.4
Crepe yarn	Special twist	6.0–9.0

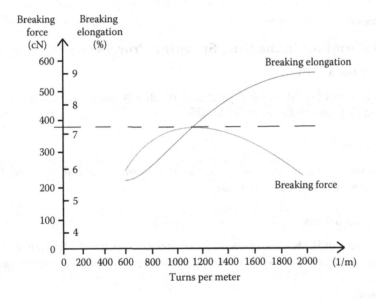

FIGURE 7.32
Effect of twist level on breaking force and breaking elongation. (Reproduced from Furter, R. and Meier, S., Measurement and significance of yarn twist, USTER ZWEIGLE Twist Tester 5—Application Report, Uster Technologies AG, USTER®, Uster, Switzerland, September 2009. With permission.)

strength, and elongation (Figures 7.32 and 7.33). The reduction of twist increases the hairiness because the number of protruding fibers increases. Higher twist C.V% significantly affects fabric appearance, strength, elongation, and dye uptake. The variation in spindle speed due to slack or too tight spindle tape will give rise to higher twist variations. A reduction of the yarn twist increases the yarn diameter and decreases the density. Twist C.V% should not exceed 3.5% to avoid quality problems, which can be recognized by naked eye.

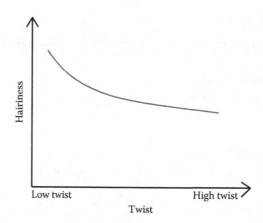

FIGURE 7.33
Effect of twist on yarn hairiness. (From Furter, R., Physical properties of spun yarns—Application Report, SE 586 Uster Technologies AG, USTER®, Uster, Switzerland, 3rd edn., June 2004. With permission.)

7.6 Nonconformities in the Ring Spinning Process

7.6.1 Hard Twisted Yarn

The various causes of hard twisted yarn and its effects and remedial measures taken in ring spinning process are listed in Table 7.25.

7.6.2 Unevenness

The various causes of yarn unevenness and its effects and remedial measures taken in ring spinning process are listed in Table 7.26.

7.6.3 Soft Twisted Yarn

Yarn that is weak indicates lesser twist. The various causes of soft twisted yarn and its effects and remedial measures taken in ring spinning process are listed in Table 7.27.

7.6.4 Hairiness

Protrusion of fiber ends from the main yarn structure. The various causes of yarn hairiness and its effects and remedial measures taken in ring spinning process are listed in Table 7.28.

7.6.5 Undrafted Ends

The various causes of undrafted ends and its effects and remedial measures taken in ring spinning process are listed in Table 7.29.

7.6.6 Higher Thick and Thin Places

The various causes of higher thick and thin places and its effects and remedial measures taken in ring spinning process are listed in Table 7.30.

TABLE 7.25

Hard Twisted Yarn: Causes and Remedies

Causes	Effects	Remedies
Count becoming very coarse because of interdoubles or lashing-in Worn out threads in fluted roller joints Failure of delay drafting mechanism in long ring frames Loose timer belt driving the front roller in long ring frames Traverse going out of drafting area, etc.	End breakage in winding Loss of productivity in winding Affects fabric appearance Shade variations in dyeing	Prevent interdoubling and lashing-in by adopting correct work practices Proper maintenance of delay drafting mechanism Maintain proper tension of timer belt that drives front roller Set traverse within drafting area

TABLE 7.26

Unevenness: Causes and Remedies

Causes	Effects	Remedies
Uneven feed material Improper settings in drafting zone Worn-out cots and aprons Improper selection of spacer Low pressure in top arms Eccentric fluted rollersand cots Jammed arbors, vibrating spindles, jammed bobbin holders Improper distribution of drafts between break draft zone and main zone Improper cleaning of drafting zone, lapping in adjacent spindles Traverse going out of drafting area, etc.	Higher count variation Affects yarn and fabric appearance Leads to shade variations in dyeing Higher end breaks in winding Loss of productivity in winding Higher hard waste in winding Higher rate of cop rejection	Effective control of feed material hank Proper selection of process parameters such as roller settings, spacer, top arm pressure, draft, and so on Check and control roller eccentricity at frequent interval Proper maintenance of cots, aprons, fluted rollers, arbors, spindles, and bobbin holders Proper follow-up of cleaning schedule; maintain the drafting zone in clean condition Set traverse within drafting area

TABLE 7.27

Soft Twisted Yarn: Causes and Remedies

Causes	Effects	Remedies
Loose spindle tapes Worn-out spindle tapes Jammed spindle bolsters Loose bobbin on spindle Jammed jockey pulley Spindle button missing, etc.	End breakage in winding Affects fabric appearance Leads to shade variations in dyeing Higher hard waste in subsequent process Higher rate of cop rejection	Maintain proper tension in spindle tapes Change worn-out spindle tapes Proper maintenance of spindle bolsters, jockey pulley Missed spindle buttons should be replaced

TABLE 7.28

Hairiness: Causes and Remedies

Causes	Effects	Remedies
Worn-out rings and travelers Worn out lappet hooks, separators Higher spindle speed Improper selection of traveler Variations in fiber lengths Lower humidity in the working area Too big a balloon Vibrating spindles	Affects fabric appearance Leads to end breaks in subsequent process More fly accumulation on the machines Poor working environment due to lint shedding	Replace worn-out parts in which the yarn passes through Optimum selection of spindle speed, traveler Length uniformity should be good Maintain proper humidity in the department

TABLE 7.29

Undrafted Ends: Causes and Remedies

Causes	Effects	Remedies
High twist in rove	End breaks in ring frame	Optimize twist level in roving
Lower break draft	Multiple breaks due to fly generation	Optimize break draft, top arm
Low top arm pressure	Productivity loss in ring frame	pressures and spacer
Higher humidity	Higher waste generation	Maintain proper humidity
Smaller spacer	Higher rate of cop rejection in winding	Proper maintenance of grooved apron
Channeled aprons and cots		and cots

TABLE 7.30

Higher Thick and Thin Places: Causes and Remedies

Causes	Effects	Remedies
Eccentric top and bottom rollers	End breakage in ring frame and	Top and bottom roller to be checked
Insufficient pressure on top rollers	winding	for roller eccentricity at frequent
Worn-out aprons and improper	Productivity loss in ring frame	interval and corrected
apron spacing	Higher soft waste generation	Optimize top arm pressure
Worn-out or improper meshing of	Poor yarn realization	Optimize draft in ring frame
gear wheels	Shade variations in dyeing	Keep roller settings as per norms
Mixing of cottons varying widely	Higher rate of cop rejection in	Proper maintenance of gear wheels,
in fiber lengths and use of	winding	cots, and fluted rollers
immature cottons		Avoid too much variation in fiber
Excessive draft		length between fibers during mixing
Wider roller settings		Maintain proper R.H% in department
Partial lapping on drafting rollers		and avoid lapping tendency
Higher stretch between bobbin		Ensure proper functioning of bobbin
holder and drafting zone		holder and arrest stretch generation
Broken roving guides		Replace broken roving guides
		Optimize roving bobbin size
		especially for low twist roving

7.6.7 Idle Spindles

The various causes of idle spindles and its effects and remedial measures taken in ring spinning process are listed in Table 7.31.

7.6.8 Slub

An abnormally thick place or lump in yarn showing less twist at that place. The various causes of slub and its effects and remedial measures taken in ring spinning process are listed in Table 7.32.

7.6.9 Neps

Neps are highly entangled fibers. The various causes of neps and its effects and remedial measures taken in ring spinning process are listed in Table 7.33.

7.6.10 Snarl

Yarn with kinks (twisted onto itself) due to insufficient tension after twisting. The various causes of snarls and its effects and remedial measures taken in ring spinning process are listed in Table 7.34.

TABLE 7.31

Idle Spindles: Causes and Remedies

Causes	Effects	Remedies
Noncreeling of bobbins in time Spindle tape and apron breakages Broken/missing spare parts, etc.	Productivity loss Higher soft waste generation Poor yarn realization Energy loss	Creeling of bobbins as per standard procedure Broken spindle tapes and apron breakages to be attended immediately Replace broken or missed spare parts immediately Implementation of TPM practices

TABLE 7.32

Slub: Causes and Remedies

Causes	Effects	Remedies
Improper mixing of fibers Too much variations in fiber lengths Improper opening Low pressure in drafting roller Inadequate drafts applied Lower setting of drafting rollers for the fiber length in use Damages in card wire points resulting in bunches of fibers Lashing in or lapping in spinning Damages in the draft gear wheels Slippage of rollers while drafting	End breakage in ring frame and winding Productivity loss in ring frame Higher soft waste generation Poor yarn realization Affects fabric appearance Shade variations in dyeing	Ensure homogeneous mixing of fibers Ensure length uniformity to be good Ensure necessary opening of fibers in blow room and carding process Draft to be kept as per norms Proper maintenance of card wire points, gear wheels, fluted rollers Ensure roller setting w.r.t fiber length

TABLE 7.33

Neps: Causes and Remedies

Causes	Effects	Remedies
Accumulation of fly and fluff on the machine parts Poor carding Defective ring frame drafting and bad piecing Improperly clothed top roller clearers	Productivity loss in winding Affects yarn and fabric appearance Higher rate of cop rejection in winding Higher hard waste in winding Poor yarn realization	Ensure proper functioning of overhead traveling cleaners in roving and ring frame Maintain drafting zone free from fly accumulation by cleaning frequently with picker gun Ensure proper opening in carding Optimize drafting in ring frame Train labor for proper piecing Top roller clearer to be cleaned frequently

TABLE 7.34

Snarl: Causes and Remedies

Causes	Effects	Remedies
Higher than normal twist in the yarn Presence of too many long thin places in the yarn Improper spindle tape tension	Entanglement with adjacent ends causing a break Affects fabric appearance Shade variation in dyeing End breakage in winding	Maintain proper tension in spindle tapes Prevent the occurrence of thin places by optimizing draft parameters Optimum twist to be used for the type of cotton processed The yarn to be conditioned

7.6.11 Crackers

Very small snarl-like places in the yarn that disappear when pulled with enough tension or yarn with spring-like shape. The various causes of crackers in the yarn and its effects and remedial measures taken in ring spinning process are listed in Table 7.35.

7.6.12 Bad Piecing

Unduly thick piecing in yarn caused by over-end piecing. The various causes of bad piecing in the yarn and its effects and remedial measures taken in ring spinning process are listed in Table 7.36.

7.6.13 Kitty Yarn

Presence of black specks of broken seeds, leaf bits, and trash in yarn. The various causes of kitty yarn and its effects and remedial measures taken in ring spinning process are listed in Table 7.37.

TABLE 7.35

Crackers: Causes and Remedies

Causes	Effects	Remedies
Mixing of cottons of widely differing staple length	More end breaks in winding	Avoid too much variation in fiber length between fibers during mixing
Closer roller settings	Higher rate of cop rejection	
Eccentric top and bottom rollers	Poor yarn realization	Optimize roller setting w.r.t fiber length
Poor temperature and relative humidity in the spinning shed	Higher hard waste generation	Rectify eccentric drafting rollers
Over spinning of cotton	Productivity loss in winding	Maintain proper R.H%
	Affects fabric appearance	Optimize draft and twist w.r.t fiber properties

TABLE 7.36

Bad Piecing: Causes and Remedies

Causes	Effects	Remedies
Wrong method of piecing and over-end piecing	End breaks in winding	Train the labor to follow proper method of piecing
	Affects fabric appearance	
	Productivity loss in winding	Minimize the end breakage rate in ring frame
	Higher rate of cop rejection in winding	
	Higher hard waste	Separators to be provided
	Poor yarn realization	

TABLE 7.37

Kitty Yarn: Causes and Remedies

Causes	Effects	Remedies
Ineffective cleaning in blow room and cards	End breaks in winding	Improve cleaning efficiency in blow room and carding process
	Affects fabric appearance	
Use of cottons with high trash and too many seed coat fragments	Productivity loss in winding	Select cotton with minimum amount of trash
	Higher hard waste	
	Poor yarn realization	Maintain proper R.H%
	Production of specks during dyeing	
	Needle breaks during knitting	

7.6.14 Foreign Matters

Foreign matters such as metallic parts, jute flannels, and so on are spun along with yarn. The various causes of kitty yarn and its effects and remedial measures taken in the ring spinning process are listed in Table 7.38.

7.6.15 Spun-In Fly

Fly or fluff either spun along with the yarn or loosely embedded on the yarn. The various causes of spun-in fly and its effects and remedial measures taken in ring spinning process are listed in Table 7.39.

7.6.16 Corkscrew Yarn

It is a double yarn where one yarn is straight and other is coiled over it. The various causes of corkscrew yarn and its effects and remedial measures taken in the ring spinning process are listed in Table 7.40.

TABLE 7.38

Foreign Matters: Causes and Remedies

Causes	Effects	Remedies
Improper handling of travelers Improper preparation of mixing Improper cleaning methods Poor work practices	Affects fabric appearance End breaks in winding Higher hard waste generation Higher rate of cop rejection Formation of holes and stains in cloth	Ensure correct method of work practices Avoid usage of jute bags for carrying cotton during mixing

TABLE 7.39

Spun-In Fly: Causes and Remedies

Causes	Effects	Remedies
Accumulation of fluff over machine parts Fanning by workers with pad or jute sacks Failure of overhead cleaners Malfunctioning of humidification plant Poor cleaning methods (usage of compressed air for cleaning)	Affects fabric appearance End breaks in winding Higher hard waste generation Higher rate of cop rejection	Maintain machine in clean condition Avoid fanning by workers Ensure correct functioning of overhead cleaners Machine should be stopped while running stopped overhead cleaners Ensure proper method of cleaning Ensure correct functioning of humidification plant

TABLE 7.40

Corkscrew Yarn: Causes and Remedies

Causes	Effects	Remedies
Feeding of two ends (instead of one) in ring frame Lashing-in ends in ring frame	Causes streaks in fabric End breaks in winding Higher hard waste generation Higher rate of cop rejection	Ring frame tenters to be trained for proper piecing methods Pneumafil suction pipe to be kept clean and properly set Ensure proper placement of separator to avoid lashing-in

7.6.17 Oil-Stained Yarn

Yarn stained with oil or grease present in the machine components. The various causes of oil stained yarn and its effects and remedial measures taken in the ring spinning process are listed in Table 7.41.

7.6.18 Slough-Off

Coils of yarn coming out of the ring cops in bunches at the time of unwinding (Figure 7.34). The various causes of slough-off and its effects and remedial measures taken in the ring spinning process are listed in Table 7.42.

7.6.19 Low Cop Content

Yarn content in the ring cop is less than expected (Figure 7.35). The various causes of low cop content and its effects and remedial measures taken in the ring spinning process are listed in Table 7.43.

TABLE 7.41

Oil Stained Yarn: Causes and Remedies

Causes	Effects	Remedies
Careless oil in the moving parts, overhead pulleys, etc. Piecing made with oily or dirty fingers Careless material handlings	Affects fabric appearance End breaks in winding Productivity loss in winding Higher rate of cop rejection Higher hard waste generation	Appropriate material handling procedures Ensure proper method of lubrication Material handling equipment to be maintained in clean condition

FIGURE 7.34
Slough-off.

TABLE 7.42

Slough-Off: Causes and Remedies

Causes	Effects	Remedies
Improper ring rail movement Worn builder cam Loose package and excessive coils in the package Soft build of cops Improper empties fit on the spindles and slack tapes	End breaks in winding Productivity loss in winding Higher rate of cop rejection Higher hard waste generation Poor yarn realization	Ring rail movement to be set right Optimum ratio of winding: binding coil and optimum chase length to be maintained

FIGURE 7.35
Low cop content.

TABLE 7.43

Low Cop Content: Causes and Remedies

Causes	Effects	Remedies
Under-utilization of bobbin height	Productivity loss in ring frame	Optimum chase length, coil spacing, and wall thickness of empty cops tube to be ensured
Lower number of coils/inch		
Higher chase length	Efficiency loss in ring frame	Ratchet/pawl movement to be properly set
Cop bottom bracket properly not set		
Improper selection of ratchet	Drop in winding efficiency	Free space of only 7.5 mm to be maintained at the top and bottom of the cop
Ratchet pawl pushing number of teeth/movement in the ratchet wheel is improper	More knots for a given length of wound yarn	Free space of 0.75 mm only to be maintained between full cops and the ring
Spinning empties wall thickness is high		

FIGURE 7.36
Improper build.

TABLE 7.44

Improper Build: Causes and Remedies

Causes	Effects	Remedies
Improper combination of ratchet and pawl	Slough-off during doffing or winding	Ratchet and ratchet/pawl movement to be accurately arrived by considering count of yarn, ring diameter, and chase length
Jerky ring rail movement (poker rod movement to check)	More breaks during unwinding (due to slough off)	Lubrication of poker rods at appropriate intervals to be carried out
	Higher hard waste in winding	

7.6.20 Improper Cop Build

Step-like appearance in the ring cop as shown in Figure 7.36. The various causes of improper cop build and its effects and remedial measures taken in the ring spinning process are listed in Table 7.44.

7.6.21 Ring Cut Cops

Damaged layers on the surface of the ring cops (Figure 7.37). The various causes of ring cut cops and its effects and remedial measures taken in the ring spinning process are listed in Table 7.45.

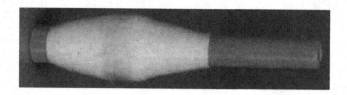

FIGURE 7.37

Ring cut cop.

TABLE 7.45

Ring Cut Cops: Causes and Remedies

Causes	Effects	Remedies
Count being coarse	End breakage in ring	Maintain count C.V% under control
Less number of teeth on ratchet wheel	frame and cone winding	(especially on coarser side)
being pushed each time	Loss of production in ring	Correct selection and setting of
Incorrect ratchet wheel (leads to low	frame and cone winding	ratchet wheel and number of teeth
package density)	Higher hard waste	pushed by it
Wobbling of spindle or empty cops	Leads to higher cop	Spindle maintenance to be done
Nonalignment of rings in the center of	rejection in automatic	properly
spindle axis	cone winding machine	Ring centering should be checked
Use of a lighter ring traveler (leads to low		Use of right traveler w.r.t yarn count
package density)		Poker bar should be polished as per
Jammed poker bars		schedule
Insufficient pressure on top rollers		Optimum setting of top roller loading
resulting in coarser count, etc.		Proper fit of empty cops with
Improper fit of empty cops with spindles		spindles should be ensured

TABLE 7.46

Lean Cops: Causes and Remedies

Causes	Effects	Remedies
Very fine count	Productivity loss in	Count C.V% should be within limits
Excessive breakages on a particular spindle	ring frame	End breakage rate should be kept
High chase length	Poor yarn realization	under control
Smaller ratchet wheel, etc.	Higher pneumafil	Correct selection of ratchet wheel,
Wobbling spindles and noncentering of lappet	waste	traveler, and chase length
hooks	Higher rate of cop	Traveler clearer setting to be kept as
Loaded traveler running on the ring	rejection in winding	per norms
Worn out rings	Higher hard waste in	Proper maintenance of rings,
Cracks in the apron	winding	traveler, spindles, lappet hooks,
Less TPI in roving		ABC rings, spacer, etc.
Scratches in the lappet hook and separators		Fit the empties correctly onto the
Missing spacer		spindle
Loosely fit empty tube on spindle		Feeding good quality roving with
Close traveler clearer setting		optimum TPI and imperfections
Wrong traveler running for that particular ring		Change traveler as per schedule
Thick places and thin places in the roving		Ensure trouble-free rotation of
Loose roving guide		bobbin holder
Lack of enough quantity of oil in the bolster of		
spindle		
Improperly set lappet hook		
Traveler burn outs		
Improper following of traveler change schedules		
Bobbin holder not rotating properly, etc.		

7.6.22 Lean Cops

If yarn breaks several numbers of times in one particular spindle, the cop made on this spindle looks deshaped and appears lean due to low yarn content on the cop. This is called as lean cop. The various causes of lean cops and its effects and remedial measures taken in ring spinning process are listed in Table 7.46.

7.7 Package Size or Cop Content

Package size is important for both technical and economic reasons. In order to minimize the yarn tension, progressively smaller cop formats and ring diameters have been adopted with increasing spindle speed.[37] The result is an enormous reduction in full package weight, thereby greatly shortening the spinning cycle. Yarn content on the ring cop or bobbin affects the efficiency of ring frames and winding machinery. Hence, it is essential to check of package weight very carefully and obtain optimum values for spinning and winding. Amount of yarn spun onto ring tubes mainly depends on the volume of yarn and yarn packing density (g/cc). The various factors influencing the package content in the ring spinning are listed in Table 7.47.

The efficiency of the ring spinning process is related to the number of doffs per shift. The increase in cop content decreases the number of doffs per shift and subsequently the efficiency increases. This necessitates the optimum shaping of spinning bobbins. The quality and formation of spinning cops have an enormous effect on the efficiency of winding and work load. Especially unattended spinning cops due to end breaks cause lower efficiency in winding. Optimum package weight is determined by volume and packing density.

TABLE 7.47

Various Factors Influencing the Cop or Ring Bobbin Content

Ring tube dimensions
Full bobbin dimensions
Conicity of tube
Clearance between tube and spindle
Spindle diameter
Yarn winding starting position
Yarn winding ending positions
Bottom stroke length
Ring rail lift
Winding length per stroke
Yarn twist
Type of raw material
Yarn count
Spinning yarn tension
Traveler weight
Yarn diameter
Fiber packing density in yarn formation

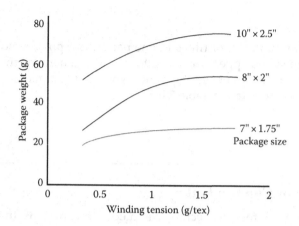

FIGURE 7.38
Effect of winding tension on package weight.

Optimum winding tension is necessary to make a package sufficiently firm to withstand handling (Figure 7.38). Winding tension (T) is given by

$$T = \frac{pD}{2}$$

where
  p is the pressure on the yarn
  D is the diameter of the package

Theoretical prediction of package weight is considerably complicated by the fact that the pressure and therefore the density decrease from the inner layer of a package to the outer layer.[37]

It can be noted that only a small gain in weight is obtained for a considerable increase in tension.

7.7.1 Coil Spacing

Lay is the measure of the closeness of the spacing of the yarn coils in any one layer of the bobbin. It is normally given in number of coils per inch measured parallel with the axis of the bobbin. The number of coils per ring rail stroke, front roller delivery rate, and yarn diameter determines the coil spacing. Coil spacing should be seven to eight times of yarn diameter. Since yarn diameter is computed for a given type of raw material and yarn count, optimum regulation of coil spacing is achieved.

$$\text{Coils per inch} = 8.69 \times \sqrt{\text{Count in Ne}}$$

7.7.2 Cop Bottom or Base Building Attachment

In an effort to achieve the maximum cop content, many spinners use a "cop bottom" or base building attachment on the ring frame especially conventional machines. This attachment facilitates reduction of ring rail traverse to about one half of the normal traverse. The reduced traverse tends to close the coil spacing of yarns.

7.8 Count C.V% and Evenness

The count variation of the yarn spun in the ring spinning process is a critical quality issue. Higher level of count C.V% deteriorates the quality of the yarn and subsequently final fabric appearance. The productivity of winding is also affected due to higher count variations. Higher count variability invariably leads to higher strength variability. The weak patches in the yarn tend to create frequent end break in further processing, which often reaches annoying levels, leading to rejection of ring bobbins. Higher count variability especially of medium to long length range results in moire-like appearance in fabric and increases warp way streaks and weft bars. Ring cuts and soiled ring packages are another problem with higher count C.V%. To overcome this, wider clearance is kept between ring diameter and full package leading to lower doff weights. The various factors influencing the count C.V% and yarn unevenness are relative humidity, fiber fineness, short fiber content, twist level, and cut length taken during testing (Figures 7.39 through 7.43). Yarn evenness is a measure of the level of variation in yarn linear density or mass per unit length of yarn. When there are large variations in yarn linear density, there will be many thin spots in the yarn, which are often the weak spots.

Ishtiaque et al.[14] reported that the yarn U% decreases with the increase in top roller pressure and traveler mass due to the fact that the increase in top roller pressure consolidates fibers strand in the drafting zone and fibers move in a more controlled manner so that the erratic movement of floating fibers is restricted. Furthermore, with the increase in traveler mass, the twist flow increases in the spinning zone that may lead to better binding of edge fibers in the yarn body and they do not eject out from the spinning triangle, resulting in

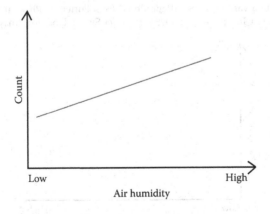

FIGURE 7.39
Effect of humidity on count C.V%. (Reproduced from Furter, R., Physical properties of spun yarns—Application Report, SE 586 Uster Technologies AG, USTER®, Uster, Switzerland, 3rd edn., June 2004. With permission.)

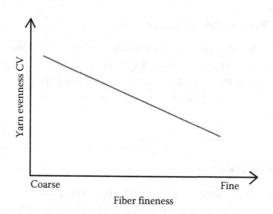

FIGURE 7.40
Effect of fiber fineness on yarn evenness. (Reproduced from Furter, R., Physical properties of spun yarns—Application Report, SE 586 Uster Technologies AG, USTER®, Uster, Switzerland, 3rd edn., June 2004. With permission.)

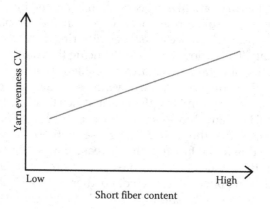

FIGURE 7.41
Effect of short fiber content on yarn evenness. (Reproduced from Furter, R., Physical properties of spun yarns—Application Report, SE 586 Uster Technologies AG, USTER®, Uster, Switzerland, 3rd edn., June 2004. With permission.)

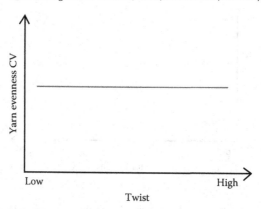

FIGURE 7.42
Effect of twist level on yarn evenness. (Reproduced from Furter, R., Physical properties of spun yarns—Application Report, SE 586 Uster Technologies AG, USTER®, Uster, Switzerland, 3rd edn., June 2004. With permission.)

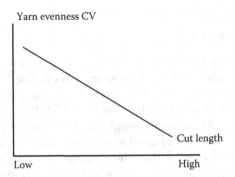

FIGURE 7.43
Effect of cut length on yarn evenness. (Reproduced from Furter, R., Physical properties of spun yarns—Application Report, SE 586 Uster Technologies AG, USTER®, Uster, Switzerland, 3rd edn., June 2004. With permission.)

better yarn evenness. The study also concluded that the imperfections first increase and then decrease with the increase in top roller pressure. The yarn imperfections initially decrease and then increase with the increase in spindle speed.

7.8.1 Drafting Waves

Floating fibers, short fibers or fibers, which are clamped neither by the pair of delivery roller nor by the pair of take-off roller of a drafting plane, can cause drafting waves or peaks (Figure 7.44). The height of such a peak depends on the number of uncontrolled fibers and the drafting value. Such a fault can be due to an unsuitably set drafting system.

The average wavelength of the drafting peak always corresponds to approximately 2.5 times the medium staple length.

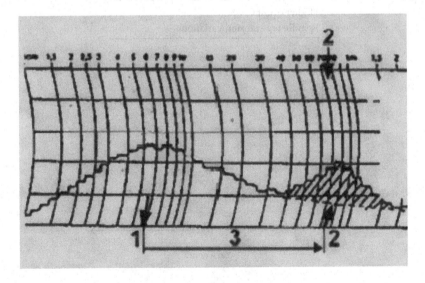

FIGURE 7.44
Spectrogram of normal and faulty drafting. (Reproduced from Furter, R., Physical properties of spun yarns—Application Report, SE 586 Uster Technologies AG, USTER®, Uster, Switzerland, 3rd edn., June 2004. With permission.)

7.9 Tenacity and Tenacity C.V%

A basic requirement for any yarn is that it can stand up to downstream processes without causing stoppages or affecting production efficiency. The tenacity of the yarn is an important property requirement in fabric manufacturing process especially weaving where the yarn has to withstand the stress imposed by the various machine elements under dynamic conditions. Minimum strength and elongation properties are needed to prevent a yarn breaking or being damaged in downstream operations as well as avoiding damage to the end products in weaving. The various causes of higher tenacity C.V% are listed in Table 7.48. The higher tenacity C.V% will tend to create more end breakages and severely affect the productivity of the weaving preparatory process and weaving process. The tenacity depends on the twist and the fiber strength. The tenacity varies between 15 and 26 cN/tex, the elongation between 4% and 10%, depending on the fiber strength, the yarn count, and twist. The tenacity variation is higher for fine yarns because the probability of weak places in fine yarns is higher (Figure 7.45).

Ishtiaque et al.[14] reported that the yarn tenacity increases with the increase in top roller pressure and traveler mass due to the fact that the higher top roller pressure increases the normal force over the fibers and this may reduce the fiber slippage and facilitates better control over the movement of fibers during their sliding. These factors are responsible for greater yarn strength.

TABLE 7.48

Various Causes of Higher Tenacity C.V%

Twist variation
Variations in yarn evenness
Humidity fluctuations inside the department
Higher count C.V%
Spindle tape tension variation

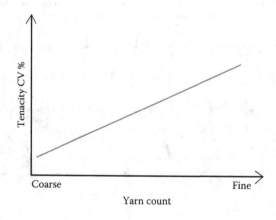

FIGURE 7.45

Effect of yarn count on tenacity C.V%. (Reproduced from Furter, R., Physical properties of spun yarns—Application Report, SE 586 Uster Technologies AG, USTER®, Uster, Switzerland, 3rd edn., June 2004. With permission.)

7.10 Roller Lapping

Roller lapping is a severe problem that every spinner faces in the ring frame process. It significantly affects the productivity and quality. The economic implications of roller lapping in the ring spinning process are listed in Table 7.49.

Lapping occurs immediately after an end break due to loose drafted strand of fibers delivered from front roller strongly adhered to the roller surface (top or bottom).[38] The adhesion[39] between fibers and roller surface arises due to the following factors:

- Higher level of honey dew in the cotton
- Negative air pressure build up around the roller creating a suction effect
- Improper humidity control in ring frame department

7.10.1 Factors Influencing Roller Lapping

Finer fibers would show a higher tendency to form laps due to their lower bending rigidity and large surface area of contact. Shiffler[40] reported that shorter fibers tend to show a greater lapping tendency, whereas results reported by Hariharan and Seshan[38] show that long and fine fibers show a higher incidence of lapping. Excessive moisture in air or improper mix-up of mixing oil or water during stack mixing may also be prone to lapping. Shiffler[41] stated that a high roving twist has a positive influence on lapping behavior. The increase in top roller load increases lapping propensity due to fibers are pressed more strongly against roller surface. Shiffler[41] also reported that a cot with large diameter significantly reduces the lapping propensity due to reduction in fiber wrap angle. Chattopadhyay[39] reported that the softer cots are likely to form more laps due to increased adhesion forces created by its longer surface area of intimate contact with the fibers.

7.10.2 Measures to Prevent Lapping Tendency

Acid treatment and berkolization has been reported to reduce lapping propensity. Due to ageing, cracks developed on cots, surface. Grinding is normally carried out to rectify the cracks developed in the cots. The distance of the pneumafil suction pipe from the front roller nip affects lapping frequency. If the distance is more, the force with which loose fibers are sucked in reduces so that the lapping frequency is likely to increase. The negative pressure in the pneumafil suction pipe also affects roller lapping. The negative pressure in the suction pipe should not fall below 70 mm of water. Balasubramanian and Trivedi[42]

TABLE 7.49

Economic Implications of Roller Lapping

Production loss
Increase in soft waste due to lapping
Increase in idle spindles
Possibility of cot damage while clearing the laps
Increase in fly liberation while clearing
Increase in end breakage rate
Poor yarn quality

reported that the provision of stationary clearer roller over top roller in the drafting zone reduces the lapping tendency due to greater frictional resistance. Incidence of lapping on top roller is due to high R.H% and that on bottom roller is due to low R.H%. The maintenance of correct level of R.H% is needed for trouble-free processing.

7.11 Yarn Quality Requirements for Different Applications

Enhanced fabric quality, in spite of higher speed, is not only the result of increased process dependability and optimization of process functions but also of continuous improvements in the quality level of the yarns to be woven or knitted. The progress is best confirmed by the Uster statistics concerning fiber yarns. Since the raw material basis for natural fibers remained fundamentally unchanged, these improvements in the quality of ring spun cotton yarns, for instance, do not touch the basic properties, but rather those characteristic quality variables that can be influenced by the spinning technological means. The cost to repair a yarn failure is much less if it occurs prior to the weaving process. In addition, a yarn failure during weaving also increases the chances for off-quality fabric. Most of the quality problems encountered during fabric forming are directly related to mistakes made during yarn manufacturing or yarn preparation for weaving. The yarn breakage is directly proportional to the speed of the machine for the same quality of yarn being used. The stringent quality requirements of spun yarn for high-speed weaving and knitting process are discussed here.

7.11.1 Weaving

In the last two decades, tremendous progress has been made in the field of weaving technology and the most significant being the replacement of convectional looms by shuttle-less looms for increasing the productivity and quality of the end product. With ever-increasing weaving speeds, the requirements on weaving preparation are also getting more stringent. The weaving process is still to be blamed for disturbances of the warp flow. Warp yarn must have uniform properties with sufficient strength to withstand stress and frictional abrasion during weaving. Tension experienced by the warp is higher in high-speed shuttleless weaving compared to conventional weaving.[43] The speed of shuttleless loom is usually three to four times faster. If the quality of warp remains the same, warp breaks will increase three to four times, resulting in low production. The weft yarns are not subjected to the same type of stresses as are the warp yarns and thus are easily prepared for the weaving process. The various quality requirements of cotton spun yarn for the high-speed weaving process are summarized in Table 7.50.

Minimum strength of yarn should not be lower than 70%–75% of average yarn strength. Minimum strength is influenced by strength C.V%.

7.11.2 Knitting

Knitted fabrics rejection should be less than 1%. Yarn faults contribute to 25% of the rejections. The requirement of yarn strength for knitting process is secondary compared to weaving process. The mechanical stress experienced by the yarn in the knitting machine is lower than that with a high-speed weaving machine. However, the yarn must have enough

TABLE 7.50

Quality Requirements of Cotton Spun Yarn for High-Speed Weaving

Count C.V%	<1.4%
TPI C.V%	<5%
Unevenness U%	<8%
Tenacity	>16.5 g/tex
Tenacity C.V%	<7.5%
Elongation %	>5.5%
Elongation C.V%	<8%
Hairiness index	<4
Hairiness C.V%	<1.5%
Imperfections/1000 m	<70/1000 m
Objectionable CLASSIMAT faults	<1/100 km
Shade variation under UV exposure	Nil

elongation and elasticity. Elastic recovery of yarns is influenced by the breaking elongation. The higher the breaking elongation of yarns, the better will be the elastic recovery and vice versa. Thick places and thin places in the yarn should be very low as these create problems such as stop holes in the knitted fabric and broken needles, respectively. The yarn meant for high-speed knitting operation should have a low friction value as the yarn has to pass freely through the various guide elements of the machine[44] (Table 7.51).

The tendency of a yarn to shed lint during abrasion with machine parts under running condition is termed as lint shedding.[45] Higher hairiness in the yarn is one of the reasons for lint shedding tendency. The lint accumulations are taken away by the incoming yarn and obstruct the passage by blocking needles and subsequently create yarn or needle breakage and deteriorate fabric appearance. Fly liberation should be less during knitting and it should be less than 25 mg/kg of yarn knitted. Nearly, 25% of all the faults occurring during knitting process are due to incidence of lint/fiber fly, which includes faults such as needle stripes and dropped stitches. The major yarn faults responsible for knitted fabric rejections are listed in Table 7.52.

Another defect in the knitted fabric called "patchiness" is the result of uneven yarn, which becomes pronounced as the cover factor is decreased. The cloudy appearance of the knitted fabric after dyeing/finishing is mainly because of random or periodic hairiness

TABLE 7.51

Quality Requirements of Cotton Spun Yarn for High-Speed Knitting

Count variation (cut length 100 m)	<1.8%
Count variation (cut length 10 m)	<2.5%
Breaking tenacity	>10 cN/tex
Elongation at breaking force	>5%
Yarn twist factor	3.1–3.6
Paraffin waxing/surface friction value	0.15 μm
Yarn irregularity	<25% value of Uster® Statistics
Hairiness H	>50% value of Uster Statistics
Hairiness variation between bobbins	<7%
Seldom occurring thick and thin place faults (CLASSIMAT values)	<A3/B3/C2/D2 or D1 or more sensitive
Remaining yarn faults (CLASSIMAT values)	A3 + B3 + C2 + D2 = <5/1,00,000 m

TABLE 7.52

Major Yarn Faults Responsible for Knitted Fabric Rejections

Unevenness and periodic irregularities
Stiff yarn—higher TPI
Contamination (especially of length more than 20 mm)
Thick and thin places
Higher friction
Higher hairiness variation
Neps
Shade variation due to cotton color variations
Black spots or kitties (vegetable matters and dust content)
White specs (immature fibers)
Lower elongation and elasticity
Higher 10 m C.V% of yarn (affects fabric appearance)

variation in the yarn. Higher twist variation in the hosiery yarn tends to create fabric barre. Objectionable fault measured by CLASSIMAT system affects yarn breakage in high-speed knitting process.[46] The tiny trash particles present in the yarn may get transferred to the knitted fabric as black spots or kitties, which will spoil the fabric appearance and create problems in dyeing. These tiny foreign matters may also increase the wear of needles and create holes in the knitted fabrics.

7.12 Technological Developments in Ring Spinning

The technology behind ring spinning has remained largely unchanged for many years, but there have been significant refinements. In the last five decades, major developments witnessed in the ring spinning machine are in the areas such as creel, drafting zone, spindle, spindle drive, ring, traveler, and driving mechanisms. Modern ring spinning machines can incorporate push-button draft and twist changes, automatic doffing (also without under winding), sliver/roving stop motions, thread break indicators, electronic speed and package building programs, and automatic piecing, online monitoring, data collection, ring cleaning, and can also be linked to the winders, with a cop steamer stage between spinning and winding. The producers of the modern spinning machines have been developing the machines with improved construction of different working elements and optimal spinning geometry, with a ring diameter of 36 mm, a tube length of 180 mm, and a spindle speed of up to 25,000 rpm. Some of the major technological advancements in ring spinning machines manufactured by different ring frame manufacturers are discussed here.

7.12.1 Developments in Rieter® Spinning Machines

1. *SERVOgrip*: The yarn has to wind several times around the lower end of the spindle to hold it in the spinning position at the time of doffing.[47,50] These under windings often cause multiples end down and lead to fiber fly when machine is restarted after doffing. SERVOgrip as shown in Figure 7.46 is a system of doffing ring cop without the under winding threads. The main element of the SERVOgrip is a patented clamping crown. At the time of doffing the ring rail moves downward and the clamping

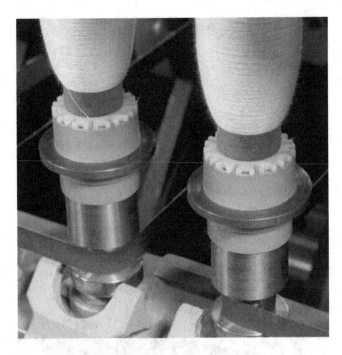

FIGURE 7.46
Rieter SERVOgrip. (From Rieter Textile Machinery Company, Rieter ring frame manuals, Winterthur, Switzerland, 2011. With permission.)

crown opens while the spindle is still revolving slowly. The yarn gets inserted in the open crown and the crown gets closed afterward. When the cop is replaced, the length of the yarn remains firmly clamped, enabling piecing after machine is started.

2. *Individual spindle monitoring (ISM)*: ISM as shown in Figure 7.47 is a quality monitoring system. This system reports faults and anomalies by means of a three-level light guidance system thus enabling personnel to locate the problem spindles without unnecessary searches. Signal lamps at the end of the machine indicate the side of the machine on which the ends down rate has been exceeded (level 1). An extra-bright LED on each section guides the operator to the location of the fault (level 2). The indicator on the spindle itself signals ends down with a continuous light and slipping spindles with a flashing light.

3. *Ri-Q-Draft system*: Ri-Q-Draft drafting system with pneumatically loaded guide arm and the Ri-Q-Bridge provides improved and consistent thread guidance in the main drafting zone at all critical points of yarn production through optimum coordination of cage lengths with draft distances. The incorporation of a new bottom apron guide bridge, the Ri-Q-Bridge guarantees optimum apron running. It is so accurately coordinated with the operation of the top apron so that absolutely precise fiber guidance is achieved in the main draft. This results in a striking improvement in yarn quality and IPI values.

4. *FLEXIdraft*: FLEXIdraft flexible drive, equipped on Rieter G 33 ring spinning machine, features separate drives for the drafting system and the spindles. This system enables change in the yarn count, twist and twist direction (S/Z) via

FIGURE 7.47
Rieter—individual spindle monitoring (ISM). (From Rieter Textile Machinery Company, Rieter ring frame manuals, Winterthur, Switzerland, 2011. With permission.)

the control panel of the machine. The drafting rollers are split in the center of the machine to ensure smooth running of drafting operation. On the basis of FLEXIdraft, each drafting system drive can be started or stopped individually via FLEXIstart system. Thus, depending on machine length, one-sided or two-sided drafting system drives are used. FLEXIdraft has a further advantage of noise level reduction due to elimination gear of wheels.

7.12.2 Developments in Toyota® Spinning Machines

1. *Optimized spinning geometry*: The reduction in stretch length and higher spinning angle on Toyota RX240 NEW ring frame results in high speed due to better twist propagation and stable ballooning with reduced yarn breakage.[48,50] Similarly, balloon control ring that moves together with the lappet at the start of winding and then with the ring from about 40% cop winding leads to stable balloon form.

2. *Toyota ElectroDraft® System*: The Toyota ElectroDraft® System (optional) features independent servo motors drive for front and back rollers. The spindles are also driven by separate tangential drive system where 1 motor drives 96 spindles. Thus the required draft and yarn twist can be set via control panel.

3. *Toyota Servo motor–driven positive lifting mechanism*: Toyota's proprietary screw shaft positive lifting mechanism is used in RX240 NEW ring for ring rail lifting motion.

This eliminates disparity in the ring rail motion during long periods of continuous operation. The different cop parameters such as chase length, cop diameter, winding start position, bobbin diameter (bottom and top), total lift, and so on can be fed via key operation of the machine panel.

7.12.3 Developments in Zinser® Spinning Machines

1. *Zinser optimized spinning geometry*: Zinser has claimed that the yarn path has been made short, thus causing fewer disturbances to twist transmission as compared with a long thread run.[49,50] The program-controlled motor gives drive to yarn guiding element; consequently, bobbin build can also be preprogrammed from the control panel by setting winding length, traverse length, and ring rail upward displacement. The traversing yarn guide remains at standstill at the start of the bobbin build operation for effective adaptation to the spinning tension that results in constant yarn tension.

2. *Zinser GUARD system (RovingGuard and FilaGuard)*: The individual yarn monitor FilaGuard monitors the rotation of the steel ring travelers on each spindle and detects any yarn break immediately. Optical signals indicate the specific yarn break, directing the operating personnel to the spindle of yarn break to rectify the problem. The automatic roving stop RovingGuard, which responds within milliseconds, interrupts the roving feed in case of yarn break, thereby preventing material loss and minimize lapping tendency.

3. *Zinser SynchroDrive, SynchroDraft, and ServoDraft*: Zinser SychroDrive is a multimotor tangential belt drive system. The system employed several motors arranged at defined positions to drive spindles through tangential belt. The consistency in spindles speed relative to each other minimizes the twist variation apart from reduction in noise level and minimum power requirement. SynchroDraft transmission is for long machines to drive the middle bottom rollers from both ends, consequently minimizing twist variation between gear end and off end of the machine. Zinser ServoDraft system employs individual motors for driving bottom rollers of the drafting system.[50] Hence, yarn count and twist change can be done by simply feeding required parameters at the control panel of the machine that adjusts the motors speed accordingly.

7.12.4 Compact Spinning System

The yarns produced in ring spinning have good strength and unique structure but the integration of many fibers is poor, and such fibers tend to generate hairiness that does not contribute to yarn strength. Compact spinning offered the potential to create a near-perfect yarn structure by applying air suction to condense the fiber stream in the main drafting zone, thereby virtually eliminating the spinning triangle.[16] The purpose of a genuine compact spinning process is to arrange the fibers in a completely parallel and close position before twist is imparted. Compacting takes place in the compacting zone following the main drafting zone of the drafting system. With the invention of compact spinning, for the first time a new spinning process was not aimed at exclusively achieving higher production, but at better yarn utilization and yarn quality. Compact spinning systems were first presented at the International Textile Machine Fair ITMA'99. At present, the Rieter, Suessen, and Zinser companies have produced compact ring spinning frames.

Yarn manufactured by means of the compact spinning system compared with classical yarn is characterized by

- Better smoothness
- Higher luster
- Abrasion fastness better by 40%–50%
- Hairiness lower by 20%–30%, as measured with the use of the Uster apparatus
- Hairiness lower by 60%, as measured with the use of the Zweigle apparatus
- Tenacity and elongation at break higher by 8%–15%.
- Smaller mass irregularity

References

1. Gordon, S. and Hsieh, Y.L., *Cotton: Science and Technology*, Woodhead Publications, Cambridge, U.K., 2007.
2. Klein, W., *A Practical Guide to Ring Spinning, Manual of Textile Technology*, Vol. 2, Textile Institute, Manchester, U.K., 1987.
3. Autotex - ASSOCIATED AUTOTEX ANCILLARIES (P) LIMITED, What is the main yarn fault caused by the bobbin holders?, http://www.autotex.net/ans/16.htm, Accessed Date: May 1, 2013.
4. Balasubramanian, N., A study of irregularities added in apron drafting. *Textile Research Journal* 39: 155–165, 1969.
5. Sujai, B. and Sivakumar, M., Effect of spinning rubber cot shore hardness on yarn mass uniformity and imperfection levels, http://www.inarco.com/pdfs/05%20technical%20information/publications/Effect%20of%20spinning%20rubber%20cot%20shore%20hardness%20on%20yarn%20mass%20uniformity%20and%20imperfection%20levels.pdf.
6. Binz, H., Yarn Spinning Innovation, Technology, *Evolution—Business and Technology Magazine from SKF*, May 15, 1997, http://evolution.skf.com/yarn-spinning-innovation/, Accessed Date: May 1, 2013.
7. Caveny and Foster, The irregularity of materials drafted on cotton, spinning machinery and its dependence on draft, doubling and roller setting—Part I - Speed frames, *JTI*, 46, T529–T550, 1955.
8. Bannot, B.N. and Balasubramaniam, N., Roving twist and apron spacing upon yarn quality and ring frame end breakages. *Journal of Textile Association* 35(4): 143–146, 1975.
9. Basu, A. and Gotipamul, R., Effect of some ring spinning and winding parameters on extra sensitive yarn imperfections. *Indian Journal of Fibre & Textile Research* (South India Textile Research Association) 30: 211–214, June 2005.
10. Tang, Z.-X., Barrie Fraser, W., Wang, L., and Wang, X., Examining the effects of balloon control ring on ring spinning. *Fibers and Polymers* 9(5): 625–632, 2008.
11. Senthil Kumar, R., Role of ring and traveller in the ring spinning, http://www.scribd.com/sen29iit, Accessed Date: May 1, 2013.
12. Reiners + Furst GmbH u.Co. KG, Riener—360° Performance, Riener traveller—Product information brochure, Reiners + Fürst GmbH u. Co. KG, Mönchengladbach, Germany, 2011, http:// www.reinersfuerst.com, Accessed Date: May 1, 2013.
13. Bracker, Short staple spinning, Bracker Technical Information Manual, http://www.bracker.ch, Accessed Date: May 1, 2013.

14. Ishtiaque, S.M., Rengasamy, R.S., and Ghosh, A., Optimization of ring frame process parameters for better yarn quality and production. *Indian Journal of Fibre & Textile Research* 29: 190–195, June 2004.
15. Barella, A., Torn, J., and Vigo, J.P., Application of a new hairiness meter to the study of sources of yam hairiness, *Textile Research Journal* 41: 126, 1971.
16. Gowda, R.V.M., *New Spinning Systems*, NCUTE Publication, IIT Delhi, Delhi, India, 2003.
17. Lunenschloss, *Textil-Praxis* 22: 649, 1967.
18. Stalder, H., New spinning process comforspin. *Melliand International* 6: 26, 2000.
19. LRT Engineering Division, *LRT Handbook*, http://www.lrtindia.com/downloads/rt_handbook_en.pdf, Accessed Date: May 1, 2013.
20. Liang, R., A new idea to reduce ring spinning yarn hairiness—Mismatch spinning. *Shang Hai Textile Science & Technology* 37(3): 16–17, 25, 2009.
21. Liang, R., Jingkun, C., Guangcai, F., Qinjie, S., and Guoqi, L., Mismatch spinning to reduce ring spinning yarn hairiness. *Progress in Textile Science & Technology* 2: 39–40, 2009.
22. Haghighat, E.A., Johari, M.S., and Etrati, S.M., A study of the hairiness of polyester—Viscose blended yarns; Part I. Drafting system parameters. *Fibres & Textiles in Eastern Europe* 16(2): 67, April/June 2008.
23. Lord, P.R., *Handbook of Yarn Technology: Technology, Science and Economics*, Woodhead Publishing Ltd., Cambridge, U.K., 2003.
24. Sokolov, G.V., *Theory of Twisting of Fibre Materials*, Legprombytizdat, Moscow, Russia, 1977.
25. Walton, W., Use of a yarn hairiness meter and results showing the effect of some spinning conditions on yarn hairiness. *Journal of Textile Institute* 59: 365, 1968.
26. Veejay Fineplast, Spinning tubes, http://www.veejaygroup.com/ring_spinning_tubes.html, Accessed Date: May 1, 2013.
27. Senthil Kumar, R., End breakage rate—A performance indicator in the spinning mill, http://www.scribd.com/sen29iit, Accessed Date: May 1, 2013.
28. Salhotra, K.R., Mechanism of end breakage in ring spinning, *NCUTE Programme Series—Ring Spinning, Doubling and Twisting*, NCUTE, IIT Delhi, Delhi, India, March 2000.
29. Biradar, M.M., End breakage rate—A major performance indicator of the spinning mill, http://www.fibre2fashion.com/industry-article/35/3422/end-breakage-in-spinning-a-major-performance-indicator-of-spinning-mill2.asp, Accessed Date: May 1, 2013.
30. Ghosh, A., Ishtiaque, S., Rengasamy, S., and Patnaik, A., The mechanism of end breakage in ring spinning: A statistical model to predict the end breaks in ring spinning. *AUTEX Research Journal* 4(1): 19–24. March 2004.
31. End breakage in ring spinning, *Textile Encyclopedia*, http://wwwsen29iitcom.blogspot.com/.../end-breakage-in-ring-spinning.html, Accessed Date: May 1, 2013.
32. Das, A., Ishtiaque, S.M., and Kumar, R., Study on drafting force of roving: Part II—Effect of material variables. *Indian Journal of Fibre & Textile Research* 29: 179–183, June 2004.
33. Das, A., Ishtiaque, S.M., and Kumar, R., Study on drafting force of roving. Part IV—Correlation between drafting force, roving strength and yarn quality. *Indian Journal of Fibre & Textile Research* 29(4): 313–317, 2004.
34. Palaniswamy, K. and Mohamed, P., Yarn twisting. *AUTEX Research Journal* 5(2): 87–90, June 2005.
35. Furter, R. and Meier, S., Measurement and significance of yarn twist, USTER ZWEUGKE Twist Tester 5—Application Report, Uster Technologies AG, Uster, Switzerland, September 2009.
36. Chellamani, K.P., Kumar, V.J., and Vittopa, M.K., Twist for maximum strength & twist contraction in compact spun yarns. *SITRA Focus*, 27(3): 2–4, September 2009.
37. Ishtiaque, S.M. and Alagirusamy, R., Optimizing package size for ring spinning, *NCUTE Programme Series—Ring Spinning, Doubling and Twisting*, NCUTE, IIT Delhi, Delhi, India, March 2000.
38. Hariharan, R. and Seshan, K.N., How to reduce roller lapping at ring frame. *The Indian Textile Journal* 98: 140, December 1987.

39. Chattopadhyay, R., Roller lapping in ring spinning, *NCUTE Programme Series—Ring Spinning, Doubling and Twisting*, NCUTE, IIT Delhi, Delhi, India, March 2000.

40. Shiffler, D.A., Roll wraps in ring spinning. Part I—Kinetics and incremental costs. *Textile Research Journal* 8: 479–487, 1993.

41. Shiffler, D.A., Roll wraps in ring spinning. Part II—Effect of fibre and spinning variables. *Textile Research Journal* 9: 515–522, 1993.

42. Balasubramanian, N. and Trivedi, G.K., Stationary clearer will reduce roller lapping. *Indian Textile Journal* 89–91, February, 1979.

43. Dhandhani, V. and Sawant, S., Yarn quality requirement for high speed weaving machines, http://www.fibre2fashion.com, Accessed Date: May 1, 2013.

44. Yarn requirements for knitting, *Technical Bulletin*, Cotton Incorporated, Cary, NC, 2–4, TRI 2006.

45. Bhowmick, N. and Ghosh, S., Role of yarn hairiness in knitting process and its impact on knitting room's environment. *WSEAS Transactions on Environment and Development* 4(4): 360–362, April 2008.

46. Chellamani, K.P and Vittopa, M.K., Quality requirements for hosiery yarns, http://www.fibre-2fashion.com/.../quality-requirements-for-hosiery-yarns1.asp, Accessed Date: May 1, 2013.

47. Rieter, Ring spinning, G36 ring spinning machine—Technical information brochures, Rieter Machine Works, Winterthur, Switzerland, Accessed Date: May 1, 2013.

48. Toyota Industries Corporation RX240 NEW, *Toyota Textile Machinery Bulletin*, 12, http://www.toyota-industries.com/textile/whatsnew/bulletin_maintenance/pdf/vol12e.pdf, Accessed Date: May 1, 2013.

49. Zinser, Zinser Ring71—Technical information brochure, Saurer, Oerlikon Schlafhorst, http://schlafhorst.saurer.com/fileadmin/Schlafhorst/Bilder/Produkte/Zinser_Ringspinnen/ZinserRing_71/ZinserRing_71_RZ_130305_EN_i.pdf, Accessed Date: May 1, 2013.

50. Singh, R.P. and Kothari, V.K., Developments in comber, speed frame & ring frame, *The Indian Textile Journal*, May 2009, http://www.indiantextilejournal.com/articles/fadetails.asp?id=2069, Accessed Date: May 1, 2013.

51. Furter, R., Physical properties of spun yarns—Application Report, SE 586 Uster Technologies AG, Uster, Switzerland, 3rd edn., June 2004.

52. Rieter ring frame manuals, Rieter Textile Machinery Company, Winterthur, Switzerland, 2011, Accessed Date: May 1, 2013.

8

Process Control in Winding

8.1 Significance of the Winding Process

Yarn winding is the final stage of the yarn-forming process and the starting point for various subsequent processes, from weaving or knitting to textile finishing. Yarns manufactured and packaged from ring spinning are not in the optimum condition to be used to form fabrics. Package size, build, and other factors make it necessary for the yarn to be further processed to prepare it to be handled efficiently during fabric formation. Yarn winding can be viewed as a packaging process, forming a link between the last few elements of yarn manufacturing and the first element of fabric manufacturing process. This interface function of winding is what makes the winding process so important. The ring-spinning operation produces a ring bobbin containing just a few grams of yarn, which is unsuitable for the efficiency of further processing, such as warping, weaving, and knitting. The winding process converts the ring bobbin of several grams into dense yarn package of several kilograms, which can unwind in the subsequent operations without interruptions.

Due to the ever-increasing emphasis on better quality of fabric for the highly competitive market and process performance, yarn has to meet the standards in respect of yarn faults, hairiness, lea count C.V%, etc., besides traditional quality standards. The yarn faults present in the spinning bobbin are removed in winding process in order to improve the quality of the final fabric and process efficiency of weaving preparatory processes and weaving process. Improper utilization of the features of the winding machine can not only cost heavily to the spinning mills but also lead to loss of good customers permanently.

8.1.1 Objectives of Winding Process

1. To remove objectionable faults from yarn
2. To build packages with dimensions compatible with requirements of the subsequent processes, which are equally important both for high-speed warping and shuttleless weaving

8.1.2 Types of Wound Packages

Wound packages are available in different types to meet the requirements of the knitting, weaving, and dyeing. The wound package is classified here in terms of how the yarn is laid on the package as

- *Parallel wound package*: yarn is laid parallel to one another so that each yarn is wound perpendicular to the package axis.
- *Near-parallel wound package*: one or more yarns are laid nearly parallel to one another owing to the slow traverse used to fill the full length of the flanged bobbin.
- *Cross-wound package*: yarn is laid on the package at an appreciable angle so that the layers of yarn one another giving stability.

Cross-wound packages have two types such as cone and cheese.

8.1.3 Cross-Winding Technology: Terminologies

- *Wind angle*: It is defined as the angle between the directions of the yarn laid on the package surface and any plane perpendicular to the package axis.
- *Wind ratio*: It is the number of revolutions made by the package while the yarn guide makes a single traverse from one end of the package to the other.
- *Traverse ratio*: It is the number of coils laid on the package during a double traverse of the yarn guide. It is twice the wind ratio.
- *Traverse length or traverse*: It is defined as the distance between extreme positions of a reciprocating thread-guide in one cycle of its movement.
- *Gain*: It is the angular displacement of the yarn at the beginning of a double traverse, with respect to the corresponding position of the previous double traverse.
- *Ribboning or patterning*: If the yarn is repeatedly laid on the top of or along the same path as the previously wound yarn, this duplication of yarn path on the package creates a defect known as Ribboning or patterning.

8.2 Demands of Cone Winding Process

In today's competitive market, the customers are becoming more and more sensitive regarding quality. Every parameter regarding yarn and package quality is of prime importance and must be met out in a close tolerance. To meet both qualitative and quantitative demands, one has to exploit all the available features of winding machine in a judicious manner. The control of the quality characteristics of a yarn during winding can be done by offline or online testing.

New generation of high-speed looms and knitting machines place increasingly more stringent demands on the quality and processability of the yarn (Figure 8.1). Companies that are best able to respond to the challenge of economically improving the quality of the yarn to effectively meet the requirements of high-speed fabric producing machines, are going to be successful in a competitive market.[1]

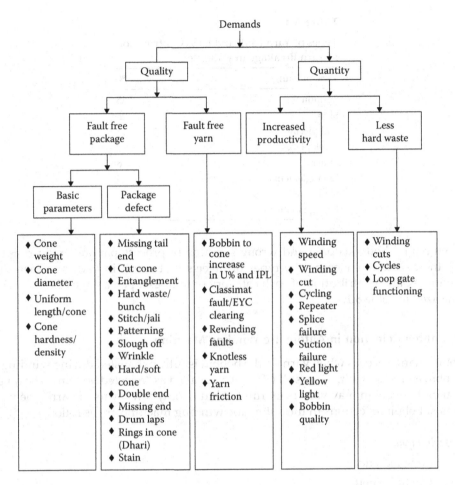

FIGURE 8.1
Demands of the cone winding process.

8.2.1 Quality Requirements of Ring Bobbins for Winding Operation

The high demands placed on quality and economic efficiency in the textile industry are subject to continuous change.[1] The significance of yarn package winding as a quality filter in the textile production chain is growing in the same measure. Demands that are still crucial in yarn winding are the conservation of the specific quality of the yarn, the further optimization of the structure and build of the yarn packages. Yarn quality and package quality of ring bobbin produced in ring spinning are vital for trouble-free winding operation. The inferior quality of yarn substantially affects the productivity of the cone winding machine due to higher clearer cuts.

The yarn quality is decided by appropriate selection of raw material and optimum process parameters from mixing to ring spinning departments. The yarn quality parameters such as unevenness, count C.V, strength, imperfections, and hairiness should be under control. Bobbins that exceed the selected limits of quality characteristics have to be ejected at the winding machine. Package quality predominantly influences the productivity and hard waste generation in the winding process. The various bobbin parameters to be kept under control were discussed elaborately in the previous chapter. The yarn faults occurred

TABLE 8.1

Types of Yarn Fault and Its % Contribution
to Yarn Breakage in Winding

Type of Fault	%
Spun-in fly	35
Slub	25
Bad piecing	13
Weak places	9
Others	8
Entanglements	5
Slough-off	3
Foreign matter	2

due to the poor housekeeping and wrong selection of process parameters in ring frame is also the matter of concern. The types of defects that result in yarn breakage during cone winding and contribution of each category of fault as a percentage of total faults are summarized in Table 8.1.

8.2.2 Bobbin Rejection in Automatic Winding Machine

A bobbin change occurs when yarn on the bobbin is fully exhausted during winding.[1] But if a bobbin is changed with yarn still left on it, it is called as "rejected bobbin." The quantity of yarn on the bobbin may vary from full bobbin to only few layers of yarn. The various reasons of bobbin rejections in the automatic winding machine are as follows:

1. *Bobbin quality*:
    a. Long tail end
    b. Kirchi/Lapetta
    c. Deshaped bobbin
    d. Overfilled bobbin
    e. Bottom spoiled bobbin
    f. Ring cut bobbin
    g. Soft bobbin
    h. Sick bobbin
2. *Bobbin feeding in magazine*:
    a. Presence of under-winding and back-winding while feeding the bobbins in the magazine leads to rejection
3. *Top bunch transfer failure*:
    a. Top bunch position is lower with respect to bobbin tip.
    b. Blowing device does not come down to concentrate blow at the bobbin tip.
    c. Very few numbers of coils at the bobbin tip.
    d. Removal of top bunch due to fault in cutter at the bobbin preparatory or any other reason. Very few numbers of coils at the top bunch.

TABLE 8.2

Acceptable Deterioration of Yarn Properties from Ring Bobbin to Cone

Yarn Properties	Acceptable Deterioration (%) from Ring Bobbin to Cone
Unevenness U%	3–5
Thin places (−50%)	0–0.5
Thick places (+50%)	15–20
Neps (+200%)	5–10
Hairiness	25–30

4. *Fault in winding unit and/or splicing failure*
5. *Yarn quality*:
   a. High degree of objectionable fault
   b. Count variation
   c. High hairiness bobbin

8.2.3 Acceptable Deterioration in Quality from Ring Bobbin to Cone

During clearing and winding the yarn, it has been practically experienced by industry that there is deterioration of certain yarn characteristics like strength, elongation, hairiness, etc. Irregularity can adversely affect many of the properties of textile materials. There is deterioration in terms of U% and IPI values and hairiness from ring frame bobbin to cone due to abrasion of yarn with various contact points in yarn path (Table 8.2).

Thus, a yarn with higher unevenness directly affects the costs of production, the likelihood of rejection of a product, and the profit. The selection of process parameters in winding such as winding speed, auto speed, tension, etc., should be done in such a manner that the final yarn would attain the satisfactory quality. Rust and Peykamian[2] revealed that fibers migrate even during the winding process, increasing yam hairiness afterwards and a higher winding tension and/or higher yam velocity leads to more fiber migration and hence more severe yarn hairiness. Tarafder[3] studied the influence of the winding process on yam hairiness; he observed that the increased hairiness at the bottom of the bobbin was greater than that at the top after winding.

8.3 Factors Influencing Process Efficiency of Automatic Winding Machine

The achievement of expected process efficiency in the winding department is a challenging task for any spinning technician. The process efficiency of winding is influenced by several factors mentioned as follows:

- Winding speed
- Number of yarn splicing or knotting per 10,000 m
- Rate of yarn breakage
- Spinning bobbin mass

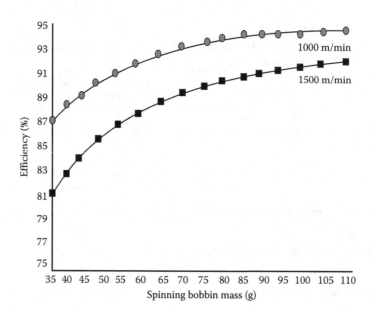

FIGURE 8.2
Effect of spinning bobbin mass and winding speed on winding efficiency.

- Waiting time
- Time for manual doffing

The spinning bobbin mass has a significant influence on the efficiency of the winding machine (Figure 8.2).

8.4 Winding Speed

The winding speed has a significant impact on the quality of the yarn as well as productivity of winding process. Higher winding speed put more stress and strain on the yarn and also increases degree of abrasion with different machine parts. The winding speed has a predominant effect on yarn properties such as imperfections, hairiness, tenacity, etc. A study conducted by SITRA[4] reported that the increase of winding speed from 1000 to 1400 m/min increases the imperfection by 40%–60% in combed counts and 65%–85% in carded counts. The aforementioned study also concludes that the yarn hairiness increases with the increase in winding speed from 20% to 35% in combed counts and 30% to 45% in carded counts. The loss in efficiency due to yarn breakage is higher at winding speed of 1500 m/min compared to 1000 m/min (Figure 8.3).

8.4.1 Slough-Off in High-Speed Unwinding

The coils of yarns lie exactly on top of each other, two or three such overlaying coils can be inadvertently pulled off together during unwinding of the yarn from the bobbin, which is termed as "slough-off." The occurrence of slough-off leads to entanglement of the coils, which then has to cut-away generates hard waste. Slough-off in the ring bobbin during

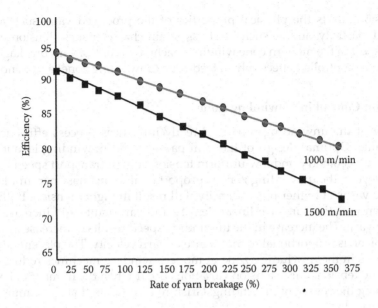

FIGURE 8.3
Effect of winding speed and yarn breakage rate on winding efficiency.

TABLE 8.3

Causes of Slough-Off and Their Frequency of Occurrence

Causes of Slough-Off	Frequency (%)
Number of coils (chase length)	42
Winding speed	17
Height of balloon breaker	13
Hardness of spinning bobbin	12
Relation between balloon breaker height and number of coils	4.5
Relation between winding speed and number of coils	4
Others	7.5

unwinding is a serious problem in cone winding that substantially affects the productivity and generates huge amount of hard waste. For high-speed winding, it is essential to minimize slough-off. It is difficult to remove sloughing completely with a yarn clearer and this can adversely affect the quality of final yarn packages. The various causes of slough-off and their frequency of occurrence are listed in Table 8.3.

The ring bobbin package dimensions should be as per recommendations in order to overcome the occurrence of slough-off.

8.5 Yarn Tension

Yarn tension plays a very important role in deciding the quality and efficiency of textile processes. Niederer[5] stated that with precise tension control, many processes can run at least 30% faster and have the added benefit of quality improvement at the same time. Variation in yarn tension affects the physical properties of the produced yarn, such as its tensile strength and elasticity, and as such its stress–strain characteristics. Tension control is an important aspect in the modern cone winding machine to control yarn breakages. Yarn tension needs to be controlled effectively in feed zone (unwinding zone) and winding zone.

8.5.1 Tension Control in Unwinding Zone

The stability of the unwinding process directly influences process efficiency and final product quality. Optimal design of the yam package and unwinding layout allows reasonable balloon geometry and low uniform tension at high transport speed. Yarn tension at the guide eye in the unwinding zone is proportional to the mass per unit length of the yarn and the yarns of higher linear density will result in higher tension. Both short-term and long-term variations in yarn linear density and yarn faults will increase the tension variations in yarn. The increase in the unwinding speed results in increase in yarn tension and yarn tension is proportional to the square of yarn velocity. The placement of unwinding accelerators in the unwinding zone in modern winding machines reduces and levels the unwinding tension. Large tension variations occur when the number of balloon loops changes during the course of unwinding. Grishin[6] has shown that the number of balloon loops (n) is given by

$$n \leq \left(\frac{\varepsilon Z}{\pi}\right) + \frac{1}{2}$$

$$n \leq \left(\frac{Z}{\pi}\right)\left(\frac{V}{a}\right)\left(\frac{\sqrt{T_o m_o}}{T_o - m_o V^2}\right) + \left(\frac{1}{2}\right)$$

where
  Z is the balloon height
  V is the yarn speed
  T_o is the yarn tension at guide eye
  m_o is the mass per unit length of the yarn
  a is the package radius at the unwinding point

During unwinding, the yarn tension is highest at the nose and lowest at the base of the spinning bobbin resulting in tension variations. Fluctuations in the unwinding angle at the point of unwinding on the package tend to generate tension variation in the yarn. Fraser[7] found that including elasticity in the theoretical unwinding model leads to a decrease in both the balloon tension and the balloon radius.

8.5.2 Tension Control in Winding Zone

The variation of yarn tension, especially in the formation of a package, has a negative effect on its technological properties, and therefore it is necessary to use suitable tension-stabilizing devices on the winding machines. Tension control is very important for uniform wound package and decides the productivity of preweaving processes. Many processes such as knitting and weaving that are feeding cone package can run at least 30% faster with better quality of package by precise tension control in winding process. To keep the yarn density even at all sections of a wound package, the automatic winders are equipped with devices

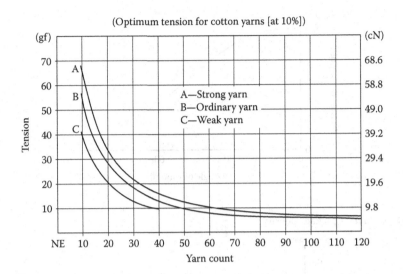

FIGURE 8.4
Yarn tension level for different counts.

that control the yarn tension uniformly, right from start to end of the package and also try to keep the tension levels lower. Tension level should be 8%–12% of single yarn strength. The tension levels for different yarn counts are shown in Figure 8.4. A study conducted by SITRA[4] reported that the yarn imperfections increased by 12%–75% in 40sCW when winding tension varied between 8% and 15%. Varying the yarn traverse speed in a predetermined manner as a function of the package diameter would also result in continuous variation in winding speed. In order to compensate for the resultant fluctuation in yarn tension, it would be desirable to appropriately vary the speed of the drive to package continuously.

The function of such devices, called tensioners, is to ensure that the tension of the yarn is maintained at a preset mean level and that its dynamic component is reduced to certain limits. Modern machines should be outfitted with active computer-controlled tensioners with the possibility of programming yarn tension at the desired level independently of the motion parameters of the yarn and its initial tension.

8.6 Package Density of Cone

Package density is a key parameter in winding process. The desired package density depends on the next process after winding and its requirements. Yarn meant for package dyeing needs soft packages of uniform density and diameter to ensure perfect dyeing results with assured consistency. The proper density and homogeneity of the package allows the dye to penetrate between the various layers of yarn, producing a uniform flow inside the batch. The low-density packages are produced by effective control of winding tension and the pressure between the package and the winding drum.

Angle of wind has a significant effect on the package density of yarn package (Figure 8.5). Package density in winding is directly influenced by the parameter "yarn tension" (Figure 8.6). Package density can be indirectly measured by "Durometer." It measures the

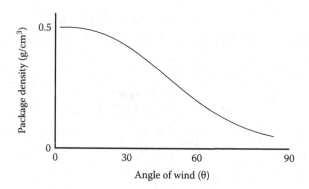

FIGURE 8.5
Effect of angle of wind on package density.

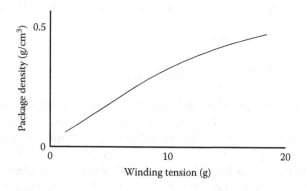

FIGURE 8.6
Effect of winding tension on package density.

TABLE 8.4

Shore Hardness of Yarn Package for Different
End-Use Application

End Use	Shore Hardness (°)
Knitting (4°20′ cone: 2.52 kg)	48–52
Weaving (4°20′ cone: 2.52 kg) kg)	55–58
Dyeing (cheese: 1.45 kg)	28–32

shore hardness of the package. Depending on the subsequent process, shore hardness of the yarn package is maintained as given in Table 8.4.

Package density of a cone can be calculated by referring Figure 8.7.

$$\text{Volume of cone } V = \left(\frac{\pi h}{12}\right)\left\{\left(D^2 + Dd + d^2\right) - \left(D'^2 + D'd' + d'^2\right)\right\} \text{cm}^3$$

$$\text{Net weight of cone} = W \text{ g}$$

$$\text{Then, package density } p = \frac{W}{V} \text{ g/cm}^3$$

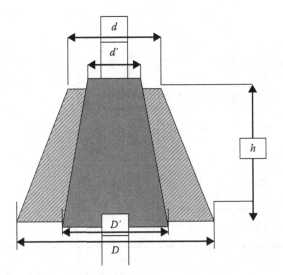

FIGURE 8.7
Determination of package density using package dimensions.

8.7 Yarn Clearing

Yarn clearing is the process of detection and elimination of yarn defects, which can lead to difficulties in subsequent production stages or defects in end product. Elimination of yarn defect always interrupts winding process and thus a loss in production. The removal of faults must be carried out in winding department due to low cost compared to subsequent weaving preparatory operations. The cost of a yarn breakage in yarn winding is lower compared to weaving or knitting. Yarn clearing is therefore always a compromise between maximum possible number of yarn defects that can be removed and minimum possible production loss.

8.7.1 Yarn Faults

The spun yarns have objectionable faults such as thick and thin places. The appearance of a fault in the finished product, that is, a woven or knitted fabric, is largely determined by its size. A yarn fault classification according to the cross section and length is therefore the basis for the assessment of yarn faults (Figure 8.8). Depending on the size or dimension and the frequency of the faults, these are divided into

- Frequent yarn faults
- Seldom-occurring yarn faults

Frequent yarn faults are also called as imperfections, which can be measured by evenness tester. Imperfections are measured per 1000 m. This type of faults does not adversely affect subsequent production process and quality of end product. The natural variability existing in the raw material and inherent variation in production process are the main reasons for the frequently occurring faults.

Seldom-occurring yarn faults differ above all in their larger mass or diameter variation and size. This faults are measured per 1,00,000 m in Classimat tester. Due to some fault

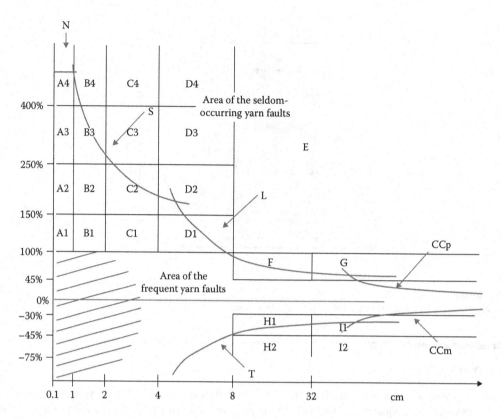

FIGURE 8.8
Setting of clearing curve in automatic cone winding machine.

in spinning process, seldom-occurring faults are created. Such types of defects adversely affects the fabric appearance and they must be removed by the yarn clearer. The causes of seldom-occurring yarn faults can be divided into three groups according to their origin:

- *Raw material and carding*: Physical characteristics such as the count, length, and short fiber content, whereas the quality of chemical fibers depends on the opening of the individual fibers. Inadequate opening in carding also results seldom-occurring faults.
- *Spinning preparation*: Inefficient drafting in draw frame or speed frame.
- *Spinning*: Spun-in flies in ring frame department, incorrect draft selection.

8.7.2 Yarn Clearers

Yarn clearers are classified into three types based on the working principle as

1. Mechanical
2. Capacitance
3. Photoelectric

Conventional manual-type winding machines are normally equipped with mechanical-type clearers. Mechanical clearers basically consist of an adjustable slot through which the yarn passes. It traps only thick places that are greater than the slot setting. The device incorporates a cutting edge that breaks the yarn fault. Mechanical clearers are rarely used nowadays as they are not efficient in removing thin places and contaminants. The capacitance clearer system measures the mass per unit length and generates a voltage that is compared with a set reference value for the mean yarn thickness. The yarn is automatically cut and the fault is removed whenever the voltage difference exceeds a given maximum value. In the photoelectric system, a beam is projected from a light source laterally across the yarn, and a photocell measures the intensity of the light passing by the yarn. By this way, variations in the yarn thickness can be monitored. The fault is detected and removed whenever the change in intensity is greater than a set level.

8.7.3 Yarn Clearer Setting in Automatic Winding Machine

The setting of yarn clearer is crucial in winding process in terms of quality and productivity. The various settings kept in the clearer are dependent on customer requirements. The following are the various channels available in the clearer system of the modern winding machine:

1. *Reference length*: Length of the yarn over which the fault cross section is to be measured.
2. *Sensitivity*: This determines the activating limit for the fault cross-sectional size.
3. *Yarn count*: The setting of yarn count helps to provide mean value of the material being processed to which clearer compares the instantaneous signals for identifying the seriousness of faults.
4. *Material type*: Input of fiber type being processed.
5. *Fault channels*:
   a. Short thick places
   b. Long thick places
   c. Long thin places
   d. Neps
   e. Count
   f. Splice
6. *Contamination clearing*: This setting removes the contaminants present in the yarn such as color thread, PP thread, jute thread, etc.

8.8 Waxing

Waxing is the process of applying minute wax particles onto the surface of yarn in winding specially for knitting purpose. In knitting process, the yarn passes through a series of reciprocating needles at a high speed. It is of extreme importance to keep the yarn tension minimum and uniform throughout the knitting process. This is achieved by sufficiently waxing the yarn and thus reducing the friction coefficient of the yarn against the thread guides and needles. Waxing should be uniform along the length of the yarn within cone

TABLE 8.5

Coefficient of Friction of Waxed and Unwaxed Yarn

| Fiber Type | Coefficient of Friction | | Wax Consumption (g/kg) |
	Unwaxed Yarn	Waxed Yarn	
Cotton	0.24–0.36	0.12–0.16	0.8–1.2
Synthetic	0.20–0.35	0.12–0.19	0.6–0.9

and also from drum to drum. Insufficient and nonuniform waxing may lead to frequent yarn breakage during knitting, too much fly generation, and needle breakage. A study conducted by SITRA[4] reported that there is no increase in imperfections after winding if the yarn is waxed. Wax consumption in the yarn is determined by the formula

$$\text{Wax consumption} = \frac{w_1 - w_2}{W} \text{ g/kg}$$

where
w_1 is the average weight of new wax (g)
w_2 is the average weight of wax after one full doff (g)
W is the average net package weight (kg)

The coefficient of friction of waxed and unwaxed yarn made of cotton and synthetic fibers are given in Table 8.5.

8.8.1 Factors Influencing Correct Waxing

1. Sufficient and uniform contact pressure between wax roll and yarn.
2. Free rotation of wax roll against the yarn.
3. Correct dimension of the paraffin wax.
4. Quality of wax:
   a. Melting point: The wax applied on the yarn surface should not melt during yarn conditioning. The optimum melting temperature of wax should be around 61°C–65°C.
   b. Penetration: It is a measure of softness of the wax. Its measuring unit is the penetration depth of a needle (in 1/10 mm) into the wax at a particular temperature and pressure. At a temperature of 25°C–35°C, the penetration should be between 9 and 16 (i.e., 0.9–1.6 mm).

8.9 Knotting and Splicing

The joining of the two yarn ends as a result of end breakage or supply bobbin exhaustion or fault removal in the winding machine is accomplished by means of knotting or splicing operation. Manual winding machines use knotting or splicing for joining

TABLE 8.6

Advantages of Splicing over Knotting in Winding Process

No higher mass variations

Only slight increase in its normal diameter

Visually unobjectionable

No mechanical obstruction

High breaking strength close to base yarn

Almost equal elasticity in the joint and basic yarn

Dye affinity is unchanged at the joint

Widest application range

the broken yarns depending upon the customer requirements. Automatic winding machine adopts only splicing for joining the yarns. The ideal yarn joining should not impair the final yarn quality and can have sufficient strength to withstand the subsequent processes. The size of the knot depends on the type of the knotter and count of the yarn, would normally be two to three times the diameter of the single yarn. The advantage of a knot is that its strength is several times that of the strength of the yarn. Knots have a detrimental effect on quality as well as process. The size of the knot can disturb its passage in subsequent processes and affect quality and productivity. A big knot in the yarn is prone to introduce a fault in fabric. The knots are responsible for 30%–60% of stoppages in weaving.

Splicing satisfies the demand for knot-free yarn joining. The advantages of splicing over knotting in winding process are listed in Table 8.6. The principle for splicing two yarn ends is to untwist a short length at the ends and then intermingle and retwist together the fibers of the two ends. The appearance of the spliced zone should be more or less similar to the yarn appearance, which is not spliced. A well-spliced joint has a diameter 20%–30% greater than the yarn over a length of approximately 15–20 mm, and an average strength of around 80% of the yarn strength with a low C.V% of strength. Kaushik et al.[8] reported that the breaking strength, breaking elongation, and work of rupture of spliced yarn were lower than for normal yarn. The number of fibers in the two yarn ends should be same or about 105 lower than in the normal cross section. The spliced joint must have as much twist as possible compared to mean twist of the normal yarn in order to attain requisite strength.

One important parameter that can adversely affect the strength of splices is insufficient compressed air pressure. To eliminate this possibility completely, the autoconer is fitted with a pressure monitor. If the air pressure in the system falls too below the predetermined level, the pressure monitor stops the splicing carriages. Splicing cannot continue until the problem has been corrected. The twisting pressure takes part in giving strength to the splicing zone, while joining the untwisted tapering ends of breaking yarn. Kaushik et al.[9] reported that the quantitative contribution of splice element found that twist contributed most to the strength of spliced yarn and there was a progressive decrease in breaking strength and elongation of spliced yarn. Splicing length plays a greater role in giving strength to the splice so the splice length control lever setting is very sensitive as well as important. Cheng and Lam[10] narrated that splice strength increased for longer fibers and decreased with increasing twist. Kaushik et al.,[9] while studying splicing process, narrated that the intermingling/tucking contributes the most to the strength of spliced yarn (52%).

8.9.1 Factors Influencing Quality of Knot

1. *Resistance of slippage*: Degree of slippage depends on fiber type, blend composition, yarn count, twist, and flexural rigidity.
2. *Size of the knot*: A bigger knot leads to uneven tension and results in end breakage in the process.
3. *Knot tail ends*: The tail length should basically be judged from the yarn count and flexural rigidity. The tail length should be of about 5–10 mm depending on knot slippage resistance.

8.9.2 Quality Assessment of Yarn Splicing

Quality of splice can be assessed by methods like load, elongation, work of rupture, % increase in diameter, and evaluation of its performance in downstream process. Splice appearance can be assessed by simple visual assessment.

1. *Splice appearance*:

   Splice appearance grade developed by ATIRA is one of the methods for grading the splice quality. Like the grading of yarn appearance, splice appearance is also rated on a numerical scale ranging from 1 to 7 according to its appearance. The following points are to be taken care while checking the splice appearance:

   a. Splice must be of equal length.
   b. The center of the splice must be well mixed.
   c. No damages must be visible in the splicing area.

2. *Splice strength*:

   Splice strength is measured by "splice strength tester." The test incorporates a comparison between splice strength and yarn strength and expressed as percentage value.

$$\text{Splice strength}\% = \frac{\text{Splice strength}}{\text{Yarn strength}} \times 100$$

   The following points are to be noted during evaluation of splice strength:
   a. At break, the yarn must jump away.
   b. The break must not always be in the splice area.
   c. The splice must not slide apart.
   d. Splice strength should be at least 80% of the single strength.

3. *Splice breaking ratio (SBR)*:

   The SBR is computed by expressing the number of breaks in the splice zone as a percentage of the total tests. One has to note down that the spliced yarn has broken in the spliced zone or elsewhere. The lower the value of SBR, the higher is the splice strength and the better the splicing quality. A splice with 40 SBR can be considered as a good quality splice.

8.9.3 Factors Influencing Properties of Spliced Yarn

1. *Fiber factors*: Fiber properties such as torsional rigidity, coefficient of friction, and breaking strength affect splice strength and appearance. The lower torsional rigidity and higher breaking strength permit better fiber intermingling. The higher coefficient of friction of fibers generates more interfiber friction to give a more cohesive yarn. Coarser yarn has a higher breaking strength and provides better fiber intermingling due to more number of fibers.

2. *Yarn factors*: An increase in twist significantly increases the breaking load and elongation, even at higher pneumatic pressure. Ring spun yarn lent best splicing compared to other spun yarns. Due to the presence of wrapper fibers, rotor spun yarns are difficult to untwist and the disordered structure is less ideal for splicing. Air-jet spun yarn is virtually impossible to splice due to low tensile strengths and elongation values.

3. *Splicing factors*: Optimum selection of opening pressure is essential for attaining required tensile strength. Higher opening pressure impairs the strength. The increase in splicing duration increases the breaking strength of the spliced yarn because of increased cohesive force resulting from an increased number of wrapping coils in the spliced joint. Splices made on longer lengths and for longer period of time have more uniform strength.

8.10 Package Defects in Winding

Drum winding machines rotate forming package through surface contact with a cylindrical drum, and the yarn is traversed either by an independent traverse, typically by a wing cam or by grooves in the drum. In this winding operation, packages are sometimes being produced with different kinds of defects in it. Defect in the package leads to cone rejection and subsequently a large amount of hard waste is generated. The various package defects associated with the winding process are discussed in this section.

8.10.1 Missing Tail End

Tail end is wound at the base of the cone at the start of new doff. In subsequent knitting or weaving process, the tail end of one cone is tied up with the free end of the next cone to maintain the continuity of the yarn in the process. There should be at least five coils of tail end at the base. Tail end length should be at least 75 cm. Tail end miss may occur due to the following reasons:

1. Tail end groove is not in the proper position of the cone.
2. Tail end groove width is too small to hold the tail end coils.
3. Cone length is short.
4. The base of the cone gets pressed during packing.
5. Tension of tail end is low.
6. Insufficient number of coils are applied at the tail end.
7. Problem with auto-doffer.
8. Workers negligence while applying transfer tail.

8.10.2 Cut Cone

This refers to cut in yarn inside the cone. This may occur due to the following reasons:

1. Cradle is not lifting from the drum surface after yarn break, resulting abrasion between the package surface and drum.
2. Brake is not functioning. As a result, the package is still rotating when the cradle lowers the package on the stopped drum.
3. Suction arm mouth touching the package when it picks up the upper end.
4. Drum surface is rough.
5. Also, improper material handling and damage during transportation of the cones may cause cut on the outer surface of the cone.

8.10.3 Yarn Entanglement

Free end of yarn gets entrapped within the layers of the package and does not come out during unwinding. This is called entanglement. Entanglement may occur due to the following reasons:

1. Suction arm setting is too close to package, causing successive layers to be sucked by the suction arm mouth.
2. Also, high subpressure of suction may cause the same.
3. Dull EYC cutter. As the yarn is not cut, suction arm mouth will not get the free end of yarn from package and will try to suck yarn from the outer layer.
4. Higher cycling intensifies the aforementioned problems.
5. Long protruding fiber of one layer catching yarn of the next layer.

8.10.4 Hard Waste/Bunch

Entangled yarn found inside the cone is called hard waste or bunch. This may occur due to the following reasons:

1. Cradle does not lift off the yarn guide drum at yarn break, causing abrasion between drum and package, and thus disturbing the layers that get entangled among themselves.
2. Yarn layers coming off the package during upper end suction due to too close setting or high suction pressure. These loose layers get entangled among themselves.
3. Extra yarn being caught from outside.

8.10.5 Stitch/Jali Formation

Yarn slips out from the reversal point of the drum and reenters in the package at a different point. This length of yarn is clearly visible at the base or nose of the yarn package. If this happens several times in the package, "jali" is formed at the nose or base of the cone.

The main cause of stitch or jali formation is uncontrolled tension variation that may occur due to the following reasons:

1. Loop gate malfunctioning, that is, it does not come back after yarn joint and always remains in touch with the bobbin surface. It hinders unwinding from bobbin.
2. Scratches inside drum guide groove.
3. Defective drum guide plate.
4. Damaged surface of tension disc.
5. Tension disc is not rotating freely.
6. Blunt EYC cutter.

8.10.6 Patterning/Ribbon Formation

Consecutive yarn layers are laid side by side causing visible pattern in the package. This may occur due to

1. Insufficient intensity of ribbon breaker.
2. Jam in package adopter. Thus, package speed will not match the speed of yarn guide drum.
3. Softer packages due to insufficient cradle pressure or low yarn tension are more prone to ribbon formation.
4. Improper setting of cone taper.
5. Improper choice of drum pitch w.r.t. package requirement.

8.10.7 Sloughing Off

Consecutive yarn layers slips toward the nose of the cone and ultimately comes out together:

1. Low cradle pressure
2. Low yarn tension
3. Yarn running out of tensioner
4. Patterning/ribbon formation as stated earlier
5. High unwinding speed

8.10.8 Wrinkle/Cauliflower-Shaped Cone

Inside yarn layers at the nose of the package pushed out and thus tend to curling. This happens due to high pressure from the outer layers. These types of defects occur due to

1. Improper package alignment of package with the drum.
2. Variation in cradle pressure. Low pressure at package starts and higher pressure as the package diameter increases.
3. Too low yarn tension.
4. Higher degree of cone tapering.
5. Deformed shape of empty cone.

8.10.9 Hard/Soft Cones

Package is too hard or too soft compared to normal packages. This may occur due to

1. Improper setting of cradle pressure
2. Rough surface of disc tensioner
3. Tension disc not rotating freely
4. Yarn running out of disc tensioner

8.10.10 Double End

Multiple yarn strands being wound on the yarn package. This may occur due to the following reasons:

1. Suction arm mouth draws two or more ends from the package as a result of extra ends on package flank, loops or loose yarn layers.
2. In case of bobbin rejection, the tension cutter should cut the yarn of the rejected bobbin. If the tension cutter fails, the yarn from the rejected bobbin and the yarn from the new bobbin, both may pass to the cone, one through the yarn clearer and the other outside the yarn clearer.
3. In case of piece bobbin (i.e., bobbin with two or more free ends), both the ends may pass to the package, one through the yarn clearer and the other outside the yarn clearer.

8.10.11 Missing End/Cob Web

A loose length of yarn exits from nose or base, may wrap around the cone tip, and reenter the package. This may occur due to the following reasons:

1. Insufficient cradle pressure, causing excessive slippage between yarn guide drum and delivery package.
2. Cone runs erratically over the drum due to damage or eccentricity. Only dimensionally stable cones should be used.
3. Defect or unevenness at the reversal points of the drums guide groove.
4. Damaged deflector plates result yarn end to lay on the flank.
5. Higher degree of yarn tension variation.
6. Inaccurate guide elements in the yarn path in the yarn tension sensor, resulting in improper tension regulation.
7. High degree of traverse displacement.

8.10.12 Drum Lap

A yarn end break takes place in the guide groove drum and the yarn is wound around the drum. In this case, the yarn may not be wound around the whole package, but on a small width. This may occur due to

1. Soiled or damaged yarn guide groove of the drum
2. High winding speed
3. High yarn tension

8.10.13 Ring in Cone

Ring may be visible from the top or bottom side of the cone especially under UV light. This may occur due to

1. Wrong bobbin passed into the cone (i.e., different count, blend, twist, etc.)
2. Mixing up of old and new material
3. Tint variation in the mixing
4. Tension variation within cone

8.10.14 Oily/Greasy Stains on Cone

The cones must be checked for oily/greasy stains on it. The yarn path must be checked for the source of such faults.

8.11 Determination of Shade Variation in Yarn Package

Shade variation is a serious problem in the final yarn package that arises due to the different material or lot mix-up at any stage of textile production process. In yarn form, shade variation may be detected on ring tube, cone, beam, and pirn. Shade variation in the spinning mill may arise from basic mixing problem in raw material in terms of color or micronaire value and incorrect blending. The properties of cottons being mixed should be compatible and within recommended tolerance levels to overcome the shade variation in yarn package. Also, the mix-up of different lap or sliver cans or roving bobbins or ring bobbins in the production process may give rise to shade variation in the final yarn package. If the cones having shade variation are processed further, it will lead to fault "barre" in the fabric. The properties that are the causes of barre are given as follows:

1. Fiber micronaire variation
2. Fiber color variation
3. Yarn linear density variation
4. Yarn twist variation
5. Yarn hairiness variation
6. Improper mixing of cotton from different origins
7. Improper mixing of cotton from different varieties
8. Improper mixing of cotton grown in different seasons

The cones produced in the cone winding should be checked for any shade variation in the UV light. Due to subtle differences in the yarn reflectance by the yarns of different lot that got mixed up in the cone, the shade variation is clearly noticed by the naked eye under UV exposure.

8.12 Control of Hard Waste

Waste generated in the yarn manufacturing process can be classified into soft waste and hard waste. Reusable wastes such as sliver, lap bits, roving ends, and pneumafil waste are normally termed as soft waste. Yarn waste obtained from ring frame and winding department is not reusable, hence it is called as hard waste. The occurrence of hard waste must be controlled as it affects the productivity of the spinning mill. Hard waste generation in winding department must be less than 1%. The various causes of hard waste generation are

- Over end piecing of yarns while attending end breakage during doffing
- Control of end breakage immediately after doffing
- Reducing the cop rejection % in winding
- Adopting proper material handling procedures
- Proper work practices in ring frame and winding

8.12.1 Practices to Be Adopted to Control Hard Waste

- Quality of ring spun supplied to winding should be maintained as per requirements.
- Proper functioning of splicing or knotting equipments should be ensured.
- Ensure proper control of package quality in ring spinning operation.
- Housekeeping and machine maintenance should be done in a better way.
- Proper material transportation using material handling equipments.
- Minimizing end breakages in ring frame during start of the doff as it leads to the practice of over-end piecing that generates huge hard waste in winding.
- Usage of knife to clean yarn bits wound on spinning empty bobbin should be prohibited.
- Control of package defects in winding by setting optimum process parameters.

8.13 Wrong Work Practices in the Winding Department

8.13.1 Poor Work Practices in Manual Winding Process

1. Making more hand waste.
2. Pulling more length of yarn from cone.
3. Leaving hand waste with the yarn.
4. Trying to find end from the nose of the fresh cop.
5. Using knife for stripping the cops.
6. Pulling too much of yarn after knotting.
7. Transferring cop end to left hand, taking end from cone, pulling out and then getting end from left hand for knotting.

8. Leaving loose end.
9. Leaving snarl knot.
10. Following groups doffing method.
11. Carrying manually four to seven cones for weight checking and dropping one or two cones over floor.
12. Changing the drums after checking the weight of cones and producing defective cones.
13. Not cleaning the entangled cops at a time and wasting time for every cop.
14. Using mixed color empties for the same count.
15. Running different color empties on the same side of the winding machine and making the winder to drop the empties in containers.
16. Waiting till a couple of cops are exhausted and then starting replenishing.
17. Failing to pass the yarn properly through the yarn passage in the first attempt and making additional attempts.
18. Dropping hand waste in alley ways/hopper/over cops, etc.
19. Not cleaning the thread path.
20. Not keeping the hopper clean.
21. Not cleaning the drum brush.
22. Not cleaning the light drum lapping.
23. Not informing the concerned regarding the defects like stop motion fault, cone holder problem, etc.
24. Pulling knotted yarn up to ear or chest while leaving for winding (after starting the drum).
25. Dropping full cops/empties in alley ways or underneath the machine.
26. Winder pulling more length of yarn even in case of short length back wind cops.
27. Winding two to three coils over four fingers unnecessarily after passing the yarn through the passage.
28. Not exchanging the empties collecting tray in time. Empties overflow and get scattered over the floor.
29. Making an attempt to mend the break or change the cop while over-head cleaner is passing.
30. Not caring about the correct tension weights.
31. Not operating the starting handle properly in the first attempt and making additional attempts.
32. Transferring cop from one hand to the other hand unnecessarily.
33. Transferring ends unnecessarily from one hand to the other hand.
34. Dropping the empties down instead of leaving it over the running conveyor.
35. Making too many attempts for creeling cop in the skewer.
36. Taking end with left hand from the cop.
37. Finding it difficult to get the end yarn from the cop due to poor eyesight, pulling out few coils and then taking end from the cop.
38. Rotating the cone with fingers too many times.

39. Not cleaning the knotter when the waste gets accumulated.

40. Not positioning ends properly in knotter and making knotter failure.

41. Using a defective knotter and making lot of knotter failures.

42. Passing the yarn below the stop motion wire after knotting, sometimes after starting the drum.

43. Missing the cone end while positioning the ends in knotter.

44. Breaking the cop end while positioning the ends in knotter.

45. Depositing hand waste in waist bag after every knot.

46. Holding the starting handle for a long time.

47. While leaving the yarn for winding, not leaving it properly inside the groove and making additional attempts.

48. Not passing the yarn below the tension weight in the first attempt and making additional attempts for this purpose.

49. Throwing the yarn over cone without knotting.

50. Throwing the yarn over empty cone after starting the drums.

51. Stripping the cop before changing the cop and starting the drum.

52. Dumping the cops at a particular place of the hopper and walking for taking every cop.

53. Breaking the yarn while leaving for winding and again knotting.

54. Not keeping the slub-catcher clean.

55. Cleaning the thread path while the yarn is running.

56. Mixing the defective/part cops with the good cops and rejecting the same cops again.

57. Not cleaning the defective portion of the cone and winding it as it is.

58. Not working efficiently after achieving the standard production.

59. Working at a very high pace and taking rest by standing at the side of the machine or disturbing other winders.

60. Standing idle till the shift end as soon as the production accounting is over.

61. Wasting time in collecting cops even when a service man is provided.

62. Using fluffs accumulated cops without cleaning.

63. Making hand waste immediately after inserting the cop.

64. Transferring the cop end from right hand to left hand after insertion, passing the yarn through left hand again transferring to right hand.

65. Making hand wastes after positioning the ends in the knotter.

66. Putting knots using fingers and not using the knotter for this purpose.

67. Not collecting the fallen empties and cops immediately.

68. Not keeping tension unit clean.

69. Using compressor air for cleaning the machine and thread path by keeping the cops and cones in open air.

70. Giving unproductive jobs like repairing empty cones, bringing cops, weighing cones and stacking at packing, disposing empties, etc., to winders and affecting machine productivity.

71. Keeping the cop containers in the machine alley and taking cop from the container by turning every time.
72. Depositing the empty bobbins inside the crates and not using the conveyor for this purpose.
73. Collecting the cones in Hessian bags and carrying them over shoulder or head.
74. Carrying only one or two cones for checking cone weights running excessive length of yarn without checking weight and not rewinding the same. Taking too much of time for cone weight checking.
75. Doffing too heavy or too light cones.
76. Providing the lesser number of balances for cone weight checking.
77. Keeping the weighing balances far away from the winding spot.
78. Using damaged weighing balances.
79. Not keeping the weighing balances clean.
80. Not inspecting the zero error on weighing balance.
81. Not checking the tare weight and using as it is.
82. Too many winders standing in queue at a time for cone weight checking.
83. Dropping the cones over floor while weight checking.
84. Not tying waist bag.
85. Moving the cop carrier along with winder while going for mending breaks.
86. Not using the cop carrier.
87. Not keeping cop carrier wheels or bearings clean and not maintaining them properly.
88. Dragging cop crates or baskets over floor or carrying them over shoulder or head.
89. Handling cops/cones and empty bobbins roughly and damaging them.
90. Using damaged plastic crates or baskets.
91. Mixing the damaged empties with good empties.
92. Working with dirty and stained hands.
93. Wearing loose full-hand shirts and working on the machine.
94. Running the machine at lesser effective winding speed than the standard speed.
95. Not passing the yarn through the slub-catcher and winding.
96. Using damaged empty cones.
97. Making count mix-up.
98. Putting bigger knots.
99. Making the finger wet either by touching the tongue or sweat for taking end from cop/count.
100. Checking the cone weights that are going for doubling.
101. Keeping the cones over uncleaned hopper.
102. Keeping the cones over hopper horizontally.
103. Holding the cone holder some time after starting the new cone.
104. Holding the running cone and feeling with fingers unnecessarily while the cone is running.

105. Making an attempt to clean the running cones/drums.
106. Using compressor air for cleaning while the cones are running.
107. Not making the wax rounds to rotate when it is not rotating.
108. Running the cone without wax that has to be waxed.
109. Mixing the winding count cops along with reeling count cops and conditioning them.
110. Winding the high-twist yarn that is to be steamed without steaming.
111. Not keeping the conveyor and conveyor pulleys clean.
112. Spending too much of time for adjusting the dress.
113. Not cleaning the cones with brush while it is running and taking more time for cleaning the cones after doffing.
114. Rejecting too much of strips and bottoms and accumulating them.
115. Distributing spoiled doffs to all winders and affecting productivity.
116. Disposing the cones without marking the ticket number.
117. Running the cone holder without empty cone.
118. Using empty cone with different count label.
119. Spoiling the cone labels while inserting it or handling it.
120. Rejecting too many strips and bottoms, cutting the yarn with knives, and making them wastes.
121. Spoiling the good wax rolls.
122. Bringing too much of wax rounds and keeping them over hoppers or window sills.
123. Using the wax rounds that are accumulated with dust and fluffs without cleaning.
124. Using knife for cutting the drum lapping and not using the hooks provided for this purpose.
125. Not cleaning the drum shaft and cone spindle lap wastes immediately.
126. Damaging the drum shafts and cone spindles while cleaning wastes wound on them.
127. Not lubricating the mechanical hand knotters in time and not maintaining them properly.

8.13.2 Poor Work Practices in Automatic Winding Process

1. Keeping the cops in armpit and creeling them
2. Keeping the container over the bigger container and dragging them
3. Not giving priority for signals
4. Starting the signal drums without rectifying the defect properly
5. Not moving the container along with the winder while creeling
6. Moving the container along with the winder while going for rectifying the signal faults

7. Keeping the container on right-hand side and turning for getting cops from container (few winders)

8. Keeping the container on the right-hand side and transferring cops from the right-hand side to the left-hand side for unwinding the backwind length

9. Leaving too much of yarn inside the suction tube

10. Not cleaning the knotter arms immediately when it is essential

11. Not inspecting the knotter's arm

12. Breaking the end while making end

13. Wasting too much of yarn while cleaning cops and cones

14. Using knives for cleaning the drums, cops, etc.

15. Creeling the rejected cops without rectifying the defect

16. Not keeping the reserve empty cones/cops readily in magazine creel

17. Rubbing the strips over machine frame and cleaning

18. Accumulating the rejected cops

19. Not attending the sorting table in time

20. Keeping the hand faraway from magazine creel while making end

21. Not holding the middle of the cop while making end

22. Dragging the empties container

23. Dragging the cop's crate

24. Not adjusting or arranging the cones while collecting them in trolley/containers

25. Not keeping the yarn clearer clean

26. Not using the given devices (e.g., Johnson bud) for cleaning the yarn clearers and cleaning them with hand or waste

27. Starting the machines after power failure without releasing the doffing carriage/splicing carriage locks

28. Starting the machine without checking the pilot spindles

29. Making an attempt to move the doffing carriage/splicing carriage when there is some struck up or jam without rotating the handle in clockwise direction

30. Not operating the top side rod immediately when there is some problem in the splicing carriage

31. Not keeping the splicing carriage wheels clean

32. Creeling different count cops in the same magazine and causing count mix-up

33. Using damaged empty cones

34. Leaving very small problematic drums idle

35. Making the doffed cone to fall down while pushing toward the conveyor

36. Dropping the cops or empties underneath the machine and not collecting them immediately

37. Not attending the conveyor drums immediately

38. Not depositing the waste inside the waist bag and dropping them over floor or leaving them inside the suction tube

39. Not attending the faulty start-up drums immediately

40. Lifting the cone holder too much and causing the automatic doffing mechanism to actuate the auto-doffer and not pressing the button immediately and causing the auto-doffer to doff the part cone

41. Doing early doffing when doffing the cone manually

42. Not winding the tail end or not leaving the sufficient length of yarn for tail end in case of export cones or weaving cones while doing the doffing manually

43. Generating too much of part and rejected cops

44. Creeling the double gaited/ring cut cops without rectifying the defects or unwinding defective portion

45. Mixing the strips and bottoms along with the empties and disposing them as they are

46. Not changing over the knotters/splicer meter knob according to the shift

47. Not collecting the suction waste as per schedule or as instructed in time

48. Not keeping the cradle holder, reversing shaft, driving shaft, etc., clean

49. Not filling water in wet splicer and not releasing the air from the tube in time

50. Running the count that is to be waxed without wax

51. Bringing too much of wax rounds and wasting them

52. Keeping the wax rounds over machine or cops or over window sills

53. Throwing or dropping the damaged wax rounds or the pieces of wax underneath the machine or above the machine or inside the plastic crates or on any other containers

54. Using the dust and fluff accumulated wax without cleaning

55. Not checking the pilot drums before switching on the machine

8.14 Yarn Conditioning

The presence of moisture in the mill atmosphere has a significant impact on the physical properties of fibers and yarns. High-speed spinning machines generate more friction, thus giving additional heat to the yarn and as a result of such heat transfer the yarn moisture content is vaporized. For quality reasons, it is absolutely important to have even distribution of this recuperated moisture throughout the entire yarn package. The two parameters such as relative humidity and temperature will decide the amount of moisture in the atmosphere. The increase in the relative atmospheric humidity causes a rise in the moisture content of the cotton fiber. The maintenance of high degree of moisture improves the physical properties of yarn. The correct follow-up of R.H% in different departments of spinning mill is very crucial, which influences the trouble-free running and dust-free environment in subsequent process. Higher R.H% tends to increase roller lapping tendency in many departments. Lower R.H% tends to drop in strength value and increase in dust liberation. The fiber strength and elasticity increase proportionately with the increase in humidity. If the water content of the cotton fiber is increased, the fiber is able to swell, resulting in increased fiber-to-fiber friction in the twisted yarn structure.

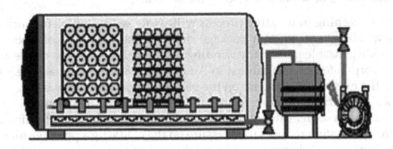

FIGURE 8.9
Yarn conditioning machine.

The ultimate concept of the yarn conditioning process is to supply the yarn with increased strength and elongation, reduced snarling of yarn, improved working at post spinning processes like warping, weaving, knitting, etc. Yarn conditioning machine as shown in Figure 8.9 is basically closed vessel capable of producing saturated steam at very low pressure. The machine produces vacuum that facilitates even penetration of steam into the layers of yarn package.

8.14.1 Benefits of Yarn Conditioning

- Increase in moisture content of yarn that enhances the mechanical properties of yarn
- Reduction in fluff and fly generation in subsequent processes such as weaving and knitting up to 30%–50%
- Reduction in needle breaks in knitting process up to 35%
- Improvement in machine efficiency in weaving and knitting
- Improvement in uniformity and appearance of the yarn
- Reduction of electrostatics effects
- Reduces the visible loss to the spinner by increasing yarn weight
- Reduces the wear and tear of needle at knitting machine
- Reduction in end breakage in weaving and knitting
- Increases softness of cloth

8.15 Technological Developments in Winding Machine

The basic objectives of the winding machine remain same since its inception. The major developments occurred in the winding machine in the recent decade have been directed toward higher output, efficient fault removal, better wound package, and minimum damage to the yarn. The major developments[11] witnessed by several winding machinery manufacturers are discussed in this section.

8.15.1 Antiribboning or Ribbon Breaker Mechanism

Ribboning or patterning in winding process will create several problems such as slough-off, poor unwinding, and uneven package density. The provision of ribbon breaking device is an indispensable part of any winding machine. Propack® is a cradle antipatterning system incorporated by Schlafhorst in Autoconer 338 winding machine.[12] Patterning occurs when the number of coils laid on the package per double traverse is a whole number. Propack system constantly determines the ratios between the drum and the package rotational speeds. As soon as the critical speed ratio that produces patterning is about to reach, the propack system reduces the pressure on the cradle by a predetermined amount. Thus, the package runs at slower speed below the critical patterning speed till the package diameter is adjusted to a value above the pattern zone.

Muratec Mach Coner® Automatic winder uses Semi-conductor device (TRIAC) for pattern breaking.[11] Antipatterning is achieved by adjusting the speed change cycle. As the drum and, therefore, the package is constantly being accelerated and decelerated, the ratios of their speeds never reach critical value to produce patterning. The multiple-groove antipatterning mechanism is incorporated on "Pac21" model of Murata automatic cone winding machine. The Pac21 monitors the critical parameters that cause ribboning and switches the yarn from one groove to another groove of the same drum to avoid patterning. Computer-Aided Package (CAP)® is an electronic-type modulation offered as an optional device on "Orion" model automatic winder by Savio.[11] Each winding head computer controls the yarn deposited on the package and automatically intervenes with a servomotor to modify the drive ratio between package and the drum only at the critical diameter. It is claimed that with CAP Orion winders produce perfect packages without patterns.[11]

8.15.2 Hairiness Reduction

Winding normally increases the hairiness due to rubbing of yarns against different parts of the winding machine. Since modern winding machines operate at high speeds, any method to reduce the yarn hairiness at the winding stage will not only reduce the processing cost in the downstream, but will also produce high-quality fabrics. The tendency of yarn hairiness is controlled by special attachment in some winding machines. Murata[13] winder offered hairiness reducing device with "Perla-A/D" model of automatic winder and is used in combination with the tension manager. This enables packages to be wound with less hairiness. The hair reducing device is placed just below the gate tensioner. When the yarn passes through this device during winding, a vortex of air (or by using discs) is applied in the yarn path that wraps the hairs on the yarn and thus reduces protruding hairs from the yarn. The attachment of hairiness reduction device in winder will help to achieve the following as claimed by the manufacturer:

- Less stoppage of weaving machine due to reduction in improper shedding caused by tangling of warp yarn
- 30% less pick-up of size
- Less fly waste at subsequent process

8.15.3 Tension Control

Yarn tension and contact pressure are the two process parameters with the greatest influence on the attributes of a package of staple fiber yarn. The integration of these parameters into a closed control loop is a prerequisite for precise adherence to the preset values and maintaining uniformly high package quality. Bal-Con® is the balloon controller equipped on Murata automatic winder.[14] Bal-Con maintains a constant balloon throughout the unwinding of the cops hence maintains a constant tension. Bal-Con is used in combination with tension manager that enables package to be wound at high speed. The function of tension manager is to sense the position of the yarn on the ring bobbins and sends signal to computer. Computer sends the command to solenoid of gate tensor to adjust the pressure on the yarn accordingly. Gate Tensor adjusts the pressure on the yarn suitably and winding tension is kept constant. The Tension Manager System controls tension fluctuation at drum start after yarn joining and at around the end of winding, based on Bal-Con detected supply bobbin data and by the pressurized tension controlled by the Gate Tensor on individual spindle. During unwinding of supply bobbin, the "Bal-Con" minimizes the contact between the yarn separated from the bobbin and the yarn layer on the bobbin.

"Boosters" are equipped on Savio automatic winder to reduce the unwinding tension.[11] This device adjusts itself at a constant distance from the nose of the tube, combines its action with the square section balloon breaker, and significantly modifies the shape of the balloon, reducing unwinding tension. This effect, also combined with the Tensor action, contributes to a considerable reduction in winding tension. The Tensor Tension Sensors are incorporated on Savio automatic winder for control of yarn tension. This device continuously detects actual winding tension and is positioned immediately before the drum. Autotense® is yarn tension control system incorporated on Schlafhorst automatic winder.[11] The yarn tension sensor continuously measures the yarn tension. The measured values are transmitted in a closed control loop to the tensioner, the pressure of which is increased or reduced as required. Thus, yarn tension is maintained at constant level throughout the complete winding process. This regulation system not only prevents an increase in yarn tension toward the end of the bobbin but also compensates for the lower yarn tension during acceleration by increasing the applied pressure thus helps in uniformity of the package build from one package to the next.

References

1. Senthil Kumar, R., Quality requirements of ring cop in modern cone winding. *The Indian Textile Journal*, May 2008, http://www.indiantextilejournal.com/articles/FAdetails.asp?id=1179, Accessed Date: November 12, 2013.
2. Rust, J.P. and Peykamian, S., Yarn hairiness and the process of winding. *Textile Research Journal* 62(11): 685–689, 1992.
3. Tarafder, N., Effect of post spinning operation on the hairiness of the ring spun cotton yarn. *Indian Journal of Fibre & Textile Research* 17: 119, 1992.
4. Chellamani, K.P., Pasupathy, R., and Vittopa, M.K., Factors influencing yarn quality change during high speed winding. *SITRA* 52(1): 3–6, May 2007.
5. Niederer, K., Achieving tension control in yarn processing. *International Fibre Journal (Quality Control Instrumentation)* 15: 46–52, October 2000.
6. Grishin, P.F., Balloon Control, Parts I & II, Platt's Bulletin 8 No. 6, 161–191.
7. Fraser, W.B., The effect of yarn elasticity on an unwinding balloon. *Journal of the Textile Institute* 83: 603–613, 1992.
8. Kaushik, R.C.D., Sharma, I.C., and Hari, P.K., Effect of fibre/yarn variables on mechanical properties of spliced yarn. *Textile Research Journal* 57: 490–494, 1987.

9

Process Control in Rotor Spinning

9.1 Significance of Rotor Spinning

Open-end spinning is a process in which fibrous material is highly drafted, ideally to the individual fiber state, creating a break in the continuum of the fiber mass. The individual fibers are subsequently collected onto the open end of a yarn that is rotated to twist the fibers into the yarn structure to form a continuous yarn length. Rotor spinning, usually referred to as open-end spinning, was commercially introduced during the late 1960s and is second only to ring spinning in terms of short staple yarn production. The rotor spinning machine is fed with draw frame sliver, which converts it into yarn and subsequently into yarn package, eliminating the roving process. With this spinning method, twisting and package building are separated by employing the false-twist principle. The major components of the rotor spinning machine are shown in Figure 9.1. The first rotor spinning machine suitable for industrial use was presented in 1967 by the name BD200 at the ITMA exhibition.[1] The current share of rotor-spun yarn is around 20% of total staple fiber yarn production, and it is increasing steadily. In summary, open-end spinning has the following major advantages compared with ring spinning:

- Elimination of roving stage
- High productivity and low energy consumption
- Large package size

Open-end spinning systems are designed to overcome some of the problems associated with ring spinning. Twist insertion in ring spinning requires the rotation of the whole yarn package. In open-end spinning, only an end of the yarn is rotated to insert twist, which consumes much less energy than rotating a yarn package. In open-end spinning, the fiber supply is reduced, as far as possible, to individual fibers, which are subsequently carried forward by an air stream as free fibers. This gives rise to the technique called "free-fiber spinning." These fibers are then progressively attached to the tail or "open end" of the already formed yarn. Hence, this is called as "open-end spinning." The interruption in the fiber supply accounts for the alternate name to the process, called "break spinning." The characteristic advantages of rotor-spun yarns are listed in Table 9.1.

Rotor spinning has certain disadvantages too. The main disadvantages of rotor-spun yarns compared with ring-spun yarns are their lower strength and the presence of wrapper fibers, which adversely affect their handle.[1] The spinning limits are also lower, generally taken to be between 100 and 120 fibers in the yarn cross section. Brandis[2] calculated the

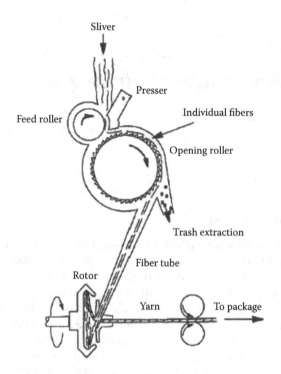

FIGURE 9.1
Parts of the rotor spinning machine.

TABLE 9.1

Characteristic Advantages of Rotor-Spun Yarns

Lower defect levels compared with the other spinning systems, particularly fewer yarn long thick and thin places
Superior knit fabric appearance
Lower fiber shedding at knitting or weaving than ring-spun yarn
Less torque than ring spun yarn
Less energy required than for ring-spinning
Less floor space required compared with ring and air jet spinning
Sophisticated real-time quality and production monitoring on each spinning position
Superior dyeability compared with ring-spun yarn

theoretical spinning limits of rotor spinning as well as the centrifugal forces (F) acting on the end of the yarn using the following formula:

$$F = 1.25 \times 10^{-6} \times \frac{n^2 D^2}{g} \times \text{linear density (tex)}$$

where
  n is the rotor speed (rev/min)
  D is the rotor diameter (m)
  g is the acceleration due to gravity (m/s²)

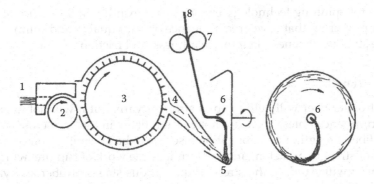

FIGURE 9.2
Fiber deposition in the rotor groove. 1, feed sliver; 2, feed roller; 3, opening roller; 4, transport tube; 5, rotor groove; 6, withdrawal tube; 7, delivery rollers; 8, yarn arm inside the rotor.

9.1.1 Tasks of Rotor Spinning Machine

The basic tasks of the rotor spinning machine are as follows:

- Opening and attenuating the feed sliver to individual fibers (fiber separation)
- Cleaning
- Homogenizing through back doubling
- Combining, that is, forming a coherent linear strand from individual fibers
- Ordering (the fibers in the strand must have an orientation as far as possible in the longitudinal direction)
- Improving evenness through back doubling
- Imparting strength by twisting
- Winding

In rotor spinning, the individualized fibers carried by air current are deposited continuously on the internal peripheral surface of a rapidly rotating drum, called the rotor, to form a fibrous ring (as seen in Figure 9.2). The rotation of the rotor imparts twist to the fibrous ring, which is then peeled off and withdrawn along the axis of the rotor.

9.2 Raw Material Selection

The careful selection of raw material by the spinner is now gaining greater significance than ever for two reasons: fiber cost and product quality.[3] An important consideration in selecting the correct raw material is the quality of the end product, coupled with the processing requirements for creating a given yarn. The selection of good quality fiber leads to good quality yarn, but the interaction of fiber with machine technology differs from one spinning system to another. It is essential to know how fiber parameters affect both yarn quality and spinning performance. The real success of spinning combines the selection of correct spinning components and machine setups with engineered fiber selection principles. Rotor spinning technology enjoys great flexibility in the choice of raw materials. The fiber properties that have great influence on yarn quality and spinnability of rotor spinning are strength, fineness, length, cleanliness, and friction.

9.2.1 Fiber Strength

The use of stronger fiber will help to spun stronger yarn, but it will not help to improve spinning performance unless there are not enough fibers in the yarn cross section. Along with higher fiber strength, fiber elongation also has to be taken into consideration, as the combination of strength and elongation determines the work of rupture, which is the best measurement of withstanding the stress of the various stages of fiber and yarn processing.[3] The strength yield that relates fiber strength to yarn strength ranges from 40% to 60% depending upon yarn count. Louis[4] reported that fiber strength has more influence on yarn strength than fiber length. Finer fiber used for producing a yarn needs lower twist level with subsequent higher spinning productivity. The increased number of fibers per yarn cross section combined with increased fiber strength and length contributes to this effect.

9.2.2 Fiber Fineness

In all the open-end spinning systems, the fiber fineness, that is, number of fibers per yarn cross section, has a crucial effect on yarn properties. Fiber fineness expressed in micronaire or in denier decides the spinning limit in the open-end spinning systems. The stable spinning performance and better yarn strength and evenness can be ensured by maintaining a minimum of between 90 and 110 fibers in the cross section of a rotor yarn. Steadman[5] reported that the yarn appearance is adversely affected by low micronaire cotton though end breakage rate is reduced.

9.2.3 Fiber Length

The importance of fiber length in rotor spinning process is less compared to the ring spinning process. The finer the yarn count, the higher the contribution of fiber length. Length uniformity is an important parameter influencing yarn evenness and spinning performance in rotor spinning of finer counts, even though rotor spinning is less sensitive to short fiber content than ring spinning. Louis[4] found more wrapper fibers with longer cotton than shorter cotton. Steadman et al.[5] reported that the long staple cottons have poorer utilization of fiber strength because of higher incidence of wrapper fibers. Manohar et al.[6] found that improvements in yarn irregularity and imperfections are more marked in waste mixing and the fall in yarn strength is also lower.

9.2.4 Cotton Cleanliness

Cotton cleanliness refers to cotton free from trash, contaminations, and dust. Rotor spinning process is very sensitive to the trash content of cotton. The feed sliver with superior cleanliness is a prerequisite for trouble-free spinning in the rotor spinning machine. Vegetable matters present in the cotton sliver will get caught in the rotor groove and obstruct the yarn formation process. This can in turn lead to end breakages or to fiber agglomeration at this point. The obstruction at rotor groove due to trash particles further tends to generate imperfections too. Very fine trash particles in the sliver lead to kitties or black spots in the delivered yarn, which deteriorates the fabric appearance produced from it. The residual

TABLE 9.2

Recommended Residual Trash Content in Feed Sliver for Rotor Spinning Process

Count (Ne)	Trash% in Feed Sliver (Should Be Less Than) (%)
Up to 6[s]	0.3
7[s]–20[s]	0.2
21[s]–30[s]	0.15
31[s]–50[s]	0.10

trash content present in the feed sliver should not exceed the values given in Table 9.2 for the specific counts as recommended by Rieter.[7]

A study by Das and Ishtiaque[8] shows that with the increase in residual trash content in draw frame sliver, the end breakage rate increases, which is due to deposition of more microdust and trash particles inside the rotor groove. Barella[9] stated that irregularity, hairiness, abrasion resistance, and resistance to abrasion deteriorate with yarns produced without periodic cleaning of rotor. Honey dew content in the cotton fiber makes the fiber to stick to machine parts and obstructs the passage of fibers, tend to create end breaks, and affects yarn quality also. The cotton selected for the rotor spinning process should be free from honey dew.

9.2.5 Fiber Friction

Fiber friction is an important parameter that determines the extent of fiber individualization, twisting efficiency, and process performance in rotor spinning. In rotor spinning, higher fiber friction levels lead to fiber damage, deposits, increased spinning breaks, reduced yarn quality, and lower spinning speeds. Cotton fibers can be spun at higher speeds than man-made fibers in rotor spinning due to the presence of natural waxes and a convoluted ribbon-like cross section, which together result in lower fiber to metal friction.

9.3 Sliver Preparation

The production of good quality rotor-spun yarn is highly possible by feeding the sliver of high uniformity. The quality requirements of draw frame sliver for the rotor spinning process are listed in Table 9.3. The control of sliver count C.V% within 3.0% ensures better yarn evenness (both short term and long term) and low yarn tenacity variations. The procedure of adopting two processes of drawing and keeping autoleveler draw frame as finisher draw frame helps to attain the lower count C.V%. Combed rotor yarns and coarser

TABLE 9.3

Quality Requirements of Draw Frame Sliver for the Rotor Spinning Process

Sliver should not contain trash of weight more than 0.15 mg

Total trash% in feed sliver should not exceed 0.3%

Nep content should not exceed 70–80 neps per gram of sliver depending on quality requirements

Count C.V% of the sliver should be under control

rotor-spun yarn normally require one draw frame between carding and rotor spinning. The required sliver count is determined by the yarn count to be spun and the draft to be given in the rotor spinning. Draft should be selected in such a way that both yarn quality and spinnability should be taken into account.

Sliver defects such as uneven sliver, split sliver, singles, thick and thin places, and undrafted ends in the feed sliver result in yarn faults. More than 80% of all yarn defects in rotor yarns are feed sliver related. The trash present in the feed sliver significantly affects the yarn quality and spinning performance. The trash particles in the sliver impede twist propagation in the rotor groove and cause spinning breaks and yarn quality deterioration, especially in finer counts.[8] By equipping adequate number of cleaners in blow room and efficient carding, maximum cleaning efficiency is achieved so that the maximum amount of trash present in the bale can be removed. Cleaning efficiency in blow room and carding should be checked at frequent intervals with the help of trash analyzer in order to keep the trash in sliver at a lower level. Rakshit and Balasubramanian[10] showed that lower production rate in carding reduced trash level in sliver and brought down the opening roller waste in rotor spinning. Simpson[11] and Louis[4] found that double carding reduces trash accumulation in rotor, brings down end breakages, and improves yarn quality.

9.4 Influence of Machine Components on Rotor Spinning Process

The rotor-spun yarn quality and spinning performance are highly influenced by the proper selection and setting of machine components in the rotor spinning machine.

9.4.1 Opening Roller

Opening roller permits much more intensive fiber individualization than roller drafting as the latter is restricted by the mechanical draft and inability of drafting rollers to run at high speeds. Action of opening roller is similar to that of licker-in in card but is more intense because of higher order of speeds. Fibers are almost individually removed from the fringe held by feed plate and feed roller. A sliver may have more than 20,000 fibers in its cross section, whereas a yarn may have approximately 100 fibers per cross section, so a total draft of around 200 will be given. The overall draft ratio is calculated from the ratio of linear densities of input sliver and output yarn. The correct selection of opening roller wire profile and opening roller speed is crucial in influencing the fiber individualization, fiber rupture, and yarn quality.

The various factors governing the selection of opening roller speed and its wire profile are listed in Table 9.4. The condition of opening roller housing decides the consistency of fiber flow through the system. The feed plate should be in clean condition and not

TABLE 9.4

Factors Governing the Selection of Wire Profile and Speed of Opening Roller

Fiber type
Fiber length
Fiber fineness
Sliver thickness or sliver linear density
Production speed

worn-out. The build-up of sticky substances (honey dew) from cotton on the feed plate should be cleaned frequently. The settings in the opening zone should be kept as per the recommendations for trouble-free processing.

9.4.1.1 Opening Roller: Wire Profile

Performance of the opening roller is influenced apart from its speed by the wire parameters such as teeth type, teeth specifications, and condition of the wire. The opening roller can be clothed with pin or saw-toothed wire, but the latter has established itself firmly in the market. A study conducted by Siersch[12] concludes that saw-toothed wire gave lower C.V% in the yarn than did comparable needle arrangements. Wire-wound clothing is recommended for cotton and cotton blends running at high throughputs, whereas pinned combing rolls are recommended for fragile fibers such as acrylics and rayon running at moderate output rates.[3] Pin type has a higher wear resistance, longer life, and further results in fewer fiber breakages. Fiber removal from opening roller to transport tube is much smoother with saw tooth wire even at higher speeds.

Important teeth specifications are rake angle "α," distance between adjacent teeth t and width of the base of the teeth b and height of wire h. t and b determine the number of teeth per unit area. Rake angle is the angle front edge makes with horizontal. The action of opening roller will increase as the rake angle reduces, but it increases fiber rupture. The rake angle of the opening roller has a significant influence on rotor yarn properties and ends down of the rotor spinning process (as shown in Figure 9.3).

A research conducted by Simpson and Murray[13] found that the lower rake angle "α" leads to more fiber breakages and poor yarn quality with cotton yarns from medium staple. Wire with 90° rake angle resulted in inferior yarn quality because of poor fiber individualization while 60° rake angle gave more fiber breakages. Rake angle of 75° produce the best yarn quality. Trash removal was better with a lower rake angle. The increase in tooth spacing of the opening roller

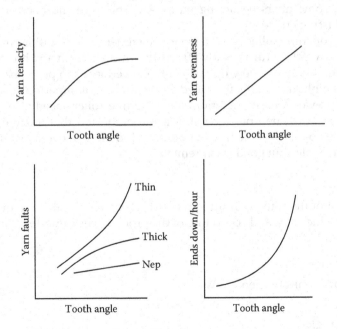

FIGURE 9.3
Effect of opening roller tooth angle on yarn properties and ends down.

TABLE 9.5

Opening Roller Rake Angle for Cotton and Man-made Fibers

Fiber Type	Rake Angle, α (°)
Cotton	66–80
Man-made fibers	90

wire will improve yarn imperfections and reduce the level of end breaks. The rake angle of opening roller recommended for processing cotton and man-made fibers is given in Table 9.5. The increase in tooth height is recommended for coarse trashy cottons, which improve the intensity of combing action and ensures better individualization and trash removal. If gentler action on cotton is needed, then lower tooth height is preferred, but fiber individualization will be inferior. Worn or damaged opening roller wire causes yarn quality problems and high ends down. Tooth density is determined by the number of wire points per unit length and the number of rows of teeth per unit width. Higher tooth density in the range of 12–18 points per square cm is recommended for processing short staple cottons and waste cottons.

The lifetime of the opening roller wire point is measured by the throughput rate in pounds or kilograms as a function of raw material used. Increasing spinning breaks, rising yarn unevenness, and imperfections are the indicators of opening roller wear. To increase the life of the opening roller wire, coatings primarily consisting of diamond particles in a nickel base were developed and coated on wire points.[3] The coating on wire point extends the lifetime by a factor of two to three times over the uncoated wire.

9.4.1.2 Opening Roller: Speed

The driving mechanism of opening roller allows a mechanical speed range from 6,000 to 10,000 rpm. The increase in opening roller speed increases fiber separation, which reduces the tendency of some fibers becoming wrappers. The lower number of wrapper fibers improves the twisting efficiency.

The increase in opening roller speed beyond recommendations will significantly increase the fiber rupture, which in turn greatly affects the yarn strength and breaking elongation (as seen in Figure 9.4). A very low opening roller speed leads to poor fiber individualization, presence of high trash in the fiber feed, more number of imperfections, increased end breakage rate, and yarn quality deterioration. Opening roller speed also affects the yarn hairiness and can affect nep production. Marino[14] reported that a decrease in opening roller speed tends to increase the number of neps. Too high an overall draft ratio can give high end breakage rates but good trash removal.

9.4.2 Rotor

Rotor is the heart of the rotor spinning process that influences the yarn quality, spinning performance, productivity, and economics of the process. The important parameters of the rotor are as follows:

1. Rotor diameter
2. Rotor groove configuration
3. Rotor speed
4. Slide wall angle

The various factors governing the selection of rotor are listed in Table 9.6.

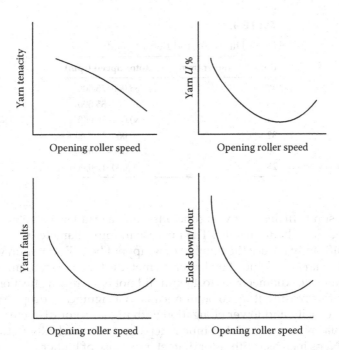

FIGURE 9.4
Effect of opening roller speed on yarn properties and ends down.

TABLE 9.6

Factors Governing the Selection of Rotor

Type of application—knitting or weaving or denim
Count of yarn
Fiber type
Fiber properties

9.4.2.1 Rotor Diameter

It is an important factor that has a significant influence on centrifugal force and the draw-off tension during spinning. The yarn draw-off tension is proportional to $C(N \cdot D)^2$, where C is the yarn linear density (tex), N is the rotor speed, and D is the rotor diameter. As a guideline, the draw-off tension acting on the yarn during spinning should be 10%–20% of the average yarn strength. The centrifugal force acting on the fibers at the sliding wall is proportional to N^2D. Rotor diameter plays a predominant role in deciding spinning stability. The speed range of rotor with respect to rotor diameter is given in Table 9.7.

A study was conducted by El-Hawary[15] on the effect of rotor diameter on cotton yarn tenacity with 45 and 55 mm rotor diameters. The results from this study exhibit that the yarn spun with the rotor diameter of 45 mm exhibits higher tenacity and breaking extension, and lower unevenness, which was attributed to increased binding-in zone of fibers at the collecting surface. The increase in rotor diameter reduces the yarn imperfections due

TABLE 9.7

Rotor Diameter and Rotor Speed

Rotor Diameter (mm)	Rotor Speed (rpm)
40	65,000–75,000
36	75,000–85,000
34	80,000–90,000
32	90,000–1,00,000
30	1,00,000–1,20,000
28	1,20,000–1,50,000

to improved twisting efficiency. It was concluded that a smaller diameter rotor with optimum rotor speed gives better results. The increase in rotor diameter with respect to fiber length may significantly reduce the number of wrapper fibers. Rotor-spun yarns of flexible and softer handle can be produced by larger diameter rotors. Balasubramanian[16] reported that an increase in rotor diameter up to 46 mm did not have much effect on strength with tandem card, but increasing it up to 56 mm leads to a significant drop in strength. Koc[17] observed lower tenacity and lower elongation with higher rotor diameter. While $U\%$ and thick and thin places are not affected much, neps increase markedly with an increase in rotor diameter. Neps increase with rotor diameter because of higher wraps per unit length and strip backs at navel as shown by Kampen.[18] Barella[19] reported that rotor diameter affects tenacity and regularity both linearly and quadratically.

9.4.2.2 Rotor Groove

The fibers reaching the rotor surface will be finally deposited at the rotor groove (as seen in Figure 9.5) in the form of layers. The geometry of the rotor groove has a significant effect

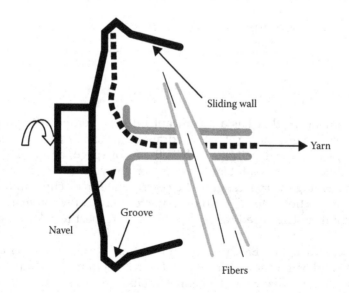

FIGURE 9.5
Rotor groove.

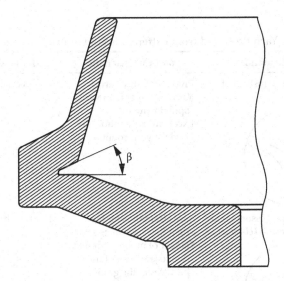

FIGURE 9.6
Rotor groove angle.

on yarn characteristics such as fiber compactness, hairiness, and handle. The groove configuration determines whether a yarn is bulky or compact, weak or strong, more or less ring yarn like, etc. The number of doublings taken place in the rotor groove tends to even out the short-term irregularities and improve the evenness of the resultant yarn. The number of doublings[20] can be calculated using the following equation:

$$\text{No. of doublings (in rotor groove)} = \frac{\text{yarn twist per meter (tpm)} \times \text{rotor circumference (mm)}}{1000}$$

The important factors influencing yarn characteristics are material of the groove, surface finish of the groove, and the groove angle. The rotor groove angle (as shown in Figure 9.6) may vary from 30° to 60°. A smaller radius rotor groove tends to trap trash particles that can lead to moire or rotor loading. The use of rotor groove with a large angle reduces the number of wrapper fibers due to the incoming fibers sliding under the yarn being formed. This may be also affected by the orientation of the groove relative to the sliding wall of the rotor. For processing denim and coarse-knitting yarns, large-angle open grooves are used due to its effectiveness of spinning. The characteristics of the yarns produced by the different rotor groove configurations and its end-use applications are summarized in Table 9.8.

9.4.2.3 Rotor Speed

The speed of rotor in connection with the speed of the opening roller has an effect on fiber configuration in the yarn structure. It also has an effect on twist level in the yarn withdrawn at a certain speed. A study conducted by Das and Ishtiaque[8] had found that the increase in rotor speed from 40,000 to 60,000 decreases the spinning-in coefficient. The reduction in spinning-in coefficient and the higher incidence of formation of wrappers have resulted in lower strength and extension of the yarn. The speed of the rotor generates centrifugal forces at the rotor groove. The centrifugal force will increase with the increase in rotor speed and fiber mass.

TABLE 9.8

Characteristics of Yarn Produced from Different Rotor Groove Configurations

Rotor Groove Configuration	Yarn Characteristics	End-Use Application of Yarn
Rotor with sharp edge instead of groove (*S-rotor*)	Produces bulky yarn Yarn is weaker than yarn spun in any rotor Excellent yarn uniformity Used for yarn counts <16 Ne	Napped fabrics
Wide groove rotor possessing good self-cleaning properties (*U-rotor*)	Yarn strength is higher than that from an S-rotor Danger of moire is present due to jamming of trash particles in the groove For yarn counts <5 Ne	Denim fabrics
Narrow rotor groove (*G-rotor*)	Yarn strength is higher than that from S-rotor and U-rotor Yarn has a low torque Danger of moire is present when sliver with high microdust content is used Used for yarn counts >6 Ne	Suitable for producing knitting and weaving yarns
Narrow, recessed grooved rotor (*T-rotor*)	Produce strongest yarn compared with all previous types Produces leanest, more compact, low hairy yarn Yarn torque is higher Uniform yarn properties Unsuitable for trashy cottons Used for yarn counts >10 Ne	Suitable for all applications except denim warp yarns

The increase in rotor speed gradually increases the yarn tension also. Maximum yarn tension directly influences the number of yarn breakages, which in turn decrease the rotor spinning machine's efficiency and raise yarn manufacturing costs. An increase in yarn tension during spinning causes a decrease in the elongation of yarn over stretching to break. Rotor speed affects yarn tenacity, elongation, and regularity in a linear manner. Lotka and Jackowski[21] found a good correlation between spinning tension and yarn count. Further, there is a good correlation between CV of yarn tension and CV of yarn count and CV of feed sliver count. Tension variations are more pronounced with finer yarns than coarser yarns. Manohar et al.[6] found that rotor speed has an insignificant or marginal effect on yarn strength, but elongation is brought down steeply with an increase in rotor speed. Kampen[18] found that while imperfections increase with rotor speed in cotton, no such effect is found with polyester and polypropylene. Manohar[6] reported that the nep content increases rapidly with an increase in rotor speed and rotor diameter. Simpson and Patureau[11] found an increase in yarn strength up to 40,000 rpm and reduction thereafter with 49 tex yarn and continuous reduction in strength with rotor speed in 25 tex yarn. This is partly because of deterioration in fiber parallelization at a higher speed. The relationship between yarn tension, rotor diameter, rotor speed, and linear density of yarn is given by the Grosberg and Mansour[22] equation:

$$T = 0.72q\omega^2R^2$$

where
   T is the yarn tension in the rotor, cN
   q is the linear density of yarn, tex
   ω is the rotational velocity of the rotor, s^{-1}
   R is the maximum diameter of the rotor, mm

The increase in rotor speed leads to an increase in winding tension too. The winding tension may need adjustment to compensate for the change of residual yarn contraction. Rotor speed is the speed of the twist insertion where each rotation of the rotor imparts one turn of twist.

$$\text{Yarn twist (tpm)} = \frac{\text{Rotor rpm}}{\text{Yarn delivery speed (m/min)}}$$

Productivity of the rotor spinning can be improved by using smaller rotors that run at higher speeds up to 140,000 rpm. However, small rotors can only be used to spin medium and fine yarns. Brandis[2] arrives at an equation (called the Krupp formula) for the maximum attainable speed (n_{max}) in rotor spinning where the yarn strength is assumed to equal the draw-off force. The equation is simplified to

$$n_{max} = \frac{2700}{D}\sqrt{\frac{B}{S}} \text{ (rev/min)}$$

where
   D is the rotor diameter (m)
   S is the ratio between the theoretical and actual draw-off forces
   B is the yarn strength (g/tex)

9.4.2.4 Slide Wall Angle

The slide wall angle in the rotor has a significant influence on the sliding of arriving fibers into the groove. The slide wall angle may vary from 12° to 50° depending upon the fiber type, rotor speed, and rotor diameter. The greater slide wall angle may delay the fiber sliding into the rotor, while a relatively smaller angle causes quicker sliding of the fiber into the groove that results in better orientation of the fiber in the groove due to the increased fiber residence in the groove.

9.4.3 Navel or Withdrawal Tube or Draw-Off Nozzle

As the spun yarn is removed from the rotor, it is pulled through the stationary nozzle. The yarn entering the navel rolls on the inside surface and false twists the yarn. The various factors influencing false twisting rolling action are listed in Table 9.9. The false twist extends back to the yarn peeling point in the rotor groove. The false twist is in addition to any real twist created by the rotor. The false twist does not stay in the yarn as it leaves the navel.

The function of the navel, which must be centered exactly to the rotor, is very important because it determines the amount of false twist, bulk, hairiness, and torque of the yarn produced. The real twist and false twist in together determine the amount of spinning-in twist at the peeling point. The navel is important in maintaining a stable spinning process and in the production of a desired yarn. The various configurations of the navel are shown in Figure 9.7.

Navel is grooved to increase the false twist. Grooved navel tends to make the yarn weaker, bulky, neppy, and hairy, particularly at higher rotor speeds. The yarn produced from the smooth surfaced navel is smoother and stronger, and has low pilling tendency and lower C.V% (as shown in Figure 9.8). Ceramic materials are used to increase the life of the navels. Smooth navel made of steel gives the best yarn quality in terms of evenness and imperfections. The various profiles of ProFil (Sussen) navels and their applications are tabulated in Table 9.10.

Cheng and Cheng[23] found higher yarn strength with smooth navel compared with four-groove navel. Roudbari and Eskadandamejad[24] found maximum strength and lower number of thick places obtained with smooth navel compared with spiral and grooved navels in 50/50 nylon cotton blends. Navel type affects hairiness but has no effect on abrasion resistance. Maghassem and Fallahpour[25] found best performance with spiral navel without torque stop and close setting between navel and rotor. Nawaz et al.[26] reported that finely grooved navel gives less hairiness compared with coarsely grooved navels with built-in notches. Yarn hairiness, neps, and lint generation are affected by the setting between the navel and the rotor.

TABLE 9.9

Factors Influencing False Twisting Rolling Action

Navel surface friction
Size of the navel contacting surface
Form of navel surface (grooved or spiral or shaped)
Spinning tension
Yarn count

Smooth   Grooved   Spiral

FIGURE 9.7
Profile of different navels.

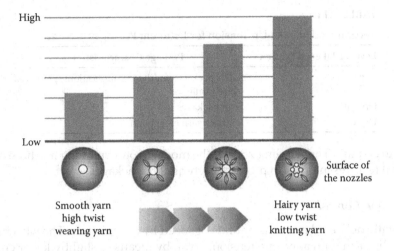

FIGURE 9.8
Effect of navel profile on yarn characteristics.

TABLE 9.10

Different Profiles of Navel and Their Applications

Type		Fiber Material	Application	Yarn Characteristics
*Pro*Fil®4				
Smooth surface 4 flutes		100% viscose	Weaving/ knitting	Universal speed range, standard hairiness, improved yarn quality
Standard diameter		100% PES	Knitting	Normal speed range, standard hairiness, improved yarn quality
		PES/cotton blends	Knitting	Normal speed range, standard hairiness, improved yarn quality
*Pro*FiL®6				
Smooth surface 6 flutes		100% cotton 100% PES	Knitting	High speed range, increased short hairiness, improved yarn quality
Small diameter		PES/cotton blends		
*Pro*FiL®S				
Smooth surface Spiral		100% cotton 100% viscose	Weaving	High speed range, low hairiness, improved yarn quality
Small diameter				
*Pro*FiL®SM				
Smooth surface Spiral		100% cotton 100% viscose	Weaving Weaving	High speed range, increased short hairiness, improved yarn quality
Soft knurled inlay Small diameter				Normal speed range, increased short hairiness, improved yarn quality

Source: Courtesy of ProFil® Navels, A new generation, Spinnovation No. 23, August 2007.

9.4.4 Winding Zone

In the winding zone, the yarn delivered from the twisting zone (spin box) is to be wound on larger packages suitable for next operation. The quality of the packages must satisfy all the requirements set by the latest winders producing fault-free packages with trouble-free unwinding properties. Optimization of the winding parameters has to be carried out

TABLE 9.11

Recommended Take-Up Tension for Different Packages

Level of Take-Up Tension	Package Characteristics
Low	Softer packages for dyeing applications; preserves yarn resilience
Optimum	Firm packages
High	Reduces yarn elongation

according to end use. The winding zone of the modern rotor spinning machine is equipped with several features in order to produce high-quality packages.

9.4.4.1 Tension Control

Yarn is usually delivered at a constant rate of speed by the spin box. The yarn package has to take up the yarn with adequate tension, given by means of slightly lower circumferential speed of the winding drum, to allow the yarn to contract. The adjustable speed differential is called "take-up tension." The recommended levels of take-up tension for different package requirements are given in Table 9.11.

9.4.4.2 Cradle Pressure

Cradle pressure alters the package density of rotor yarn package. Cradle pressure gradually decreases from beginning to end of the package build in order to compensate the growing package weight. Cradle pressure ranges between 3 and 8 lb. The recommended levels of cradle pressure in packages intended for different applications are given in Table 9.12.

9.4.4.3 Stop Motion

The correct functioning of stop motion is essential to minimize the productivity-related losses. A stop motion should function during yarn breaks, interference in spin box, a sliver run-out, rotor chokes, and yarn laps immediately in order to prevent waste generation.

9.4.4.4 Ribbon Breaker

If the yarn layers are placed in the package in a whole number synchronous ratio of speed between winding drum speed and yarn traverse motion, the so-called "ribbons" or patterns will occur wherein yarn layers are placed precisely on top of each other. These ribbons inhibit dye and steam penetration, tend to resist unwinding, or may slough-off, or even cause yarn breaks. The diameter of the patterning is dependent upon the winding angle and the stroke length. To minimize the patterning effect, the winding angle is constantly varied slightly by the ribbon breaker device.

TABLE 9.12

Recommended Cradle Pressure in Packages for Different Applications

Level of Cradle Pressure	Package Intended For
Low	Cheese for dyeing applications
Medium	Low twist yarn packages
High	Cheeses or cones

TABLE 9.13

Influence of Higher Draft in Rotor Spinning

Intensive opening of feed sliver—better fiber individualization
Minimizes imperfections level in yarn
Improves yarn evenness
Facilitates processing of coarser drawing sliver so that draw frame productivity increases
Opening roller wire prone to high wear at higher drafts

9.5 Draft

Draft is a function of delivered yarn count and linear density of the sliver fed. Very high draft is used to attenuate the feed sliver into individual fibers. The influence of higher draft given in rotor spinning is listed in Table 9.13. Drafting is performed by the opening roller (mechanical draft), which opens the input sliver followed by an air stream (air draft). Fibers coming out of the opening roller are airborne through an air duct. This zone of draft is of special significance because of its impact on fiber orientation. The amount of draft exercised by the opening roller is very high and can be determined by the following equations:

$$\text{Draft given by opening roller} = \frac{\text{Surface speed of opening roller (m/min)}}{\text{Sliver feed speed (m/min)}}$$

$$\text{Draft} = \frac{\text{Yarn count in Ne}}{\text{Sliver count in Ne}}$$

$$\text{Total draft ratio} = \frac{\text{Take-up speed}}{\text{Feed roll speed}}$$

The total draft given in rotor spinning is the combination of real draft from the feed roll to the rotor (in the order of thousands) and a condensation to accumulate the fiber groups into a fiber ring inside the rotor.

9.6 Twist

The individual fibers are collected at the yarn open end, and twist is then inserted at the yarn open end. The twist insertion is required to give stability to the yarn and facilitate the yarn to be formed and wound. The degree of fiber compactness in the rotor groove helps twist insertion and produces improved yarn strength. The degree of fiber compactness will depend on the rotor speed, rotor diameter, and tightness of the groove angle. The lowest value of twist is a function of the radius at the base of rotor groove and the type of navel in use. The various factors governing the amount of twist given in rotor spinning are listed in Table 9.14.

TABLE 9.14

Factors Governing Amount of Twist in Rotor Spinning

Strength requirements
Processing needs of weaving yarns
Fiber length
Count of the yarn
End use (knitting or weaving)

TABLE 9.15

Influence of Twist Multiplier on Rotor Yarn Properties and Production

Yarn Properties	Level of Twist Multiplier	
	Lower	Higher
Yarn strength	↓	↑
Spinning performance	↓	↑
Abrasion resistance	↓	↑
Yarn bulkiness	↑	↓
Hand	↑	↓
Production	↑	↓
Liveliness	↓	↓
Hairiness	↑	↓
Elongation	↓	↑

The impact of various levels of twist multiplier (TM) on rotor yarn properties and production of the rotor spinning machine is given in Table 9.15. The selection of optimum TM is essential to balance the various properties and production (as seen in Table 9.15).

The relationship between fiber length, yarn count, and twist level is shown in Figure 9.9. The relationship between twist and yarn count is given by

$$\text{Twist per inch} = \text{Twist multiplier} \times \sqrt{\text{Count (Ne)}}$$

Twist multiplier followed in the rotor spinning process is in the range of 2.0–8.0 TM. The relationship between twist, rotor speed, and delivery speed is given by

$$\text{Twist per meter} = \frac{\text{Rotor speed in rpm}}{\text{Delivery speed in m/min}}$$

Kampen et al.[18] showed that the twist difference (%) at the navel increases with rotor speed. As a result, elongation drops with increase in rotor speed. Twist loss in rotor

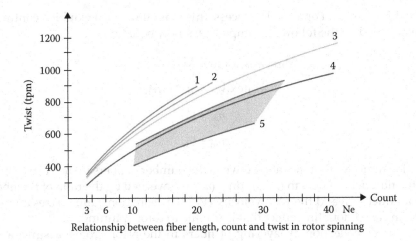

Relationship between fiber length, count and twist in rotor spinning

FIGURE 9.9
Effect of fiber length and yarn count on twist in rotor spinning. 1, comber noil (TM = 5.1); 2, cotton waste (TM = 5.0); 3, cotton 1″ to 11/8″ (TM = 4.7); 4, synthetic fiber 38 mm (TM = 3); 5, twist for hosiery yarns, raw material according to 3, 4 (TM = 3.2–4.1). (Courtesy of Rieter, Winterthur, Switzerland.)

spinning takes place because of slippage of yarn tail on rotor surface. Twist loss is given by

$$T_l = \left(\frac{\left(T_m - T_y \right)}{T_m} \right) \times 100$$

where
T_l is the twist loss (%)
T_m is the machine twist
T_y is the actual twist in yarn measured by the untwist twist method

Manich et al.[27] found increase in twist loss with linear density, twist multiplier, diameter of navel, and with grooved navel. Twist loss increases with navels that produce more false twists. Rotors that produce more friction with yarn reduce twist loss. Palamutcu and Kadoglu[28] reported lower twist loss with coarser and shorter fibers, which may be because of lower incidence of wrapper fibers. Salhotra[29] also found that twist loss increases with an increase in sheath fibers.

9.7 Doubling Effect

The mass flow per unit time of fibers in the rotor spinning system, particularly from the air duct to the take-up zone, provides an interesting insight into the contribution of fibers to the quality of rotor-spun yarn. The product of the fiber mass and fiber velocity can determine this quantity. For a stable process, this mass flow must exhibit a continuity that can be determined by the following simple mass-flow equation:

$$M_{\text{(fibers in the air duct)}} = M_{\text{(fibers in the yarn)}}$$

$$n_{\text{fd}} \text{tex}_{\text{f}} V_{\text{fd}} = n_{\text{fy}} \text{tex}_{\text{f}} V_{\text{fy}}$$

$$n_{\text{fy}} = V_{\text{fd}}$$

$$n_{\text{fd}} = V_{\text{fy}}$$

This equation indicates that the ratio between the number of fibers in the yarn cross section (n_{fy}) and the number of fibers in the air duct (n_{fd}) is governed by the ratio of the fiber velocity in the air duct (V_{fd}) and the yarn velocity (V_{fy}). The equation summarizes an important phenomenon called doubling effect that is unique to rotor spinning.

The continuous collection or layering of fibers in the rotor groove assures a uniform distribution of fibers over the rotor circumference. This action tends to even out short-term irregularities in the yarn. This doubling action contributes largely to the low irregularity of rotor-spun yarn. The number of doubling in rotor spinning can be estimated by the ratio $n_{\text{fy}}/n_{\text{fd}}$. Thus, an increase in the number of fibers in the yarn and/or a reduction in the number of fibers in the air duct can enhance the uniformity of the yarn.

9.8 Winding Angle

The winding angle is decided by the speed of the yarn traverse guide and the winding drum. Random cross-wound packages in rotor spinning change in speed as diameter increases, but maintain a constant angle of wind. The effect of increase in winding angle on rotor yarn package formation is listed in Table 9.16.

The angle of wind has predominantly influenced unwinding performance, package appearance, and optimum package weight. The winding angle recommended for different package type is given in Table 9.17.

TABLE 9.16

Effect of Increase in Winding Angle on Yarn Package Formation

Package stability increases (less bulging)
Package width is slightly decreased
Reduction in package density and package weight
Increases the tension imposed on yarn by the traverse guide, which may cause end breaks

TABLE 9.17

Recommended Winding Angle for Different Package Types in Rotor Spinning

Package Type	Winding Angle (°)
Cylindrical	30–37
Dye packages	37–40
Conical	32–37

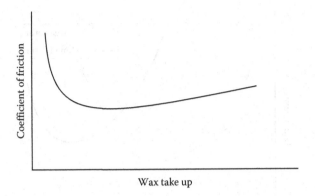

FIGURE 9.10
Effect of wax take-up on coefficient of friction of yarn.

9.9 Waxing

The yarn meant for knitting applications requires a lubricant on the surface of the yarn to reduce its friction with the knitting needles. The degree of yarn deflection over the paraffin wax roll face determines the amount of wax pickup. The wax roll should be positively driven in order to ensure a uniform wax application and to maintain a smooth wax roll surface.

The optimum level of wax take-up is essential to attain optimum coefficient of friction in the rotor yarn (as seen in Figure 9.10). The maximum reduction in the yarn coefficient of friction is normally reached with 0.5–2.0 g of wax per kilogram of yarn, depending on the wax characteristics, that is, hardness, penetration, and melting point.

9.10 End Breakage in Rotor Spinning

End breakage in the rotor spinning process not only affects the productivity but also deteriorates the quality of the yarn. The yarn structure and yarn quality are highly affected by spinning tension that is created by the centrifugal forces generated in the rotor and the frictional forces experienced by fibers and yarns from rotor groove to take-up rollers. The centrifugal forces are composed of rotor speed, rotor diameter, and fiber mass.[3] The centrifugal force can be calculated approximately by the K-factor (Schlafhorst[3]):

$$K\text{-factor} = \frac{(\text{Rotor diameter in mm})^2 \times (\text{Rotor speed in rpm} \times 1000)^2}{10^6}$$

K-factors can range between 6 and 12. The better yarn quality under a given spinning condition is possible with lower values of K-factor. Spinning stability may not be sustainable at very low K-factors. The optimum selection of rotor speed in small rotor or large rotor is necessary to control the end breakage rate (as seen in Figure 9.11).

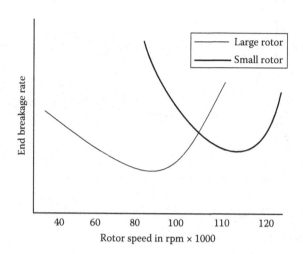

FIGURE 9.11
Rotor speed and end breakage rate.

Foreign matters such as neps, dusts, human hair, and plastics are also deposited and pressed by centrifugal force against the collecting surface of the rotor. The density of these foreign matters differs from that of the fibers; this happens when they stay on the collecting surface after the ribbon of fibers is withdrawn by the open yarn end. By the effect of centrifugal force, the increased mass of the impurity cannot be overcome by the torque of the rotating yarn end, and the continuity of fiber flow is interrupted, thereby breaking the yarn.

The different types of end breakages that occurred in the rotor spinning process are listed in Table 9.18. The yarn breaks due to tension fluctuations are found in the already spun yarn, normally in between the take-off nozzle and take-up rollers. The yarn ends on the yarn package have a blunt appearance, and short broken yarn ends are in general found in the rotor groove. On the other hand, the spinning yarn breaks occur in the yarn peel-off zone in the rotor groove when continuous fiber spin-in is interrupted. The end breakage due to sliver breaks or other interference factors outside the spinning box can be easily controlled by proper maintenance and good work practice. The ends down of the rotor spinning machine can be calculated by the formula

$$\text{Ends down per 1000 rotor hours} = \frac{\text{Number of ends down} \times 1000}{\text{Number of positions} \times \text{running time (h)}}$$

Das and Ishtiaque[8] conducted studies on the effect of various factors on end breakage rate in the rotor spinning process. They conclude that the finer the count, the end breakage

TABLE 9.18

Type of End Breakages in the Rotor Spinning Process

Yarn breaks due to tension fluctuations
Spinning yarn breaks
Yarn breaks due to sliver breaks
Yarn breaks due to interference outside spin box

rate increases drastically. As the count of the yarn becomes finer for a particular fiber fineness, there will be a lower number of fibers in the yarn cross section reaching closer to the spinning limit, which causes an increase in end breakage. The study mentioned earlier also shows that with the increase in rotor speed, the end breakage rate increases consistently. As the rotor speed increases, the proportion of end breakages with a trash particle embedded end and with a blunt end increases. With the increase in opening roller speed, the end breakage rate increases. The end breakage due to a seed coat embedded end reduces with the increase in opening roller speed, whereas the breakages due to small trash particles embedded end increase. Steadman et al.[5] report that the use of finer drawing sliver and lower draft in rotor reduces end breakage rate.

9.11 Relative Humidity in Rotor Spinning Process

The relative humidity of air in rotor spinning influences the physical properties of rotor-spun yarns as shown in Figure 9.12. The interactions between fiber, yarn, and machine elements depend on friction. Humidity in the air and moisture content of sliver and yarn predominantly influence frictional behavior; it is important that an adequate and uniform R.H% is maintained in the rotor spinning department[3] (as seen in Table 9.19).

The relative humidity should be maintained as per the recommendations in order to ensure lower end breakage rate and optimum yarn strength, especially while processing 100% cotton fibers. The correct level of R.H% in the spinning room prevents fiber damage and fly liberation and accumulations. Nield and Ali[30] reported that the mean fiber extent is affected by the relative humidity in the spinning room and as a result yarn quality is affected.

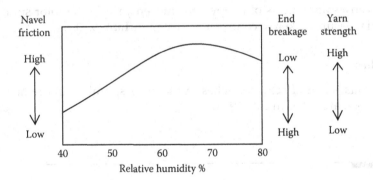

FIGURE 9.12
Effect of R.H% on navel friction, end breakage, and yarn strength.

TABLE 9.19

Recommended R.H% in the Rotor Spinning Process

Raw Material	R.H% (%)	Temperature (°F)
100% cotton	60	74
100% polyester/cotton blends	55	74

TABLE 9.20

Neppy and Uneven Yarn: Causes, Effects, and Remedies

Causes	Effects	Remedies
Dust accumulation within rotors	Productivity loss	Proper cleaning and maintenance
Surface damage in rotor groove or	Yarn quality deterioration	of rotor and opening roller
rotor cover	Affects fabric appearance	Proper maintenance of R.H%
Lapping on opening roller		
Damaged wire on opening roller		

TABLE 9.21

Stitches: Causes, Effects, and Remedies

Causes	Effects	Remedies
Lapping on the cradle sides	Productivity loss	Proper maintenance of cradle, package
Choke up in the traverse path	Yarn and package quality	holder, and traverse mechanism
Lapping on drum	deterioration	Proper maintenance of R.H%
Damaged cradle bearing		
Damaged package holder		
Snap in traverse bar belt		

9.12 Defects Associated with Rotor Spinning Process

9.12.1 Neppy and Uneven Yarn

The various causes and effects of neppy and uneven yarn in the rotor spinning process and remedial measures taken to control it are listed in Table 9.20.

9.12.2 Stitches

The various causes and effects of stitches in the rotor spinning and remedial measures taken to control it are listed in Table 9.21.

References

1. Gowda, R.V.M., *New Spinning Systems*, NCUTE Publication, IIT Delhi, Delhi, India, 2003.
2. Brandis, C., Physical limits to spinning. *International Textile Bulletin—Spinning* 2: 257, 1975.
3. Deussen, H., *Rotor Spinning Technology*, Schlafhorst Inc., Charlotte, North Carolina, 1993.
4. Louis, G.L., Some factors affecting open end cotton yarns. *Textile Research Journal* 51: 674, 1981.
5. Steadman, R.G., Gipson, J.P., Mehta, R.D., and Soliman, A.S., Factors affecting rotor spinning of fine cotton yarns. *Textile Research Journal* 59: 371, 1989.
6. Manohar, J.S., Rakshit, A.K., and Balasubramanian, N., Influence of rotor speed, rotor diameter and carding conditions on yarn quality in open-end yarns. *Textile Research Journal* 53: 497, 1983.
7. Schwippl, H., *Technology Handbook for Service Engineers*, Rieter, Winterthur, Switzerland, Edition 2006.

8. Das, A. and Ishtiaque, S.M., End breakage in rotor spinning: Effect of different variables on cotton yarn end breakage. *AUTEX Research Journal* 4(2): 52–59, June 2004.
9. Barella, A., Virgo, J.P., and Tarres, F.J., Contribution to the study of turbine cleanliness on the properties of cotton and blend open end yarns. *Textile Research Journal* 45: 160, 1975.
10. Rakshit, A.K. and Balasubramanian, N., Influence of carding conditions on rotor spinning performance and yarn quality. *Indian Journal of Textile Research* 10: 158, December 1985.
11. Simpson, J. and Patureau, M.A., Effect of rotor speed on open-end spinning and yarn properties. *Textile Research Journal* 49: 468, 1979.
12. Siersch, E., Ein Beitrag zum Mechanismus der Fasertrennung und des Fasertransportes beim OE-Rotorspinnen. *Fortschritt Berichte der VDI Zeitschriften* 3: 56, 1980.
13. Simpson, J. and Murray, M.F., Effect of combing roller wire design and rotor speed on cotton yarn properties. *Textile Research Journal* 49: 506, 1979.
14. Marino, P.N., Carpintero, J., Manich, A.M., and Barella, A., The influence of the under-pressure in the rotor on the properties of open-end spun cotton yarns at different values of the rotor speed and opening roller speed. *Journal of Textile Institute* 76: 86, 1985.
15. El-Hawary, I.A. and Sultan, M.A., Comparison of the properties of open-end-spun and ring-spun yarns produced from two egyptian cottons, *Journal of The Textile Institute*, 1/1974; 65(4): 194–199.
16. Balasubramanian, N., Rotor spinning: Effect of rotor, navel parameters, winding tension. *The Indian Textile Journal* July, 2013, http://www.indiantextilejournal.com/articles/FAdetails.asp?id=5383, Accessed Date: August 1, 2013.
17. Koc, E., Lawrence, C.A., and Iype, C., Wrapper fibres in open-end rotor spun yarn—Yarn properties and wrapper fibres. *Fibres and Textiles in Eastern Europe* 13(2): 50, 2005.
18. Kampen, W., Lunenschlose, J., Phos, T.T., and Rossbach, D., Influencing the structure of OE rotor yarns—Possibilities and limits. *International Textile Bulletin—Spinning* 3: 373, 1979.
19. Barella, A., The apparent loss of yarn twist in rotor spinning. *Melliand Textilberischte, English Edition* 13: 281, 1984.
20. Ratnam, T.V., Seshan, K.N., Chellamani, K.P., and Karthikeyan, S., *Quality Control in Spinning*, SITRA Publications, Coimbatore, India, 1994.
21. Lotka, M. and Jackowski, T., Yarn tension in the process of rotor spinning. Yarn Technology,text ile2technology.com/yarnspinning/Spinning/Rotor_spinningyarntension.htm, Accessed Date: May 1, 2013.
22. Grosberg, P. and Mansour, S.A., Yarn tension in rotor spinning. *Journal of the Textile Institute* 66: 228, 1975.
23. Cheng, Y.S.J. and Cheng, K.P.S., Selecting processing parameters that influence of rotor spun yarn formed on SDL quick spin system. *Textile Research Journal* 74: 792, 2004.
24. Roudbari, B.Y. and Eskadandamejad, S., Effect of some navels on properties of cotton/nylon66 blend (1:1) rotor spun yarn and wrapper formation: A comparison between rotor and ring spun yarn. *Journal of Textiles*, February 2013, http://www.hindawi.com/journals/jtex/2013/262635/, Accessed Date: November 1, 2013.
25. Maghassem, A. and Fallahpour, A., Selecting doffing tube components for rotor spun yarns for weft knitted fabric using multi-criteria decision making approach. *Journal of Engineered Fibres and Fabrics* 6: 44, 2011.
26. Nawaz, Sh.M., Jamil, N.A., Iftikhar, M., and Farooqi, B., Effect of multiple open-end processing variables upon yarn quality. *International Journal of Agriculture and Biology* 6(6): 256–268, 2004.
27. Manich, A., De Castellar, D., and Barella, A., Influence of yarn extraction nozzle on the apparent loss of twist on rotor open end acrylic yarns. *Textile Research Journal* 56: 207, 1986.
28. Palamutcu, S. and Kadoglu, H., Effect of process parameters on the twist of 100% polyester OE yarns. *Fibres and Textiles in Eastern Europe* 16(4): 24, 2008.
29. Salhotra, K.R., The role of sheath fibres in influencing twist loss in rotor spinning. *Textile Research Journal* 51: 710, 1981.
30. Nield, R. and Ali, A.R.A., Some aspects of humidity in open end spinning. Part II. The effect of relative humidity during spinning on fibre arrangement in the yarn. *Journal of the Textile Institute* 68(1): 10, 1977.

10

Energy Management in the Spinning Mill

10.1 Significance of Energy Management in the Spinning Mill

The need of energy conservation has assumed paramount importance due to the rapid growth of industries causing substantial energy consumptions in textile operations. Global energy crisis, as well as high cost of fuels resulted in more activities to conserve energy to the maximum extent. The textile industry retains a record of the lowest efficiency in energy utilization and is one of the major energy-consuming industries.[1] About 34% of energy is consumed in spinning, 23% in weaving, 38% in chemical processing, and another 5% for miscellaneous purposes.[1]

The energy cost is increasing at a faster pace and is the largest conversion cost accounting for about 10% of the sales and almost five times the net profit margin of a spinning mill. The three major factors for energy conservation are high capacity utilization, fine tuning of equipment, and technology upgradation. With more and more utilization of electrical energy and faster growth of industries, the impetus of energy conservation is increasing by leaps and bounds.[2] Today, energy conservation is a must for survival and it is a much more powerful imperative. As prices—for materials and finished products—are increasingly dictated by global markets, what remains within our control is conversion cost, and energy costs are perhaps our greatest opportunity for cost reduction in the country today. The conservation and efficient management of energy has become very important to keep the industry productive, profitable and competitive. The emphasis on awareness about the energy conservation is essential in the present circumstances. This chapter analyzes in detail the actual conditions of energy consumption in each department and measures energy conservation in each area of the spinning industry.

10.2 Manufacturing Cost of Yarn in Spinning Mill

In today's competitive yarn market, competing with other manufacturers depends on producing high-quality yarns with reasonable costs. The cost of yarn consists of several factors such as raw material, energy or power, labor, and capital. The cost of yarn excluding raw material is termed manufacturing cost.

The manufacturing cost of 20 tex cotton combed yarn in various countries[3] is summarized in Table 10.1. The share of the factors in manufacturing cost changes according to the yarn properties, machine operational properties, and economical situation of the spinning mill. Raw material (fiber) forms nearly half of the yarn's total cost, and other cost

TABLE 10.1

Manufacturing Cost of 20 tex Cotton Combed Yarn in Various Countries

Countries	Manufacturing Cost Factors for Chosen Countries (%)						
	India	China	Korea	Turkey	United States	Italy	Brazil
Raw material	51	61	53	49	44	40	50
Waste	7	11	8	8	6	6	7
Labor	2	2	8	4	19	24	3
Energy	12	8	6	9	6	10	4
Auxiliary material	5	4	4	4	4	3	4
Capital	23	14	21	26	21	17	32
Total	100	100	100	100	100	100	100

Source: Koc, E. and Kaplan, E., *FIBRES & TEXTILES in Eastern Europe*, 15(4), 63, October/December 2007.

factors such as labor, energy, capital cost of machines, auxiliary material cost, and waste make up the remaining part. After raw material, capital and energy costs have the highest proportions in the total. Energy is necessary for each step of spinning processes to drive machines, air conditioning, and lighting, but the highest energy consumption occurs during the spinning process in spinning machines.

10.3 Energy Distribution in Ring Spinning Process

The distribution of energy consumption among various departments of spinning mill is shown in Figure 10.1. The ring spinning system has two main problems such as low production speed and high energy consumption that cause high production costs.

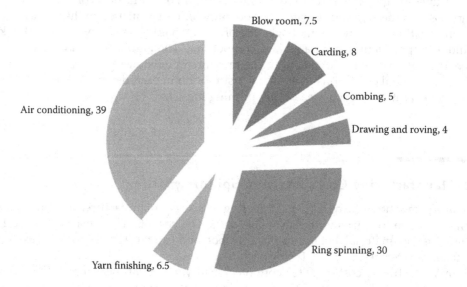

Blow room, 7.5
Carding, 8
Combing, 5
Drawing and roving, 4
Air conditioning, 39
Ring spinning, 30
Yarn finishing, 6.5

FIGURE 10.1
Energy distribution in spinning process.

Energy is generally used for operating machines, air conditioning, and illuminating the atmosphere where yarns are manufactured in spinning mills.[1] In addition to these, compressors, which provide compressed air to the spinning line, use energy. Among all departments, air conditioning plant consumes nearly 39% of the total energy consumed by the spinning mill. Ring spinning system consumes more energy than any other spinning systems. Energy consumed to drive the spindle will constitute nearly 85% of the total power consumption in the ring frame machine.[4] It will depend on the details such as yarn count, package size, and spindle speed. The remaining energy is consumed by drafting and lifter mechanism.

10.4 Calculation of Energy Consumption of Ring Frame Machines

Since the highest energy consumption occurs in spinning machines during yarn manufacturing, many studies have been carried out on the energy consumption of spinning machines. A study conducted by Krause and Soliman[5] shows that specific energy consumption in a ring spinning machine can be calculated by the following equation:

$$\text{SEC} = 106.7 \times F^{-1.482} \times \text{Dr}^{3.343} \times n^{0.917} \times \alpha_{\text{text}}^{0.993}$$

where
  SEC is the specific energy consumption (kWh/kg) in a ring spinning machine
  F is the linear density of yarn (tex)
  Dr is the diameter of the ring (m)
  n is the speed of the spindle (1000 rpm)
  α_{text} is the twist factor of the yarn

The production parameters assumed for 20 tex combed yarn are n = 17,500 rpm; α_{text} = 3,828; and Dr = 0.04 m. If these parameters are used in the preceding equation, the specific energy consumption of ring machines is found to be 1.36 kWh/kg. However, there might be a slight difference between the calculated and actual values, which is attributable to the difference in parameters such as speed, waste ratio, mechanical efficiency, and energy loss of ring spinning machines.

10.5 Energy Management Programs

A sound energy management program is required to create a foundation for positive change and to provide guidance for managing energy throughout an organization. Continuous improvements to energy efficiency, therefore, occur only when a strong organizational commitment exists.[6] Energy management programs help to ensure that energy efficiency improvements do not just happen on a one-time basis but rather are continuously identified and implemented in a process of continuous improvement.[7]

In companies without a clear program in place, opportunities for improvement may be known but may not be promoted or implemented because of organizational barriers, even when energy is a significant cost. These barriers may include a lack of communication

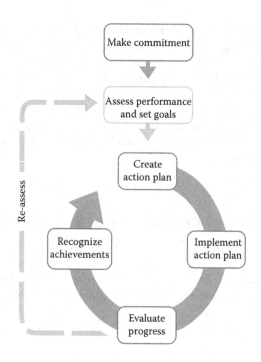

FIGURE 10.2
Energy management program. (Adapted Energy Star, Guidelines for energy management, available at: http://www.energystar.gov/index.cfm?c=guidelines.download_guidelines, 2004, accessed May 1, 2013.)

among plants, a poor understanding of how to create support for an energy efficiency project, limited finances, poor accountability for measures, or organizational inertia adverse to changes from the status quo. The major elements in a strategic energy management program are shown in Figure 10.2. It could be noted that the concept shown in this figure for energy management system builds from the ISO quality management system's philosophy of plan-do-check-act[1] (as seen in Figure 10.2).

A successful energy management program begins with a strong organizational commitment to the continuous improvement of energy efficiency. This involves assigning oversight and management duties to an energy director, establishing an energy policy, and creating a cross-functional energy team.[1] Steps and procedures are then put in place to assess performance through regular reviews of energy data, technical assessments, and benchmarking. From this assessment, an organization is able to develop a baseline of energy use and set goals for improvement. Such performance goals help to shape the development and implementation of an action plan. An important aspect for ensuring the success of the action plan is involving personnel throughout the organization.[6] Personnel at all levels should be aware of energy use and goals for efficiency. Staff should be trained in general approaches to energy efficiency in day-to-day practices. Evaluating performance involves the regular review of both energy use data and the activities carried out as part of the action plan. Information gathered during the formal review process helps in setting new performance goals and action plans and in revealing best practices.[1] Establishing a strong communications program and seeking recognition for accomplishments are critical steps to build support and momentum for future activities.[6]

10.6 Energy Conservation in the Spinning Mill

10.6.1 Spinning Preparatory Process

10.6.1.1 High-Speed Carding Machine

Carding machine individualizes the lumps of fibres that results disentangling of tufts in the "opening-and-picking" stage of primary processing and simultaneously removes impurities, neps, and short fibers, improving the orientation of fibres in the longitudinal direction and producing card sliver.[1,7] The latest generation carding machine is large, and each machine consumes considerable amounts of electricity. On the other hand, since productivity is high, one-third of the number of new machines and half the total power can produce the same production capacity as ordinary carding machines. For instance,[8] 12 conventional machines requiring 27 kW/machine can be replaced by 4 new machines requiring 41 kW/machine, thus resulting in power savings of 160 kW.

10.6.1.2 Installation of Electronic Roving End-Break Stop-Motion Detectors Instead of Pneumatic Systems

The conversion of pneumatic suction tube detector to a photoelectric stop motion system end break detector in the roving end break system of the speed frame machine saves considerable amount of energy. This measure is implemented in many textile plants around the world. The average energy saving reported is 3.2 MWh/year/machine.[9]

10.6.1.3 Ring Frame

The ring frame converts the roving made up of natural/man-made fibres into yarn. The power consumption of ring frames is the major part of the total power requirement of spinning mills. The energy cost in ring frame department will become even more critical with the modern machines having higher spindle speeds and larger packages, which consume more energy.[10] The break up of power for driving different components of a ring frame is given in the following Table 10.2. A further break-up of the spindle power is given in Table 10.3.

TABLE 10.2

Break-Up of Power for Driving Different Components of a Ring Frame

Component	Power Consumption (% of Total)
Spindle driving power	39.3
Spinning power	19.7
Tin roller/pulley shaft	12.3
Package	8.8
Drafting	6.6
Empty bobbin	4.9
Motor	8.4
Total	100.0

Source: Mishra, S.N. et al., BTRA powerspin system for estimation, monitor and control of energy consumption for ring frames in textile mill, *BTRA, 43rd Joint Technological Conference*, March 2–3, 2002.

TABLE 10.3

Break-Up of Spindle Power Consumption

Item	Power Consumption as % of Total Spindle Power	Factors Responsible
Power to drive spindles	33	Lubrication, spindle bearing, spindle weight
Bending resistance of spindle tape	42	Tape stiffness tape width
Air friction of tape surface	17	Surface finish
Jockey pulley tension	8	Tension load

Source: Mishra, S.N. et al., BTRA powerspin system for estimation, monitor and control of energy consumption for ring frames in textile mill, *BTRA, 43rd Joint Technological Conference*, March 2–3, 2002.

10.6.1.3.1 Use of Energy-Efficient Spindle Oil

In ring spinning frame, spindles constitute about 30–40% of the energy consumption. The incorporation of a dispersant additive system to the mineral-based spindle oil may result in energy savings of up to 3% when compared with conventional oils. The extent of energy savings is highly influenced by the machinery condition and their operational conditions. However, synthetic-based spindle oils (energy-efficient grades) along with certain metal compatibility additives may result in higher energy savings in the range of 5%–7% depending on viscosity.[7] The energy efficiency potential of spindle oil can be assessed in two ways:

1. The reduction in electricity consumption
2. The reduction in bolster temperature rise over ambient: energy saving oils resulting in lesser temperature increases

While selecting any energy saving spindle oil, one should carefully evaluate the important characteristics related to the service life of the oil, that is, temperature rise, thermal stability, metal compatibility, sludge-forming tendency, and antiwear/antifriction properties.[11]

10.6.1.3.2 Oil Level in the Spindle Bolster

The electricity consumption in the ring frame increases with an increase in the oil level in the bolsters because of the resistance caused by the oil.[11] Also, an excessively high oil level in the bolster may disturb the proper running of the spindle. Normally, 75% of the bolster capacity is filled with oil. The usual method of determining the depth of oil level in the bolster is by lifting the spindle and observing the oiliness of the spindle blade.[7] The correct or exact amount of oil for each type of spindle insert could be assured by using a dipstick. The dipstick has two distinct markings, that is, the bottom marking for minimum and the top marking for maximum oil levels. The oil level should, therefore, be checked with a dipstick after every topping, and the level should be maintained within the limits prescribed by the machinery manufacturers.[11] In this regard, various oil dosing equipment is available today for filling and topping spindle bolsters with the predetermined quantity of oil. As there is no excess filling of spindle oil into the bolster, it prevents wastage of oil as well as energy.

10.6.1.3.3 Light Weight Spindles

Ring frames are the largest energy consumer in the ring spinning process. Within a ring frame, spindle rotation is the largest energy consumer. The various factors influencing the power consumption of spindle in the ring frame are listed in Table 10.4. Spindles consume power to overcome the friction at the bearing, the viscosity of the oil, and the air resistance when fitted with the tube. Energy saving also can be achieved by using lightweight spindles.[7] Lighter (from 320 to 265 g) and lower wharve diameter (from 25/22.5 to 18.8 mm)

TABLE 10.4

Factors Influencing Power Consumption of Spindle

Spindle speed
Spindle size
Radial load on the spindle bearing
Quantity of oil in the spindle inserts
Viscosity of the oil

spindles consume about 10% less energy with the lower speed operation of ABS pulley shaft. A spinning plant in India replaced the conventional spindles with lighter weight ones in their ring frames and on average saved 23 MWh/year/ring frame.[12]

10.6.1.3.4 Synthetic Sandwich Tapes for Ring Frames

Synthetic sandwich spindle tapes are made of polyamide, cotton yarn, and a special synthetic rubber mix. Sandwich tapes run stable, have good dimensional stability, do not break, result in less weak-twist yarn, do not cause fiber sticking, and are made of soft and flexible tape bodies.[7] Because of these special characteristics, these tapes offer 5%–10% energy saving. The installation of energy-efficient tapes in ring frames, on average, resulted in energy savings of 4.4 MWh/year/ring frame.[2]

10.6.1.3.5 Optimization of Ring Diameter with Respect to Yarn Count in Ring Frames

Ring diameter significantly influences the energy use of the ring frame.[1] Larger ring diameters facilitate higher bobbin content with a heavier package, resulting in excess energy consumption.[7,13] The reduction of 10% in bobbin content decreases the ring frame energy intensity by about 10%. For finer yarn counts and for medium yarn counts, 38 millimeter (mm)/36 mm diameter and 40 mm ring diameter, respectively, are recommended. Before ring diameter modification is undertaken, the technical feasibility of the modification should be assessed, which may include the following items:[13]

- The yarn count range for a specific period is unpredictable. The solution to this issue is that, based on the count ranges, the ring frames should be segregated by suitable ring diameter in the way that each group of ring frames produces a specific yarn count.
- The life span of existing rings is also unpredictable due to high-speed operation.
- Overall efficiency is reduced in the postspinning phase (e.g., cone winding) due to lower bobbin content. However, the reduction in efficiency in the postspinning phase can be compensated by running the machines at a higher speed.

10.6.1.3.6 Installation of Energy-Efficient Motors in Ring Frames

As ring frames are the most energy-intensive equipment in the spinning process, it is important to make sure that the electric motors installed in the ring frames have the highest possible efficiency. Even a slight efficiency improvement in ring frame motors could result in significant electricity savings that could pay back the initial investment in a short period.[7] The following points should be taken care of in motors to conserve energy:

- Rewound motors reduce motor efficiency. No-load tests should be conducted on motors during energy audits. These no-load tests are an indication of the rewinding quality of the motors. Typically, the no-load current of a motor is about 35%–40% of the full load current.

- To improve the efficiency of motor for the given motor and load characteristics, it is advisable to bring down the motor body temperature.

- During a typical energy audit, the operating parameters of the motors such as operating current, voltage, power factor, and the running kW are measured using sophisticated instruments.

10.6.1.3.7 Provision of False Ceiling in the Ring Frame Department

The control of atmospheric conditions inside the spinning mill is vital for the yarn manufacturing process as it influences the quality and process. The humidification plant takes care of proper maintenance of relative humidity and temperature inside the ring frame department. The energy used by the humidification facility is directly related to the volume of the facility where the spinning process is carried out. The use of a false ceiling can help to reduce this volume, thereby reducing energy consumption. The electrical energy savings with the installation of a false ceiling for a 15,000 spindles capacity spinning mill is 125 MWh/year (8 kWh/spindle/year).[14]

10.6.1.3.8 Replacing Conventional Aluminum Fans with Energy-Efficient Fiberglass Reinforced Plastic Fans in the Pneumafil Suction System of Ring Frame

Ring frames have pneumafil suction fans, which are used to collect fibers when a yarn break occurs. Energy-efficient fiberglass-reinforced plastic (FRP) fans could be installed in place of conventional aluminum fans in the pneumafil suction system of ring frames.[15] The average electricity saving[12,16] from the implementation of this measure is reported to be between 5.8 and 40 MWh/year/fan.

10.6.1.3.9 Use of Light Weight Spinning Bobbins in Ring Frames

In ring frames,[1] yarn is collected on spinning tubes or empties. Spinning tubes are inserted into the spindles that are rotating at higher speeds. The rotating of spindles is the highest energy consumption activity in ring machines. The heavier the bobbins are, the greater the energy requirement for the rotation of bobbins and hence spindles. Nowadays, the use of lighter bobbins in place of conventional ones is gaining more attention. The replacement of 30–35 g bobbins (cops) with 28 g bobbins resulted in average electricity savings[17] of 10.8 MWh/year/ring frame (assuming 12 doff a day).

10.6.1.3.10 High-Speed Ring Spinning Machine

High-speed ring frame has an increased operating speed by 10%–20% with similar power consumption as compared with the conventional ring frame. As a result, the power requirement is 36.0–40.5 kW in comparison with that of 45 kW for conventional ring spinning machines for the same production.[8]

10.6.1.3.11 Installation of a Soft Starter on Ring Frame Motor Drives

The starting current drawn by an induction motor is directly proportional to the applied voltage. A soft starter is designed to make it possible to choose the lowest voltage possible (the "pedestal voltage") at which the motor can be started—the lowest voltage being dependent on the load on the motor.[7] The soft starter is also suited to situations where a smooth start is desirable to avoid shocks to the drive system or where a gradual start is required to avoid damage to the product/process/drive system and accessories. A soft starter can reduce the costs incurred by yarn breaks on a ring frame when its motor starts after each doff, as smooth starts and gradual acceleration of motors eliminate shocks during starting.[1,7] Average electricity saving reported from the implementation of this measure on ring frames is about 1–5.2 MWh/year/ring frame.[18] In addition to the electricity saving, the other advantages of this measure are a reduction in the maximum power demand and an improvement in the power factor.

10.6.2 Energy Conservation in Postspinning Process

10.6.2.1 Intermittent Modes of the Movement of Empty Bobbin Conveyors in Autoconer/Cone Winding Machines

The continuous movement of empty bobbin conveyor belts can be converted into an intermittent mode of movement. This measure results in not only substantial energy saving but also maintenance cost savings and waste reduction.[9] This resulted in electricity savings of 49.4 MWh/year.

10.6.2.2 Two for One (TFO) Twister

In two-for-one twisting machines, the balloon tension of yarn accounts for about 50% of total energy consumption.[13] The balloon diameter can be reduced with a reduction in yarn tension by providing a modified outer pot. This measure saves about 4% of total energy consumption in TFOs. It has been observed that TFOs consume less electricity at lower balloon settings.[7] Balloon size can be optimized by taking account of various studies with respect to different yarn counts. An energy saving of 250 MWh/year by optimizing the balloon setting of its TFO machines is reported in a mill.[19]

10.6.2.3 Yarn Conditioning Process

Yarn packages are kept inside the yarn conditioning machine in which required moisture is provided to the yarn. The machine produces vacuum that facilitates even penetration of steam into the layers of yarn package. The increase in moisture in the yarn improves the mechanical properties of the yarn. The water spraying nozzle type has significant influence on energy consumption in the yarn conditioning system. The replacement of jet nozzles with energy-efficient mist nozzles in yarn conditioning machine saves electricity[20] of about 31 MWh/year.

10.6.3 Energy Conservation in Humidification System

Humidification system[15] creates amiable atmosphere to the machines and operatives. Automatic humidification system consumes 25%–35% of the total mill power. The following areas in humidification system have highest possibilities for energy conservation.

10.6.3.1 Replacement of Aluminum Fan Impellers with High-Efficiency FRP Fan Impellers in Humidification Plants and Cooling Tower Fans

Axial fans are widely used for providing required airflow in cooling towers, air-conditioning, ventilation, and humidification systems. Optimal aerodynamic design of FRP fan impellers provides higher efficiencies for any specific application.[7] A reduction in the overall weight of fans also extends the life of mechanical drive systems. Fans with FRP impellers require lower drive motor ratings and light duty bearing systems.

Fans with FRP impellers consume less electricity compared with fans with aluminum alloy impellers under the same working conditions.[15] This retrofit measure[21] resulted in average savings of 55.5 MWh/year/fan.

10.6.3.2 Installation of Variable Frequency Drive on Humidification System Fan Motors for Flow Control

Temperature and humidity levels must be closely monitored and maintained for textile processes (especially spinning and weaving) so that yarns will run smoothly through the processing machines; a well-functioning ventilation system is imperative to the plant's successful operation.[22] Ventilation systems use supply fans (SFs) and return fans (RFs) to circulate high humidity air to maintain proper ambient conditions, cool process machinery, and control suspended particulate and airborne fibers. Initially, the mixture of return air and fresh air is cleaned, cooled, and humidified by four air washers. This air is then supplied to the facility by the SFs and distributed to the plant through ceiling-mounted ducts and diffusers, producing required temperatures and relative humidity levels.[22] The RFs then pull air through the processing machines into a network of underground tunnels that filter out suspended particles and fibers, usually through rotary drum filters on the inlet of each RF. The important points to be considered in supply air system and return air system for improved efficiency are listed in Tables 10.5 and 10.6, respectively.

Factors that influence the pressure, volume, or resistance of the system directly impact the fan energy requirements. Therefore, air density, changes to damper positions, system pressure and air filter pressure drops, supply and return air system interaction, and parallel fan operation all affect how much energy the fans require and must be monitored to ensure the efficient functioning of the system. VFDs can be installed on flow controls; these devices control fan speed instead of changing the dampers' position. Thus, damper control is no longer necessary, so in the use of VFDs, fan control dampers are opened 100%, thereby saving electricity use by the fans. The average electricity saving[21] reported for this retrofit in a plant is 105 MWh/year/fan. The saving and cost of the measure depend on various factors such as the size of the fan, the operating conditions, the climate, and the type of VFD.

TABLE 10.5

Points to Be Considered in Supply Air System for Improved Efficiency

Attic ventilation is to be provided to reduce transmitted heat through the roof.
False ceiling is to be insulated.
False ceiling height should be minimum 4.25 m (14 ft) from floor level to improve the uniform air circulation.
V-shaped fresh air filter can be provided for reducing fluff accumulation inside the supply air system.
Sufficient size of fresh air damper will improve plant efficiency.

TABLE 10.6

Points to Be Considered in Return Air System for Improved Efficiency

Return air trenches should be designed with minimum corners and bends and should be given a smooth finish to reduce air resistance.
Rotary return air drum filters with effective suction fan for continuous fluff removal will help to keep the return air filters clean and reduce air resistance, thereby reducing power consumption of return air fan motor.

10.6.3.3 Other Areas in Humidification System

1. *Under deck insulation*:
   a. Generally, industrial roofing is of asbestos cement sheets or metal sheets (GI/aluminum), and underdeck insulation with minimum 50 mm thick resin bonded glass wool of 32 kg/m^3 density is necessary to reduce the heat gain through the roof and minimize the attic temperature rise within the false ceiling.[1,15]

2. *Use of PVC air inlet louvers and eliminators*:
   a. PVC blades and moderate air velocity (not exceeding 600 fpm for low velocity air washer system and not exceeding 1200 fpm for high velocity air washer system) are desirable to reduce the air resistance in the air washer with consequent reduction in design static pressure for supply air fan selection so as to reduce fan motor power consumption.

3. *Replacement of ordinary V-belt with cogged V-belt*:
   a. Ordinary V-belts can be replaced with cogged V-belts to reduce friction losses, thereby saving energy. The implementation of such modification[23] on 20 V-belt drives in a spinning plant resulted in electricity savings equal to 30 MWh/year.

4. *Better maintenance of humidification system*:
   a. *Air washer maintenance*: Nozzle blockage, leakages, wear-out of nozzle, blockage and leakage in eliminators, water leakages from pump, tank over flow, etc., have to be rectified immediately to conserve energy.
   b. *Duct maintenance*: Ducts of supply air and return air and diffusers should be regularly cleaned; otherwise, there will be pressure loss.

10.6.4 Overhead Traveling Cleaners

Spinning mill usually needs to effectively manage the waste (fluff) generated during fiber processing, which affects the quality of the outgoing yarn/fabric.[24] Fluff removal and machine cleaning can be accomplished with the support of overhead traveling cleaners (OHTCs). A common waste collection system (WCS) is an independent subsystem designated to collect waste from groups of OHTCs. In modern mills, one OHTC serves every 1008/1200 spindle ring spinning frame. It moves on rails at a speed of about 16 m/min. It takes about 140 s to move from one end of the ring frame to the other end. OHTC is continuously blowing/sucking off air and waste in and around different component parts of the machine during its traverse motion. In general, one overhead traveling cleaner consumes about 17,000 kWh/year.[24] The adoption of the following features in OHTC saves electrical energy in a significant manner.

10.6.4.1 Attachment of Control Systems in OHTC

The attachment of the following control systems in OHTC gives rise to considerable energy savings:

- *Timer-based control system for OHTC*: An energy-efficient control system using timer circuits can be introduced in addition to a main contactor provided in the control box to start and stop the OHTC whenever it touches the ends of the ring frame over which it moves in a linear path. An off timer can be incorporated, with a feature of extending the delay of operation for 0–30 min in a stepped manner. This method can be adopted for plants processing fine counts in which dust liberation is less. The average electricity saving reported from a case-study was 5.8 MWh/year per OHTC.[24]

- *Optical control system OHTC*: An optical sensor to sense the position of the OHTCs on the ring frames can also be used. This system will start running the blower fan of the WCS only during the required operation time. It is reported that the energy consumption was reduced by 41% when compared with base conditions.[24]

10.6.4.2 Provision of Energy-Efficient Fan Instead of Blower Fan in OHTC

Existing blower fans of OHTCs can be replaced by energy-efficient fans with smaller diameters and less weight. An energy saving of about 20% is achievable with a quick payback period of less than 6 months. A case study implementing this measure reported average electricity saving of 2 MWh/year for each fan.[24]

10.6.5 Electrical Distribution Network

10.6.5.1 Cable Losses

The cable losses are directly proportional to the length. The losses are proportional to the square of current. Hence, for energy-efficient electrical design, the main concept in the mind should be that there will be minimum cable length.[25] Hence, the selection of panel board position and main panel room and transformer location is very important.

10.6.5.2 Power Factor

There are many electric motors in a spinning plant that can cause reactive power. Therefore, reducing reactive power by improving the power factor of the plant is an important measure in reducing energy use and costs. The replacement of low value capacitors with the addition of new capacitors reduces the system losses, which considerably saves 24.1 MWh/year. It is also possible by incorporating automatic PF control panels to achieve maximum power factor.[26]

10.7 Lighting

Lighting system constitutes about 4% of the total energy consumption in a spinning mill. Electricity saving fluorescent lamps, with 70–80 lux/W, have much higher luminous efficiency than tungsten ones with 10–15 lux/W. The effectiveness of illumination is also affected by various other factors such as the layout of the working area, the color of the interior, and the distance of the light source to the illuminated area.[1,27] Other lighting saving measures may include turning off lights, computers, and printers when not in use,

TABLE 10.7

Characteristics of Different Lamps Used in Industry

	Standard Incandescent	Full Size Fluorescent	Mercury Vapor	Metal Halide	High-Pressure Sodium
Wattages	3–1500	4–215	40–1,250	32–2,000	35–1,000
System efficacy (lm/W)	4–24	49–89	19–43	38–86	22–115
Average rated life (h)	450–2000	7,500–24,000	24,000+	6,000–20,000	16,000–24,000
Color rendering index	98+	49–85	15–50	65–70	22–85
Life cycle cost	High	Low	Moderate	Moderate	Low
Source options	Point	Diffuse	Point	Point	Point
Start-to-full brightness	Immediate	0–5 s	3–9 min	3–5 min	3–4 min
Restrike time	Immediate	Immediate	10–20 min	4–20 min	1 min
Lumen maintenance	Good/excellent	Fair/excellent	Poor/fair	Good	Good/excellent

particularly after office hours. In one plant, this has helped to reduce lighting bills by around 15% a year. The characteristics of different lamps used in industry are summarized in Table 10.7.

LT servo stabilizer of suitable capacity is recommended in the main lighting feeder with the set voltage of around 200 V. This measure ensures a saving of about 10%. Required electric power to achieve a recommended light level can be estimated as follows:

$$P = \frac{b}{\eta_e \cdot \eta_r \cdot l_s}$$

where
P is the installed electric power (W/m² floor area)
b is the recommended light level (lux, lm/m²)
η_e is the light equipment efficiency
η_r is the room lighting efficiency
l_s is the emitted light from the source (lm/W)

Light equipment efficiency (η_e) expresses how much of the light is really emitted from the light to the room. The room lighting efficiency (η_r) expresses how much of the light is absorbed by the room before entering the activity area. Light equipment efficiency and room lighting efficiency influence each other. Common values of the product $\eta_e \cdot \eta_r$ are in the range of 0.3–0.6.

10.7.1 Replacement of T-12 Tubes with T-8 Tubes

T-12 tube lights (3.8 cm dia.) are replaced with T-8 tube lights. The initial output for T-12 tube lights is high, but energy consumption is also high. T-12 tube lights also have extremely poor efficiency, lamp life, lumen depreciation, and color rendering index. Because of this, maintenance and energy costs are high. Replacing T-12 lamps with T-8 lamps approximately doubles the efficacy of the former, thereby saving electricity.[1,7]

TABLE 10.8

Standard Illumination Level in Various Departments of Spinning Mill

Department	Standard Illumination Level (Lux)
Spinning preparatory	40–50
Ring frame	75–100
Winding	100–125

10.7.2 Replace Magnetic Ballasts with Electronic Ballasts

A ballast[1] is a mechanism that regulates the amount of electricity required to start a lighting fixture and maintain a steady output of light. Electronic ballasts save 12%–25% of electricity use compared to magnetic ballast. A textile plant in India has reported energy savings of 936 kWh/ballast/year with the implementation of this measure.[7]

10.7.3 Optimization of Lighting (Lux) in Production and Nonproduction Areas

In many plants, the lighting system is not specifically designed for the process. There are lux standards for each type of textile process. The required lux in the blow room should be much lower than that of ring frame section (as listed in Table 10.8).

If the lighting provided is higher than the standard (required lux) for any part of the production, this results in a waste of electricity. Therefore, the plant engineers should optimize the lighting system based on the standard lux specific for each process step. Electricity savings of 31–182 MWh/year are reported from different textile plants as the result of optimization of plant lighting. Energy savings vary based on the plant-specific situation and preexisting lighting system.[1]

10.7.4 Optimum Use of Natural Sunlight

Most of the spinning mills do not utilize natural sunlight to an optimum level. And also, it is practically possible to use natural sunlight in every department due to the usage of humidification system. In areas such as cotton godown and blow room, the transparent roof is placed to utilize the natural sunlight. In addition to optimizing the size of the windows, a transparent sheet can be installed at the roof in order to allow more sunlight to penetrate into the production area. This can reduce the need for lighting during the day. Energy savings of 1–11.5 MWh/year are reported as the result of the efficient use of natural light.[7]

10.8 Compressed Air System

Compressed air system is an inevitable ancillary section in the modern spinning mill. This section constitutes nearly 6% of the total mill energy consumption. More than 85% of the electrical energy input to an air compressor is lost as waste heat, leaving less than 15% of the electrical energy consumed to be converted to pneumatic compressed air energy.[1] This makes compressed air an expensive energy carrier compared with other energy carriers. Many opportunities exist to reduce the energy use of compressed air systems. The following points[1,7,28] should be considered for energy savings in compressed air system:

- We can expect up to 5% of energy saving by centralizing compressor of higher capacity and higher efficiency. This is due to lesser friction losses.
- The use of required pipe size, minimum right angle bends, reduces friction loss.
- The design should ensure that the pressure drop should not be more than 0.5 kg/cm² (7 psi) in the longest line.
- Clogging of filters creates chocking of filter, leading to drop in suction pressure, which reduces compressor efficiency.
- Avoid placing the compressor in hot environment which rises the inlet air temperature that increases the energy consumption.
- Moisture in compressed air leads to corrosion in pipelines which increases internal resistance that tends to create more pressure drop which affects efficiency of compressor.
- Proper tension of the driving belt should be maintained, which saves a sufficient amount of energy.
- The components of the compressor such as piston, piston rings, bearings, and valve gears should be maintained with proper lubrication.
- Excess friction due to worn-out parts or lack of grease leads to higher power consumption.
- The oil level in the compressor should be maintained properly.
- The compressed air pipeline system should not have any air leakages.
- Pressure guns should be provided to avoid wastage of compressed air.
- When the compressed air pipelines are not maintained properly, a maximum of 15% of power loss may be noticed in the mills.

10.9 Energy Demand Control

Demand control is a follow-up analysis that is normally conducted after the development of a demand/load profile by energy auditors. Demand control is nothing more than a technique for leveling out the load profile, that is, "shaving" the peaks and "filling" the valleys.[7] The main advantage of demand control and load management is the reduction of electricity costs. The first step in demand control analysis is to analyze a plant's electricity utility tariff structure and past history of power demand. The load factor (LF) is a useful tool in demand control analysis.

10.9.1 Calculating the Load Factor

The LF is the ratio of the energy consumed during a given period (the period of an electricity bill) to the energy that would have been consumed if maximum demand had been maintained throughout the period.[1]

$$\text{Load factor}(\%) = \frac{\text{Energy used during the period (kWh)}}{\text{Maximum demand (kW)} \times \text{Time under consideration (h)}} \times 100$$

Maximum demand and total kilowatt-hours are easily obtained from past electricity bills. Normally, the LF is less than 100%. That is, the energy consumed is less than the maximum power demand at any time in the period multiplied by the total period time. In general, if the LF in a plant is reduced, the total cost of electricity will be higher. In other words, the LF is a useful method of determining if a plant is utilizing its energy-consuming equipment on a consistent basis or using the equipment for a shorter duration (lower LF), thereby paying a demand penalty. Therefore, the plant's LF should be analyzed to determine the opportunity for improvement and demand control. The simplest method for reducing peak loads is to schedule production activities in a way that the big electrical power users do not operate at the peak time at all, or at least some of them do not operate at the same time, if possible.

Machine scheduling is the practice of turning equipment on or off depending on the time of day, day of week, day type, or other variables and production needs. Improved machine scheduling is achieved through better production planning.[1] Efficient production planning that takes into account the energy aspects of production is one of the most effective ways to avoid machine idling and to reduce peak demand. The second method relies on automatic controls, which shut down nonessential loads for a predetermined period during peak times by means of some load management devices such as simple switch on–off devices, single load control devices, demand limiters, or a computerized load management system.[1]

10.10 Motor Management Plan

A motor management plan is an essential part of a plant's energy management strategy.[29] A motor management plan can help companies realize long-term motor system energy savings and will ensure that motor failures are handled in a quick and cost-effective manner. The key elements of a sound motor management plan[29] are as follows:

- Creation of a motor survey and tracking program
- Development of guidelines for proactive repair/replace decisions
- Preparation for motor failure by creating a spares inventory
- Development of a purchasing specification
- Development of a repair specification
- Development and implementation of a predictive and preventive maintenance program

10.10.1 Motor Maintenance

Motors consume almost 90% of the electricity in a spinning mill. The purpose of motor maintenance is to prolong motor life and to foresee a motor failure. Motor maintenance measures can therefore be categorized as either preventive or predictive. Preventive measures include voltage imbalance minimization, load consideration, motor alignment, lubrication, and motor ventilation. The purpose of predictive motor maintenance is to observe ongoing motor temperature, vibration, and other operating data to identify when it becomes necessary to overhaul or replace a motor before failure occurs.[30] The various motor maintenance activities to be done in spinning mills in order to prolong motor life are listed in Table 10.9.

TABLE 10.9

Various Motor Maintenance Activities to Be Followed in Spinning Mills

Motor Maintenance Activity	Frequency
Cleaning of motor body and fan cover	Daily
Observation of motor noise	Daily
Checking the condition of motors in humidification plant	Daily
Checking the motor surface temperature using IR gun	Fortnightly
Energy study of motors under load conditions	Quarterly
No-load power measurement of motors (above 10 hp)	Twice in a year
No-load power measurement of motors (below 10 hp)	Yearly
Thermography analysis	Yearly
Overhauling	Once in 3 years

Source: Shanmuganandam, D. et al., *Indian Textile J.*, February 2013, http://www.indiantextilejournal. com/articles/FAdetails.asp?id=5016, accessed May 1, 2013.

TABLE 10.10

Benchmark Values to Decide Motor Replacement

Parameter	Good	Average	Poor
Motor loading (%)	85–95	60–85	<60 or >85
Ambient temperature (°C)	<35	35–40	>40
Motor surface temperature (°C)	<85	85–100	>100
Supply voltage (near motor) (V)	415–420	400–415	<400 or >420
Supply frequency (Hz)	50	49	<49 or >50

Source: Shanmuganandam, D. et al., *Indian Textile J.*, February 2013, http://www.indiantextilejournal. com/articles/FAdetails.asp?id=5016, accessed May 1, 2013.

Normally, an induction motor is expected to work for 20 years.[30] In case there is no burnout at all even after long years of working, then the other two criteria, viz, increase in no-load power or drop in efficiency, must be considered. The benchmark values to replace a motor are given in Table 10.10, which may be taken as a broad guideline.

10.10.2 Energy-Efficient Motors

Energy-efficient motors reduce energy losses through improved design, better materials, tighter tolerances, and improved manufacturing techniques. With proper installation, energy-efficient motors can also stay cooler, may help reduce facility heating loads, and have higher service factors, longer bearing life, longer insulation life, and less vibration.[31] The choice of installing a premium efficiency motor strongly depends on motor operating conditions and the life cycle costs associated with the investment.

10.10.3 Rewinding of Motors

In some cases, it may be cost effective to rewind an existing energy-efficient motor, instead of purchasing a new motor. As a rule of thumb, when rewinding costs exceed 60% of the costs of a new motor, purchasing the new motor may be a better choice.[32] When best rewinding practices[33] are implemented, efficiency losses are typically less than 1%.

10.10.4 Motor Burnouts

If a motor is not maintained properly, it would be a source for production loss (due to motor burnout or failure and subsequent machine idle time) as well as draining of power. An intermill study conducted by SITRA[34] covering around 161 mills on motor burnouts found that the average number of motor burnouts for a 2-year period was 0.94 per 1000 spindles per year, ranging from a low of just 0.04 burnout to a high of slightly over 5 burnouts between mills. Of the total number of burnouts, winding failures account for 30%, followed by the bearing failures (26%) and protective devices failures (20%). Failures due to single phasing (8%) are the fourth major reason. The miscellaneous reasons include voltage fluctuations, switch fault, and inverter fault. The number of burnouts is highest in the small capacity motors (1–5 hp) at 37%, followed by the medium capacity motors (5–10 hp) at 26%. The ring frame department alone accounts for one-sixth of the total number of burnouts, followed by the OHTC (14%), humidification (13%), carding (12%), and blow room (9%).

10.10.5 Power Factor Correction

Power factor is the ratio of working power to apparent power. It measures how effectively electrical power is being used.[7] A high power factor signals efficient utilization of electrical power, while a low power factor indicates poor utilization of electrical power. Inductive loads like transformers, electric motors, and high intensity discharge (HID) lighting may cause a low power factor.[1] The power factor can be corrected by minimizing the idling of electric motors (a motor that is turned off consumes no energy), replacing motors with premium-efficient motors, and installing capacitors in the AC circuit to reduce the magnitude of reactive power in the system.

10.10.6 Minimizing Voltage Unbalances

A voltage unbalance degrades the performance and shortens the life of three-phase motors. A voltage unbalance causes a current unbalance, which will result in torque pulsations, increased vibration and mechanical stress, increased losses, and motor overheating, which can reduce the life of a motor's winding insulation.[1,34] Voltage unbalances may be caused by faulty operation of power factor correction equipment, an unbalanced transformer bank, or an open circuit.[1] A rule of thumb is that the voltage unbalance at the motor terminals should not exceed 1% although even a 1% unbalance will reduce motor efficiency at part load operation. A 2.5% unbalance will reduce motor efficiency at full load operation. By regularly monitoring the voltages at the motor terminal and through regular thermographic inspections of motors, voltage unbalances may be identified.[7]

References

1. Senthil Kumar, R., Energy management in spinning mill, http://www.scribd.com/doc/170513567/Energy-Management-in-the-Spinning-Mill, Accessed Date: May 1, 2013.
2. Palanichamy, C. and Sundarbabu, N., Second stage energy conservation experience with a textile industry. *Energy Policy* 33(5): 603–609, March 2005.
3. ITMF, International comparison of manufacturing costs, Spinning/weaving/knitting, International Textile Manufacturers Federation, Stockholm, Sweden, 2003.

4. Koc, E. and Kaplan, E., An investigation on energy consumption in yarn production with special reference to ring spinning—Cukurova University, Turkey. *FIBRES & TEXTILES in Eastern Europe* 15(4): 63, October/December 2007.
5. Krause, H.W. and Soliman, H.A., Energy consumption of rotor type OE-spinning machines as compared to ring spinning frame. *International Textile Bulletin* Third Quarter: 285–303, 1982.
6. Energy conservation: Time to get specific II edition, A CII study conducted by Forbes Marshall.
7. Hasanbeigi, A., Energy-efficiency improvement opportunities for the textile industry, China Sustainable Energy Program of the Energy Foundation, US Department of Energy under Contract No. DE-AC02-05CH11231, September 2010.
8. New Energy and Industrial Technology Development Organization, Japan (NEDO), 2008. Global Warming Countermeasures: Japanese Technologies for Energy Savings/GHG Emissions Reduction (2008 Revised Edition). Available at: http://www.nedo.go.jp/library/globalwarming/ondan-e.pdf, Accessed Date: May 1, 2013.
9. Energy Manager Training (EMT), 2008a. Best practices/case studies-Indian Industries, Energy-efficiency measures in Rishab Spinning Mills, Jodhan. Available at: http://www.emt-india.net/eca2008/Award2008CD/31Textile/RishabSpinningMillsJodhan-Projects.pdf.
10. Mishra, S.N., Anjali P. Pande, and A.N. Desai, BTRA Powerspin System for Estimation, Monitor and Control of Energy Consumption for Ring Frames in Textile Mills, http://www.emt-india.net/process/textiles/pdf/BTRA Powerspin System for Estimation Monitor and Control of Energy Consumption for Ring Frame in Textile Mills.pdf, Accessed Date: May 1, 2013.
11. Jha, A., 2002. Conservation of Fuel Oils and Lubricants. Available at: http://www.emt-india.net/process/textiles/pdf/Conservation%20of%20Fuel%20Oils.pdf, Accessed Date: May 1, 2013.
12. Energy Manager Training (EMT), 2008b. Best practices/case studies-Indian Industries, Energy-efficiency measures in Gimatex industries. Available at: http://www.emt-india.net/eca2008/Award2008CD/31Textile/GimatexIndustriesPvtLtdWardha.pdf, Accessed Date: May 1, 2013.
13. Chandran, K.R. and Muthukumarasamy, P., SITRA Energy Audit—Implementation strategy in textile mills, SITRA, Coimbatore, India, http://www.emt-india.net/process/textiles/pdf/SITRA%20Energy%20Audit.pdf, Accessed Date: May 1, 2013.
14. Energy Manager Training (EMT), 2008d. Best practices/case studies-Indian Industries, Energy-efficiency measures in DCM Textiles. Available at: http://www.emt-india.net/eca2008/Award2008CD/31Textile/DCMTextilesHissar-Projects.pdf, Accessed Date: May 1, 2013.
15. Roy, M.M., Humidification for textiles mills. *Air Conditioning and Refrigeration Journal*, January–March 2005, http://www.ishrae.in/journals_20042005/2005jan/article02.html, Accessed Date: May 1, 2013.
16. Energy Manager Training (EMT), 2007a. Best practices/case studies-Indian Industries, Energy-efficiency measures in RSWM Limited Banswara. Available at: http://www.emt-india.net/eca2007/Award2007_CD/32Textile/RSWMLtdBanswara/Projects_13.pdf, Accessed Date: May 1, 2013.
17. Energy Manager Training (EMT), 2008c. Best practices/case studies-Indian Industries, Energy-efficiency measures in Kanco Enterprises Ltd Dholka. Available at: http://www.emt-india.net/eca2008/Award2008CD/31Textile/KancoEnterprisesLtdDholka-Projects.pdf, Accessed Date: May 1, 2013.
18. Vijay Energy, 2009. Energy saving soft-starter. Available at: http://www.vijayenergy.com/esss.html, Accessed Date: May 1, 2013.
19. Energy Manager Training (EMT), 2007a. Best practices/case studies-Indian Industries, Energy-efficiency measures in RSWM Limited Banswara. Available at: http://www.emt-india.net/eca2007/Award2007_CD/32Textile/RSWMLtdBanswara/Projects_13.pdf, Accessed Date: May 1, 2013.
20. Energy Manager Training (EMT), 2008e. Best practices/case studies-Indian Industries, Energy-efficiency measures in RSWM Limited Banswara. Available at: http://www.emt-india.net/eca2008/Award2008CD/31Textile/RSWMLimitedBanswara-Projects.pdf, Accessed Date: May 1, 2013.

21. Energy Manager Training (EMT), 2004a. Best practices/case studies-Indian Industries, Energy-efficiency measures in Jaya Shree Textiles Rishra. Available at: http://www.emt-india.net/eca2004/award2004/Textile/Jaya%20Shree%20Textiles%20Rishra.pdf, Accessed Date: May 1, 2013.

22. United States Department of Energy (U.S. DOE), 2005. Improving ventilation system energy efficiency in a textile plant. Available at: http://www1.eere.energy.gov/industry/bestpractices/case_study_ventilation_textile.html, Accessed Date: May 1, 2013.

23. Energy Manager Training (EMT), 2008c. Best practices/case studies-Indian Industries, Energy-efficiency measures in Kanco Enterprises Ltd Dholka. Available at: http://www.emt-india.net/eca2008/Award2008CD/31Textile/KancoEnterprisesLtdDholka-Projects.pdf, Accessed Date: May 1, 2013.

24. Express Textile, Innovative energy conservation measures in overhead cleaners, available at: http://www.expresstextile.com/20051031/technext01.shtml, 2005, Accessed Date: May 1, 2013.

25. Energy Star, Guidelines for energy management, available at: http://www.energystar.gov/index.cfm?c=guidelines.download_guidelines, 2004, Accessed Date: May 1, 2013.

26. Schönberger, H. and Schäfer, T., Best available techniques in textile industry, available at: http://www.umweltdaten.de/publikationen/fpdf-l/2274.pdf, 2003, Accessed Date: May 1, 2013.

27. Easton United Nations Industrial Development Organization (UNIDO), Energy conservation in textile industry—Handy manual, available at: http://www.unido.org/fileadmin/import/userfiles/puffk/ textile.pdf, 1992, Accessed Date: May 1, 2013.

28. Centre for the Analysis and Dissemination of Demonstrated Energy Technologies (CADDET), Saving energy with efficient compressed air systems, *Maxi Brochure 06*, Sittard, the Netherlands, 1997.

29. Nadel, S., Elliott, R.N., Shepard, M., Greenberg, S., Katz, G., and de Almeida, A.T., *Energy-Efficient Motor Systems: A Handbook on Technology, Program, and Policy Opportunities*, Second Edition, Washington D.C.: American Council for Energy Efficient Economy, 2002.

30. Copper Development Association (CDA), High-efficiency copper-wound motors mean energy and dollar savings, available at http://energy.copper.org/motorad.html, 2001, Accessed Date: May 1, 2013.

31. Consortium for Energy efficiency (CEE), Motor Planning Kit, Version 2.1, Boston, MA, 2007.

32. Electric Apparatus Service Association (EASA), The effect of repair/rewinding on motor efficiency, 2003, http://www.iecex.com/dubai/speakers/Day%202C_1100-1200_John_Allen_EASA_Rewind_Study1203.pdf., Accessed Date: May 1, 2013.

33. Easton Consultants, Strategies to promote energy-efficient motor systems in North America's OEM markets, Easton Consultant Inc., Stamford, CT, 1995.

34. Shanmuganandam, D., Subash, P., and Sreenivasan, J., Towards zero burnouts in motors in spinning mills. *The Indian Textile Journal*, February 2013, http://www.indiantextilejournal.com/articles/FAdetails.asp?id=5016, Accessed Date: May 1, 2013.

11

Humidification and Ventilation Management

11.1 Importance of Maintaining Humidity in Spinning Process

Air is a vital element for every human being. The efficiency of the worker inside the mill and his health conditions may seriously be impaired by insufficient ventilation. The air quality, air temperature, and circulation of air inside the work place predominantly influence the worker's efficiency and comfort. Air free from contaminants, which is closely maintained within a fixed range of temperature and humidity, is an essential requirement of the textile mill. The generation of temperature is high in spinning mill, especially when more people are working under one roof along with more number of machines working at a higher pace. The frequent usage of water spray and provision of wide windows for fresh air are not sufficient to improve the working conditions.

Air humidity is measured as "relative humidity," which is defined as the amount of water in a sample of air compared with the maximum amount of water the air can hold at the same specific temperature.[1] Humidity plays a significant role in maintaining quality, trouble-free processing, and control of dust liberation. Cotton is a hygroscopic fiber, and it absorbs or releases moisture depending on the R.H% of the surrounding air. The cotton will give up its moisture to the air under lower humid conditions and vice versa. The moisture gain and loss occurs at every stage of the processing of fibers. The fluctuations in moisture content of cotton due to changes in temperature of department have a direct impact on the properties of fibers such as tensile strength, elasticity, fiber dimensions, and fiber friction. A drop in the equilibrium relative humidity of a textile may cause it to be weaker, thinner, less elastic, and therefore more brittle.[2] By maintaining the air humidity while processing the fibers, this loss in moisture to the atmosphere is minimized.

Humidity has a high influence on production, machinery condition, worker's efficiency, quality, and productivity. Controlled humidification helps to protect humidity-sensitive materials, personnel, delicate machinery, and equipment. The generation of static electricity during processing in spinning leads to dust and fly liberation. The proper maintenance of moisture content in the fibrous material lowers the insulation resistance and helps to carry off the electrostatic charges. Moisture loss during processing cannot be totally eliminated as the act of processing will increase the temperature of the material, which will cause it to become drier.[3] However, by increasing the humidity of the air surrounding the textile directly after processing, the material experiences "regain." Moisture is reabsorbed by the textile, thus improving the quality and performance of the fabric. This regain also has a direct impact on the weight of the textile. As textile yarns are sold by weight, if a drop in humidity leads to a 4% reduction in weight, this will require 4% more fiber to be included in the sale product. For a mill manufacturing 80 tons of textile per day, this can lead to a loss of 3200 kg of product per day due to incorrect humidity control.

11.1.1 Humidification: Terms

Humidification is simply the addition of water to air. The various terms associated with the humidification process are given as follows:

1. *Relative humidity (R.H)%*: The ratio of the vapor pressure (or mole fraction) of water vapor in the air to the vapor pressure (or mole fraction) of saturated air at the same dry-bulb (DB) temperature and pressure.

2. *Sensible heat*[4]: Heat that when added to or taken away from a substance causes a change in temperature or, in other words, is "sensed" by a thermometer. The unit of sensible heat is Btu.

3. *Latent heat*[4]: Heat that when added to or taken away from a substance causes or accompanies a phase change for that substance. This heat does not register on a thermometer, hence its name "latent" or hidden. The unit of latent heat is Btu.

4. *Evaporative cooling*: A process in which liquid water is evaporated into air. The liquid absorbs the heat necessary for the evaporation process from the air; thus, there is a reduction in air temperature and an increase in the actual water vapor content of the air.[5]

11.2 Humidity and Working Conditions

The maintenance of correct ambient conditions in the spinning mill is essential for preventing the fiber degradation during intense processing of fibers from blow room to winding. The processing of fibers in the spinning mill highly depends on the correct moisture conditions in various departments of the spinning mill. Cotton fibers tend to stick at higher moisture conditions and lead to laps formation on the machine rollers, which disrupt the production process. The removal of fibrous laps on the rollers is not only a time-consuming process but may also result in damage to top roller rubber cots. Latest spinning machines are designed to operate at higher speed; however, the increase in ambient temperature curtails the speed limits of operation. The need for proper maintenance of ambient conditions in the spinning mill is listed in Table 11.1.

TABLE 11.1

Need for Proper Maintenance of Ambient Conditions in Spinning Mill

Low moisture regain in cotton leads to poor quality and lower productivity.
Yarns with poor moisture content are weaker, more brittle and less elastic, creates more friction, and more prone to static charges.
Fibers processed at optimum regain are less prone to breakage, heating, and friction effects.
Yarns processed at optimum regain have fewer imperfections and more uniform.
Reduces static charge accumulation that makes fibers more manageable and increases machine speed.
Weight of the textile is standardized at 65% R.H and 20°C.
Reduces fly and dust liberation.
Provides healthier and more comfortable working environment.

Sengupta[2] investigated the impact of various variables on carding forces (the force required to individualize fibers from tufts of fibers) and concluded that increased relative humidity resulted in decreased carding forces, which they attributed to the reduction of the flexural rigidity of the cotton fibers as a result of an increase in the moisture content of the fibers. The reduction in flexural rigidity of fibers as a result of increased moisture also reduced the breakage rate of fibers at carding.[2] Increased relative humidity resulted in a reduced end breakage rate at spinning.

11.3 Humidity and Yarn Properties

The properties of textile fibers are in many cases strongly affected by the atmospheric moisture content. Many fibers, particularly the natural ones, are hygroscopic in the sense that they are able to absorb water vapor from a moist atmosphere and to give up water to a dry atmosphere. The amount of moisture that such fiber contains strongly affects many of their most important physical properties. The optimum humidity level in the yarn strongly influences the evenness, strength, elongation, and handle of the yarn produced (Figures 11.1 through 11.3). Hairiness in the yarn depends on the fiber quality, spinning system, speed, settings, and humidity. Lower humidity condition in the ring spinning department is one of the reasons for the increased hairiness in the yarn. The optimum humidity smoothens the hairs projecting from the yarn surface and lubricates the yarn surface. The requirement of humidity is lower at blow room at around 45%–50%, moderate at spinning processes from carding to ring spinning at around 55%.

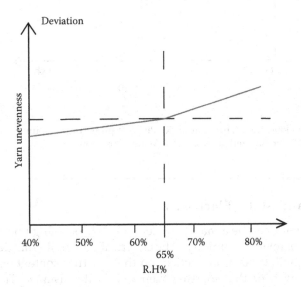

FIGURE 11.1
Effect of R.H% on yarn unevenness. (From Furter, R., Physical properties of spun yarns—Application report, Uster Technologies AG, Uster, Switzerland, June 2004. With permission.)

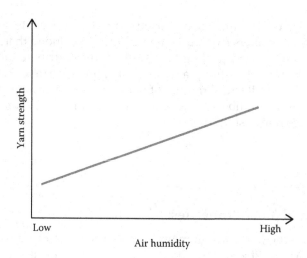

FIGURE 11.2
Effect of R.H% on yarn strength. (From Furter, R., Physical properties of spun yarns—Application report, Uster Technologies AG, Uster, Switzerland, June 2005. With permission.)

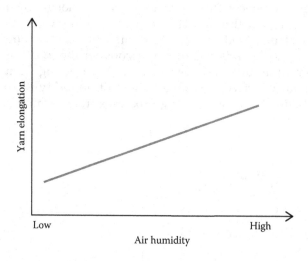

FIGURE 11.3
Effect of R.H% on yarn elongation. (From Furter, R., Physical properties of spun yarns—Application report, Uster Technologies AG, Uster, Switzerland, June 2006. With permission.)

11.4 Humidity and Static Electricity

The occurrence of static can be a major problem while processing fibers, and it is directly related to the levels of relative humidity.[6] The electrical sensitivity that determines whether static electrification will occur is dependent on the moisture content of the air and fibers. As the fibers lose moisture, they increase their electrical resistance. This means they can no longer easily dissipate the electrical charge, which is generated by the frictional contact with the machinery. In a yarn production facility with a low humidity, static discharges can jump up to 4–5 in. and, although they have a low current, can build up to several

TABLE 11.2

Electrostatic Discharge in Various Materials at Different R.H%

	Electrostatic Discharge (ESD)	
Means of Static Generation	**10%–20% Relative Humidity**	**65%–90% Relative Humidity**
Walking across carpet	35,000	1500
Walking over vinyl floor	12,000	250
Worker at bench	6,000	100
Vinyl envelopes for work instructions	7,000	600
Common poly bag picked up from bench	20,000	1200
Work chair padded with polyurethane foam	18,000	1500

hundred thousand volts. This presents a danger to staff working with the machines as it is not only very uncomfortable if they are shocked, but it can cause a person to jump and fall, which presents extreme risks when working near to textile machinery. The static discharge can also present a direct health risk to people with weak hearts or pace makers fitted. The electrostatic discharge of various materials at different relative humidity conditions is given in Table 11.2.

As well as the physical danger to staff, static electrical build-up will cause materials to stick together and be less manageable. This in turn will slow machinery, directly effecting production schedules. Also, as most machines are now microprocessor controlled, an uncontrolled electrical discharge in the wrong place can damage the electronics of the unit resulting in expensive repair bills and significant downtime. By maintaining humidity at around 50% R.H, static build-up is eliminated and all these associated problems are avoided.

11.5 Humidity and Hygiene

Hygiene is of paramount importance when releasing water into an atmosphere as any viruses or bacteria in the water could potentially be inhaled by people in the vicinity. Modern humidification systems incorporate a variety of hygiene features, but the most effective type should combine both flush cycles and a form of silver ion dosing. The flush cycles will ensure that water cannot stagnate in the pipes and allow bacteria to form.[7] Any cold water humidification system should typically autoflush at least every 24 h. The influence of relative humidity on various hygiene-related factors is shown in Figure 11.4.

Silver ion dosing is a relatively new development in hygiene control in humidifiers. As silver is effective against over 650 types of bacteria and virus, it provides added reassurance by eliminating any organisms in the water before they enter the system. Silver also has a residual effect throughout the pipe work. In the past, humidifiers typically used to incorporate UV sterilization, but this can potentially allow viruses to enter the system "shadowed" by particles in the water or allowed in by UV bulbs that have dulled with age. Regular servicing is also an important aspect of hygienic humidification. Inspections should be carried out by a competent individual from time to time to ensure optimum and hygienic performance. Some spray systems incorporate self-cleaning nozzles that can reduce maintenance to just an annual check.

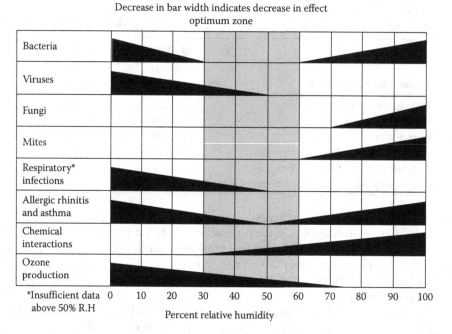

FIGURE 11.4
Optimum humidity range for human comfort and health.

11.6 Humidity and Human Comfort

Studies[6] indicate people are generally most comfortable when relative humidity is maintained between 35% and 55%. When air is dry, moisture evaporates more readily from the skin, producing a feeling of chilliness even with temperatures of 75°F or more. Because human perception of R.H is often sensed as temperature differential, it is possible to achieve comfortable conditions with proper humidity control at lower temperatures.[7] The savings in heating costs are typically very significant over the course of just a single heating season.

11.7 Humidity and Electronic Components

The microprocessors, PLCs, and modern electronic controls equipped in modern spinning machineries require controlled temperature, which should not exceed 32°C.[6] Central to all electronic circuits today is the integrated circuit (IC) or "chip." The heart of the IC is a wafer-thin miniature circuit engraved in semiconductor material. Electronic components—and chips in particular—can be overstressed by electrical transients (voltage spikes). This may cause cratering and melting of minute areas of the semiconductor, leading to operational upsets, loss of memory, or permanent failure. The damage may be immediate or the component may fail sooner than an identical part not exposed to

TABLE 11.3

Constituents of Cotton Dust

Plant matter
Fiber
Bacteria
Fungi
Soil
Pesticides
Other contaminants

an electrical transient. A major cause of voltage spikes is electrostatic discharge (ESD). Although of extremely short duration, transients can be lethal to the wafer-thin surfaces of semiconductors. ESD may deliver voltage as high as lightning, and it strikes faster.

11.8 Humidity and Dust Control

The textile mills are reminded for its cotton dust–laden environment. Cotton dust is defined as dust present in the air during the processing of cotton. The various constituents of cotton dust are listed in Table 11.3. Dust is not only a cleaning and maintenance nuisance but a common vehicle for microorganisms. It is well known that the R.H of the air will significantly affect the amount of dust in the air. A higher level of R.H (50%) will cause the particles to settle out of the air. Major amount of dust is liberated in the blow room and carding departments where opening and cleaning are accomplished. The dust exposure level in the remaining department is comparatively lower than blow room and carding.

The workers are exposed to such working environment and inhale fibrous particles and dust whole day. The health problems associated with inhaling cotton dust were discussed in Chapter 12 in detailed manner. The modern spinning mills are equipped with automatic waste evacuation system, dust filtration, and humidification plants. The dust and fly liberated by the machines are sucked by pneumatic ducts. The dust-laden air is filtered, humidified, and recirculated into the spinning department. The number of air changes per hour is optimized in each spinning department to maintain the air clean and hygienic.

11.9 Moisture Management in Ginning

Moisture content is a major factor affecting the cotton ginning process from unloading through bale packaging.[8,9] Moisture management is critical to cotton cleaning, handling, and fiber quality preservation at the gin. Cotton with too high a moisture content will not easily separate into single locks but will form wads that may choke and damage gin machinery or entirely stop the ginning process. Cotton with too low a moisture content may stick to metal surfaces as a result of static electricity generated on the fibers and cause machinery to choke and stop. When pressing and baling low moisture cotton, hydraulic pressure dramatically increases causing excessive equipment wear and problems with bale tie breakage escalate. Research[8] has shown moisture contents for seed cotton cleaning and ginning cotton is best at 6%–7% moisture content (wet basis), which allows for sufficient cleaning with minimal fiber damage. Bale packaging at these moisture contents minimizes press force, static, and bale-tie breakage. Bale storage at moisture contents greater than 8% can cause degradation to the fiber color during long-term storage. Restoring moisture to cotton fibers improves processing and adds weight to the bale.[8]

Textile mills can also have a difficult time in processing high-moisture bales. Too much moisture can reduce the bloom height of the bale at opening and cause the fibers to matt, making separation and blending difficult.[9] Many approaches have been used to restore moisture in cotton fiber using humidified air, liquid water sprays, and combinations of these systems.

11.10 Humidification Management in Spinning Mill

11.10.1 Air Washer

Air washer plants are large centralized unit in the spinning mill. The various functions of air washer are listed in Table 11.4. Spray air washers using spray water as the medium for adiabatic cooling of air are extensively useful in humidification systems for textile mills.[7] Water sprays in an air washer configuration have the ability to filter air with an efficiency of 95% for particles larger than 5 μm.

Spray air washers generally used in humidification plants consist of a chamber containing multiple banks of spray headers with spray nozzles, a tank for collecting spray water as it falls and an "eliminator section" with PVC blades having three or four bends for removing droplets of water from the air, which is humidified after passing through the curtain of spray water, before discharge to the air ducts for distribution to the humidified areas.[10] Air velocity, water spray density, spray pressure, and other design criteria are optimized by each manufacturer, depending on the air washer dimensions and spray header/nozzle sizes and configuration and eliminator design that they have developed for different applications.[7]

11.10.2 Determination of Department Heat Load

The total heat load in a department consists of internal heat load and transmitted heat load. Internal heat load is the sum of heat generated by machines, workers, and lights. Transmitted heat load is the heat flowing into the department because the outside temperature is higher than that inside the department. The heat transmission through walls,

TABLE 11.4

Functions of Air Washer

Cool the air
Humidify the air
Dehumidify the air
Clean the air of particles
Hyper ionize the air

TABLE 11.5

Approaches to Reduce Heat Load in the Department

Roof cooling
Discharge of pneumafil air to the outside
Minimization or elimination of windows area
Discharge of motor heat

windows, and roof varies considerably from season to season and even throughout a day. The heat transmission rate[11] of a department is calculated by the following equation:

$$H_T = U_w A_w [T_o - DB_i] + U_{wi} A_{wi} [T_o - DB_i) + U_r A_r [t_s] DB_i]$$

where
H_T is the transmitted heat load, kcal/h
U is the overall heat transmission coefficient kcal/h m²°C
A is the area, m²
T_o is the temperature outside, °C
DB_i is the dry bulb temperature inside, °C
t_s is the temperature of the roof top
w, wi, and r stand for wall, window, and roof, respectively

The rate of heat received by any surface depends upon a number of factors such as the time of day, color of the surface, wind velocity, dust in the atmosphere, and materials used. Variation in the outside temperature changes the actual heat load from time to time, which in turn requires adjustment in the actual supply of air from humidification plants. Some of the approaches to reduce the heat load in the department are listed in Table 11.5. These adjustments are made by means of dampers in the plants. The degree of adjustments will be minimized if the variable heat loads are reduced by various means such as good insulation or a false roof.[1] The room-sensible heat load calculations are worked out as the sum total of

1. Solar heat gain through insulated roof
2. Heat dissipation from the machines
3. Heat of air
4. Heat load from lighting
5. Occupancy heat load

11.10.2.1 Solar Heat Gain through Insulated Roof

The temperature of roof top increases much higher compared with surrounding atmosphere due to solar radiation emitted by the sun. The heat load on roof increases substantially due to the rise in temperature. The rise in surface temperature due to radiation depends on various factors such as the intensity of radiation, absorptivity of the roof surface, and wind velocity. The estimation of temperature rise is quite complicated because of the complex relationship between these factors. The maximum temperature rise of a black horizontal roof is 38.3°C whereas it is 12.8°C if it is white. The heat load caused by radiation from the sun can be reduced if roof are white instead of black. The roof is generally covered with water proofing that is coated with tar, which is black. Heat load[11] due to transmission through the roof is given by the equation

$$H_r = U_r A_r (t_2 - t_1)$$

where
 H_r is the roof heat load, kcal/h
 U_r is the overall heat transmission coefficient for roof from outside surface to air in the department, kcal/h m²°C
 A_r is the area of roof, m²
 t_2 is the temperature of outside surface, °C
 t_1 is the dry bulb temperature inside the department, °C

Some of the ways to keep t_2 low are

- The roof should be colored white to reflect radiated heat from the sun back in the atmosphere
- A false roof

11.10.2.2 Heat Dissipation from the Machines

The textile manufacturing machines generate enormous amount of heat due to metal to metal friction and metal to fiber friction. The electricity consumed by the motors is converted into heat.

The heat generated by the machine[11] is calculated by the equation

$$H_m = \frac{K_w - LF}{n} \times 860$$

where
 H_m is the heat of the machine, kcal/h
 K_w is the power of motors installed in a department, kW
 LF is the load factor
 n is the efficiency of the motor

H_m is the heat load due to the total electrical power consumed by the motors in the department. The motors convert electrical power into mechanical power to drive the production machines. A part of the total power is consumed by the motor itself and the remaining by the machine. The load factor is generally 0.85. The heat of the motor is usually about 18%–25% of the total machine heat load and can be reduced if the motors are installed over a perforated top of an underground exhaust duct and enclosed by a perforated cover.

11.10.2.3 Heat of Air

The machines such as speed frame, ring frames, and automatic cone winding machines are provided with suction arrangement for collecting broken ends and dust. The temperature of the air discharged by the suction system is more than that of the department from which the air is drawn. The air added to heat load by the suction system[11] is determined by the equation

$$H_a = K_{wa} \times 860 \times LF$$

where
H_a is the heat of the suction system, kcal/h
K_{wa} is the rated power of motor of the suction system, kW
LF is the load factor

11.10.2.4 Heat Load from Lighting

All types of light transmit heat. The heat emitted by light source is directly related to the wattage of the light source. The heat produced by fluorescent lamp is approximately 25% greater than that expected from the rated wattage.[12] This rise in heat is due to the additional electricity required by the ballast (choke). Heat emitted by the light to the department depends on

- Preferred light level in the room
- Type of lights and their construction
- Location of the light equipment

11.10.2.5 Occupancy Heat Load

The human beings working inside the plant are a source of heat generation (sensible heat and latent heat). The heat produced by a person depends upon the energy that is being exerted. The effects of different types of physical activity on heat production by the human beings are listed in Table 11.6.

11.10.3 Determination of Supply Air Quantity

The fresh air supply to a department[11,12] can be calculated as

$$q = nV$$

where
q is the fresh air supply (ft^3/h, m^3/h)
n is the air change rate (h^{-1})
V is the volume of room (ft^3, m^3)

TABLE 11.6

Amount of Heat Generation by Different Activities of Human Beings

Activity by Human Beings	Heat Generation (Btu/h)
Sleeping	291
Sitting at rest	384
Typing rapidly	558
Walking at 2 mph	761
Walking at 4 mph	1390
Walking up stairs	4365

TABLE 11.7

Air Changes in Different Departments of Spinning Mill

Department	Air Changes per Hour
Blow room	20
Spinning preparatory	15
Spinning	35

Air change rate is a measure of how many times the air within a defined space is replaced. The number of air change is an important factor to ensure that air is clean and safe. In the textile industry, the minutes per air change range from 5 to 15. Air changes are required more when the generation of heat is more due to more people working in a section or the generation of dust is more and the maximum dust level is allowed in the air.[11] Typical air changes per hour of the various departments of Indian spinning mill during summer condition is given in Table 11.7.

If air is used for cooling, the needed air flow rate[12] may be expressed as

$$q_c = \frac{H_c}{\rho c_p (t_o - t_r)}$$

where

q_c is the volume of air for cooling (m^3/s)
H_c is the cooling load (W)
t_o is the outlet temperature (°C) where $t_o = t_r$ if the air in the room is mixed
t_r is the room temperature (°C)

The required supply air quantity is calculated as follows:

$$\text{Supply air quantity (cfm)} = \frac{\text{Room sensible heat (Btu/h)}}{1.08 \times \text{Temperature rise of humidified air (°F)}}$$

11.10.4 Water Quality

The quality of water to be recirculated and sprayed in the air washer should be within the following limits:

- pH value: 7.5–8.5
- Total hardness: preferably above 100 mg/L and within 250 mg/L

Demineralized water used in boilers should not be used in air washers as this may lead to corrosion. Moderate hardness helps to build a thin protective coating that reduces harmful effects of chlorides in coater. Because of evaporation of water drops, minerals tend to concentrate during recirculation. Hence, it is necessary to bleed some water from the air washer tank and also do periodic cleaning to prevent the growth of algae.

11.10.5 Types of Humidifiers

Humidification is simply the addition of water to air. There are several methods for adding moisture to the air. The proper type of humidification equipment can help to achieve effective, economical, and trouble-free control of humidity.

11.10.5.1 Steam Humidification

Steam is readily available water vapor that needs only to be mixed with the air. Unlike other humidification methods, steam humidifiers have a minimal effect on DB temperatures. This water vapor does not require any additional heat as it mixes with the air and increases relative humidity. Steam is pure water vapor existing at 100°C. Water spray humidification disperses water as a fine mist into the air stream where it evaporates. As it evaporates, it draws heat from the air and cools it. Since the temperature of the air stream does not change, the process is isothermal.

11.10.5.2 Atomizing Humidifiers

This type of humidifier produces extremely smaller water droplets that are introduced directly to the air stream. The air's sensible heat evaporates these water droplets into water vapor. Atomizers create minute droplets by directing water onto a spinning disk or a cone or they force the water through high-pressure spray nozzles. Atomizing humidifiers are very efficient, but they require a nearly mineral-free water source.

11.10.5.3 Air Washer Humidifiers

An air washer humidifier uses a heated water spray chamber to adiabatically add moisture to the air stream. Most air washers are selected for their cooling capacity. As a humidifier, air washer relies on the air's sensible heat to evaporate water droplets and atomized water particles.

11.11 Conventional Humidification System

11.11.1 Merits

1. System and spare parts are cheap.
2. Technical expertise for critical maintenance and troubleshooting are available.
3. The system is more suitable for Indian conditions where cotton fibers are predominantly used.

11.11.2 Demerits

1. Return air system has inherent problems of fluff deposition on filters that has to be manually cleaned regularly. Deviation in cleaning schedule makes the system ineffective.

2. System is designed for peak summer conditions so manual control of humidity and temperature during other seasons becomes difficult and inaccurate.

3. The ineffective return system tends to make the shop floor atmosphere dirty with fluff and dust.

4. Due to manual cleaning of filters, air washer water tanks, there is always negligence, resulting in performance deterioration and operation problems.

5. The scope for conserving energy is limited as the fan and pump motors are always running at the same speed.

6. System requires regular maintenance, which is time consuming.

7. No possibility of measuring air washer tower efficiency on a regular basis.

11.12 Modern Humidification System

11.12.1 Merits

1. Atmospheric conditions can be met more accurately due to automation in damper control.

2. Rotary air filters perform much better due to continuous automatic cleaning system.

3. Air washer tower performance is better, but there are no cross-checks for poor efficiency.

11.12.2 Demerits

1. System requires relatively high maintenance, and cost of maintenance is higher.

2. System consumes much more energy.

3. Investment cost is much higher than the conventional system.

References

1. Hale, S., Importance of correct level of humidity for textile industry. *The Indian Textile Journal*, March 2012, http://www.indiantextilejournal.com/articles/FAdetails.asp?id=4362, Accessed Date: January 15, 2014.

2. Sengupta, A.K., Vijayaraghavan, N., and Singh, A., Studies on carding force and quality of carded sliver, Resume of Papers. p. 22–26, in Joint Technol. Conf. ATIRA, BTRA, SITRA, and NITRA, 23rd, Bombay, India, February 12–13, 1982.

3. Humidification control in the textile industry, http://www.mistingsa.co.za/humidification-control-textile-industry, Accessed Date: January 15, 2014.

4. Roy, M.M., Humidification for textile mills. *Air-Conditioning and Refrigeration Journal*, January–March 2005, http://www.ishrae.in/journals_20042005/2005jan/article02.html, Accessed Date: May 1, 2013.

5. Humidification basic concepts, http://www.devatec.com/humidity/humidification/basics. html, Accessed Date: January 15, 2014.

6. *Humidification Solution Source*, Armstrong International Inc., Three Rivers, MI, http://www. armstronginternational.com/files/commo/hvacsolutionsource/humidificationsolutionsourcebook. pdf, Accessed Date: January 1, 2014.

7. Purushothama, B., *Humidification and Ventilation Management in Textile Industry*, Woodhead Publishing Ltd., New Delhi, India, 2009.

8. Valco, T.D., Moisture management is important, USDA, http://www.extension.org/mediawiki/ files/f/ff/Moisture_PUB.pdf, Accessed Date: December 1, 2013.

9. Managing moisture at the gin is crucial for best efficiency, http://www.cottonfarming.com/ home/issues/2010-09/2010_SeptCF-Ginners.html, Accessed Date: January 1, 2014.

10. Bhattacharya, D.K., Chawla, M., and Saxena, S., Performance improvement and energy conservation through automation in textile humidification system, *ACRE Conference*, New Delhi, India, September 26–28, 2001.

11. Mushtaq Ahmed, M., Evaporative cooling, http://www.fibre2fashion.com, Accessed Date: January 15, 2014.

12. Building loads analysis, energy-models.com/building-loads-analysis-program, Accessed Date: December 1, 2013.

13. Furter, R., Physical properties of spun yarns—Application report, Uster Technologies AG, Uster, Switzerland, June 2004–2006.

12

Pollution Management in Spinning Mill

12.1 Significance of Pollution in Spinning Mill

Textile industry is the second largest industry in the world next to agriculture.[1] The textile industry provides employment to numerous people around the world. The pollutants released by the global textile industry are continuously doing unimaginable harm to the environment. They pollute land and make them useless and barren in the long run. It has become utterly necessary to reduce the pollutants emitted by the textile industry. Contamination of air, water, and land by the textile industries and its raw material manufacturing units has become a serious threat to the environment. It has endangered the life of human beings and various other species on Earth. Global warming is a direct result of the pollutants released by such industries. It also causes harmful diseases and health issues in people getting exposed to the pollutants in the long run. The emphasis on awareness about the environmental concern such as air, water, and noise pollution during the processing from fiber to fabric is essential in the present circumstances. The use of organic raw material can help in fighting the emission of pollutants by the textile units. Organic cotton is especially beneficial as the production of cotton needs a maximum amount of pesticides and fertilizers.

12.1.1 Types of Pollutant in the Spinning Process

The spinning process usually generates air pollution and noise pollution at the most, whereas the textile coloring process generates water pollution heavily. The major pollutants of the spinning mill are cotton dust and noise generation.

12.2 Cotton Dust

Cotton dust[1] is defined as dust present in the air during the handling or processing of cotton, which may contain a mixture of many substances including ground-up plant matter, fiber, bacteria, fungi, soil, pesticides, noncotton plant matter, and other contaminants that may have accumulated with the cotton during the growing, harvesting, and subsequent processing or storage periods. Any dust present during the handling and processing of cotton through the weaving or knitting of fabrics and dust present in other operations or manufacturing processes using raw or waste cotton fibers and cotton fiber byproducts from textile mills are considered cotton dust within this definition.

TABLE 12.1

Classification of Cotton Dust

Type	Size of the Particle (µm)
Trash	Above 500
Dust	50–500
Microdust	15–50
Breathable dust	Below 15

12.2.1 Classification of Cotton Dust

The microdust comprises 50%–80% fiber fragments, leaf and husk fragments, 10%–25% sand and earth, and 10%–25% water-soluble materials.[1] The high proportion of fiber fragments indicates that a large part of the microdust arises in the course of processing. Nearly about 40% of the microdust is free between the fibers and flocks, 20%–30% is loosely bound, and the remaining 20%–30% bound to the fibers. The cotton dust classified[2] according to the size of the particle is listed in Table 12.1.

12.2.2 Types of Dust

1. *Inhalable dust*: It is a term used to describe dust that is hazardous when deposited anywhere in the respiratory tree, including the mouth and nose (Figure 12.1).[1]

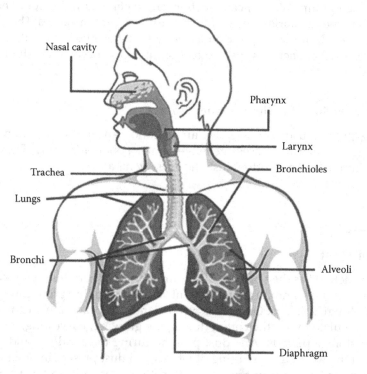

FIGURE 12.1
Human respiratory system.

2. *Thoracic dust*: It is defined as those materials that are hazardous when deposited anywhere within the lung airways and the gas exchange region.

3. *Respirable dust*: Respirable dust is defined as that fraction of the dust reaching alveolar region of the lungs.

12.2.3 Generation of the Cotton Dust during Manufacturing

- Ginning factories discharge large amounts of cotton dusts. Cotton ginning and pressing have been identified as traditional industries under the unorganized sector that functions on a seasonal basis.[1]
- Major problem of cotton dust exists in the blow room and carding section of spinning mill, whereas the exposure level in other areas is comparatively not much.
- Poor relative humidity follow-up in the department.
- Blow-down, or blow-off, is the cleaning of equipment and surfaces with compressed air.
- Cleaning of clothing or floors with compressed air.
- Improper handling of waste during transportation.
- Insufficient ventilation system.
- Improper suction system in the key areas such as blow room and carding and wherever there is a chance of dust generation.
- When materials such as laps, sliver cans, and roving bobbins are delayed in process or stored for an extended period in an area where there is a likelihood of significant dust or lint accumulation. Poor follow-up in covering the material leads to dust formation.
- Usage of spring loaded cans and carts as waste receptacles creating dust dispersion during compression of the spring loaded bottoms.
- Poor working procedures and cleaning methods.

12.2.4 Health Hazards Associated with Cotton Dust Exposure

Workers exposed to cotton dust–laden environment generally become patients of byssinosis.[1]

12.2.4.1 Byssinosis

It is a breathing disorder that occurs in some individuals with exposure to raw cotton dust. Characteristically, workers exhibit shortness of breath and/or the feeling of chest tightness when returning to work after being in the mill for a day or more. There may be increased cough and phlegm production.

Change in the levels of ESR, LDH3, and histamine may be used as indicators to assess pulmonary dysfunction in the workers those are exposed to cotton dust. It was suggested that the low hemoglobin and poor immunity against diseases may also predispose the outcome pulmonary dysfunction at an earlier stage. Cotton dust extract induces the release of histamine from samples of human lung tissue in vitro. Therefore, it is believed that histamine release is responsible for the major symptoms of byssinosis, viz., "chest tightness."

Dr. Richard Schilling, a British physician, developed a system of grading workers based on their breathing complaints on the first workday of the week.[3] Schilling's classification grades byssinosis according to how far it has progressed. Schilling's classifications are as follows:

- Grade 0 = no complaints of breathing problems
- Grade 1/2 = chest tightness and/or shortness of breath sometimes on the first day of the workweek
- Grade 1 = chest tightness and/or shortness of breath always on the first day of the workweek
- Grade 2 = chest tightness and/or shortness of breath on the first workday and on other days of the workweek
- Grade 3 = chest tightness and/or shortness of breath on the first workday and other days as well as impairment of lung function

It is believed that the degree or severity of response for individuals with symptoms of byssinosis is related to the dust level in the workplace. The beginning steps in yarn preparation generally produce more dust. Therefore, the closer to the beginning of the process, the higher will be the dust level and the more likely the pulmonary reaction or response for some workers.

12.2.5 Permissible Exposure Limits for Cotton Dust for Different Work Areas

A report[4] by occupational safety and health guideline for cotton dust stated the permissible exposure level (PEL) of various departments of spinning mill as listed in Table 12.2.

12.2.6 Medical Monitoring

Medical examinations are to be provided to prospective employees prior to their initial assignment. As a minimum, the examinations should include[1]

- A medical history to identify any existing health problems or diseases that may affect breathing
- A standardized respiratory questionnaire inquiring about such concerns as cough, chest tightness, and smoking history

TABLE 12.2

Permissible Exposure Level in Various Departments of Spinning Mill

Department	PEL ($\mu g/m^3$)
Opening	200
Picking	200
Carding	200
Combing	200
Roving	200
Spinning	200
Winding	200
Warping	200
Slashing	750
Weaving and knitting	750
Waste house	750

- A pulmonary function (breathing) test including the forced vital capacity (FVC), the amount of air one can force out after taking a deep breath and forced expiratory volume in 1 s (FEV1), the amount of air forced out during the first second of expiration

12.2.7 Environmental Exposure Monitoring

- Sampling of the workplace must be done at least every 6 months to determine the amount of cotton dust in the environment.[1]
- Measurements must be representative of all employees in the workplace.
- Sampling will be done in all work areas and on each shift.
- Sampling is done for a period equal to at least three-quarters of the shift.
- While sampling is being done, other information is collected that may pertain to the generation of cotton dust. The percent of cotton fiber in the mix; the grade of the cotton and where it was grown; types of yarn being run; and the number and types of machines operating in each area may all affect the amount of cotton dust in the workplace.

12.2.8 Vertical Elutriator

It is used to monitor employee exposure to cotton dust in the workplace. Air is drawn into the vertical elutriator (Figure 12.2) at a specified speed, and particles of 15 µm or smaller are collected on a filter.[1] The particles collected are measured to determine the amount of respirable dust (dust that can get into the lungs) there is in the work area. It is important to

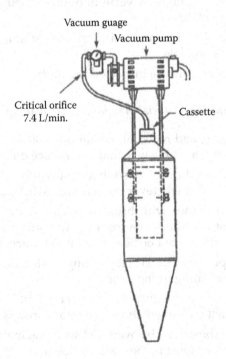

FIGURE 12.2
Vertical elutriator.

realize that other "dusts," such as starch or oil mist, are also collected on the filter and may contribute to the cotton dust levels.

12.2.9 Dust Control Measures

- Monitoring of cotton dust concentration in the occupational environment
- Provide medical surveillance to the cotton dust–exposed workers
- Establishing safe working practices to reduce the exposure level
- Training and education of workers
- Engineering controls to reduce the emission
- Use of dust masks

12.2.10 Preventive Measures to Be Followed during Manufacturing Process

12.2.10.1 General Practices

- Usage of compressed air for cleaning purposes should be prohibited when other means of cleaning are possible. Where blow-down cleaning is done (meaning general cleaning of the entire room, including the walls and ceilings ventilation ducts), employees performing the cleaning must wear respirators.
- All other employees not involved with the blow-down cleaning must leave the area.
- Cleaning of clothing or floors with compressed air is prohibited.
- Floor sweeping will be done by vacuum or other methods designed to minimize the breathing of dust.
- Waste will be handled by mechanical means. Manual handling should be limited as much as possible.
- Ventilation systems should be inspected regularly.

12.2.10.2 Work Practices during Material Handling and Cleaning

- Cotton, cotton waste, and materials containing cotton dust should be stacked or handled properly in such a way that will reduce dust level.
- Brooms should be used properly so that dust will be controlled and vacuum cleaners should be used wherever dust control is difficult.
- When cleaning machines with brushes or cloths, the individual doing the cleaning should stroke the waste from top to bottom as far from the face as possible. Surfaces should not be beaten or fanned during cleaning.
- Waste should be placed in the corresponding waste storage container immediately before accumulating in the floor.
- Waste receptacles or waste transport containers should be placed in the respective places such that creating disturbance by any means would be avoided.
- Waste receptacles should not be overfilled such that material spills to the floor during storage or transport to the waste godown.

- Spring-loaded cans and carts should not be used as waste receptacles in order to avoid dust dispersion during compression of the spring-loaded bottoms.

- When materials such as laps, sliver cans, and roving bobbins are delayed in process or stored for an extended period in an area where there is a likelihood of significant dust or lint accumulation, the materials should be covered. The storage area and the covers should be periodically cleaned to prevent lint and dust accumulation.

12.3 Pollution in Cotton Cultivation and Processing

Cotton is one of the most chemically intensive among all field crops. Cotton is grown on an estimated 3% of the total cultivated area in the world, but uses about 25% of all insecticides consumed in agriculture. Pests are such a serious threat to cotton production that economic yields are almost impossible to achieve without monitoring pests and adopting chemical controls. Plant protection operations have become the crucial aspect of production practices and pesticides that are banned for use on food crops are commonly used on cotton. In many countries, especially where cotton is machine picked, herbicides, insecticides, growth regulators, and harvest aid chemicals in addition to fertilizers are integral parts of production practices.[5] Even after harvesting, cotton fabric at textile mills is treated with a variety of chemicals for improving appearance and performance. Cotton fabrics are often processed with toxic dyes and formaldehydes before they reach end users. Growing cotton without synthetic fertilizers and other chemicals has been termed green, environment-friendly, biodynamic, etc., but organic production is the most popular name used in the cotton industry.[5] Organic cotton production is a system of growing cotton without synthetic chemical fertilizers, herbicides, conventional synthetic insecticides, growth regulators, growth stimulators, boll openers, or defoliants. It is a system that contributes to healthy soils and/or people. The organic system promotes enhanced biological activity, encourages sustainability, and commands proactive management of production.[5]

12.3.1 Impact of Chemical Used in Cotton Cultivation

1. Many chemicals used in conventional farming were first developed for warfare.[6]
2. 25 million people worldwide is poisoned by pesticides every year.[6]
3. 25% of the pesticides and fertilizers used in the world are sprayed in conventional cotton crops.[6]
4. Over 0.75 kg of toxic chemicals are used to grow the cotton needed for a conventional cotton sheet set. About 0.5 kg to make a T-shirt and pair of jeans.[6]

Among all the pesticides used, roughly 65% of the chemicals are used against insects, 20% are herbicides, and 14% are defoliants and growth regulators while fungicides and others comprise only 1% of the total toxic chemicals used on cotton.[6]

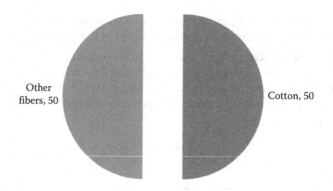

FIGURE 12.3
Fiber usage.

12.3.2 Cotton Usage

Cotton is the most widely used natural-fiber cloth in clothing today. It accounts for almost 50% of the textile market worldwide (Figure 12.3). It is used to make a number of textile products such as bath towels and robes, denim, shirts, socks, underwear, T-shirts, bed sheets, etc.

Globally, nearly 90 million acres of cotton is grown in more than 70 countries.[5] It is estimated that little over 8000 ha of organic cotton is grown in various countries, the United States being the largest producer in the world. Organic cotton is also said to be produced in Argentina, Australia, Brazil, Ecuador, Egypt, Greece, India, Nicaragua, Paraguay, Peru, Senegal, Tanzania, Turkey, and Uganda.[6]

12.3.3 Organic Cotton

Organic cotton is grown using methods and materials that have a low impact on the environment. Organic production systems replenish and maintain soil fertility, reduce the use of toxic and persistent pesticides and fertilizers, and build biologically diverse agriculture.[7] Third-party certification organizations verify that organic producers use only methods and materials allowed in organic production. Organic cotton is grown and processed without toxic chemicals that can be absorbed easily when in contact with the user's skin. Pesticides, fertilizers, and chemicals used to grow and process conventional cotton fabrics may go directly to the user's blood stream, which consequently affects the body's organs and tissues.

Organic cotton production is not simply an elimination of fertilizers and insecticides but it is a complete production system that requires equally sound knowledge of cotton production practices. With respect to insect control in particular, a thorough knowledge of nonchemical means of insect control is a prerequisite for organic production. The use of chemicals in the form of fertilizers, herbicides, insecticides, growth regulators, defoliants, and desiccants has increased the cost of production to the extent that cotton is losing its profitability against other field crops. Environmental concerns are also increasing. Organic cotton production provides an alternative to grow cotton without chemicals. Organic cotton production requires careful planning so as to realize optimum yield. It includes a number of factors such as site selection, crop rotations, variety, weed control, nonchemical means of insect control and skill to manage organic crop. Similarly, there is a need to perfect the agronomic requirements of a crop to be grown without synthetic fertilizers and pesticides. Besides the naturally soft organic cotton, fabric is a lot more comfortable to use and is available at competitive prices.

12.3.4 Necessity to Shift to Organic Production

Organic cotton production is also a consumer-driven initiative. There are many harmful chemicals that people do not know about. Twelve of these chemicals are known as persistent organic pollutants (POPs), and are the most hazardous of all man-made products or wastes that cause deaths, birth defects, and diseases among humans and animals. They are so dangerous that 120 nations agreed at a United Nations Environment Programme conference to outlaw them. Of the 151 signatories to the convention, 98 states have ratified it; sadly the United States and Russia have not yet done so.[5] There are three of those chemicals used in cotton manufacturing. The following are the main factors[5] responsible for organic cotton production:

- *Concern for the environment*—Fertilizers and pests applied to the soil, but all the chemicals are not taken up by the cotton plant. Some elements are released into the environment while others leach into the soil and also pollute water.
- *Concern for family health*—Danger of insecticide inhalation by the spray men during back-mounted manual spraying without any protective equipment.
- *Life style*—Some people were interested in insecticide-free cotton apparel due to allergies.
- *To reduce input prices*—Insecticide use changed the insect complex in many countries. Some minor insects became major and certain new insects were introduced. Consequently, there was an increase in the consumption of insecticides.

12.3.5 Comparison of Conventional Cotton and Organic Cotton Production

A comparison of production of conventional cotton and organic cotton is given in Table 12.3.

TABLE 12.3

Comparison of Production of Conventional Cotton and Organic Cotton

Criteria	Conventional	Organic
Seed preparation	Treatment of seed with fungicide of insecticide. Use of GMO (BT) seeds	Uses untreated seeds No use of GMO/BT
Soil and water	Applies synthetic fertilizers Loss of soil due to monocrop culture.	Uses biofertilizers like cow dung Strengthens soil through crop rotation
Weed control	Applies herbicides and insecticides Repeated application infecting air, water, and soil	Physical removal of weeds and biopesticides. Use of mechanical and hand methods and totally harmless
Pest control	Heavy use of insecticides. Consumes about 25% of world's total insecticides Use pesticides that are highly toxic and carcinogenic Use of spray that affects air, water, and also affects human life and nature	Use of natural predators to kill insects Uses beneficial insects to control pests Use of trap crop to control pests
Harvesting	Defoliates with toxic chemicals	Natural defoliation

TABLE 12.4

Factors Limiting the Expansion of Organic Cotton Production

Lack of suitable varieties
Pest control
Lack of production technology
Lack of information on cost of production
Price premium
Need for alternate inputs
Tied crop rotations
Nonorganic genetically engineered cotton
Certification
Marketing

12.3.6 Limitations of Organic Cotton Production

There are many reasons why organic cotton production has not extended to other countries. Nineteen countries tried to produce organic cotton during the 1990s. But many of them have already stopped, not for lack of desire or demand for such cotton, but for economic reasons. Insecticides need to be eliminated from the cotton production system because they are dangerous to apply, have long-term consequences on the pest complex, and have deleterious effects on the environment.[8] Also, heavy reliance on pesticide use has pushed many countries out of cotton production.[8] The various factors limiting the expansion of organic cotton production are listed in Table 12.4.

12.4 Significance of Noise Pollution

Noise is ubiquitous in industry.[9] Noise is a pollutant. Noise is an important cause of environmental pollution worldwide especially in urban centers, where the industries are situated. Noise has become serious environment pollution in the daily life of labor and is an increasing public health problem according to the World Health Organization's (WHO) guidelines for community noise. Development of modern automated machines in textile industries has considerably decreased the physical burden of work on workers, but one of the most undesirable and unavoidable product of these machines is noise pollution. High-level noise not only hinders communication between the workers also depending upon the level, quality, and exposure duration of noise have physical, physiological, and psychological effects on the workers. In developing countries also, there has been a great concern about the magnitude of industrial noise exposure, particularly in the textile industry.

Exposure to excessive noise is the major avoidable cause of permanent hearing impairment worldwide. The estimated costs of noise to developed countries range from 0.2% to 2% of GDP (gross domestic product). There is a serious shortage of accurate epidemiological information on prevalence, risk factors, and costs of noise induced hearing loss, especially in the developing countries. The main concern of noise-control is therefore the development, production, and preferred use of low-noise working equipment and processes. Comprehensive research is needed in technical measures for noise abatement, setting standards, improving hearing protectors, and low-cost medications for prevention.

12.4.1 Noise: Terminologies

- *Sound*: An alteration of pressure that propagates through an elastic medium such as air that produces an auditory sensation.
- *Noise*: Any undesired sound.
- *Decibel*: The intensity of sound waves produces a sound pressure level, which is commonly measured in a unit called the *Decibel*.

$$\text{Decibel} = 20 \log 10 \left(\frac{\text{Measured sound pressure}}{\text{Reference pressure}} \right)$$

Reference Pressure (P_{ref}) = 0.00002 N/m^2.

12.4.2 Ambient Air Quality Standards in Terms of Noise

The Environment Protection Act 1986 Schedule III gave the Ambient Air quality standards[10] in respect to noise for industrial, commercial, silent, and residential zones (Table 12.5).

The noise levels in industries are well above the recommended safe limits of exposure of 90 dB for 8 h daily for a 30-year working life as proposed by the International Organization for Standardization Rept. No. 1999.[11]

12.4.3 Perceived Change in Decibel Level

The perceived changes in decibel level to the human ear are listed in Table 12.6.

12.4.4 Noise in the Textile Industry

High noise levels are generally found in the textile process from fiber to fabric.[9] The noise because of clattering of gear wheels, high-speed whine of twisting and spinning

TABLE 12.5

Noise Level Limits in Different Areas

Area	Noise Level Limits in dBA (6 a.m.–9 p.m.)	Noise Level Limits in dBA (9 p.m.–6 a.m.)
Industrial area	75	70
Commercial area	65	55
Residential area	55	45
Silence zone	50	40

TABLE 12.6

Perceived Change in Sound Level by Human Ear

Change in Sound Level (dB)	Perceived Change to the Human Ear
+1	Not perceptible
+3	Threshold of perception
+5	Clearly noticeable
+10	Twice (or half) as loud
+20	Fourfold (4×) change

machinery, and noise of weaving machines have long been considered as necessary evil of the business. The ring sheds of the spinning mills are the noisiest departments.

Noise sources in any industrial process may be due to the following reasons:

- Propagation through air (airborne noise)
- Propagation through solids (structure-borne noise)
- Diffraction at the machinery boundaries
- Reflection from the floor, wall, ceiling, and machinery surface
- Absorption on the surfaces

Noise emission rises nonlinearly because of higher rotary and travelling speeds in machine parts. For example, for every doubling of the rotary speed, the noise emission for rotating print machines rises by about 7 dBA, for warp knitting looms 12 dBA, and for fans is between 18 and 24 dBA. The weaving operation on a shuttle loom consists of shuttle picking, checking, and beating up. Noise emitted during shuttle picking and checking is of a very high intensity. Despite the fact that spinners and weavers have been found to have significantly greater hearing loss than a controlled unexposed population, little progress has been made in reducing the noise in textile industries. The trend toward greater speeds has resulted in higher noise levels, often exceeding 110 dB (A) in some operations.

Continuous exposure to noise levels above 90 dBA can produce adverse auditory and nonauditory health effects. Studies carried out by NIOH (National Institute for Occupational Health, India) showed that the sound pressure levels were very high in textile industries ranging from 102 to 114 dBA. Hearing acuity of textile weavers aged 25–39 years, exposed to a noise level of 102–104 dBA, was found to be poor.

12.4.5 Method of Noise Evaluation

Personal noise exposure histories were calculated as time-averaged equivalent noise exposure levels (L_{eq}) using previously described methods as follows[9,12]:

$$L_{eq}, \quad T = A \log \left[\left(\frac{1}{T} \right) 6 \left(t_1 610 \left(\frac{L_1}{A} \right) + t_2 610 \left(\frac{L_2}{A} \right) + \cdots t_n 610 \left(\frac{L_n}{A} \right) \right) \right]$$

where
  T is the time duration over which the equivalent level is being determined
  A is the exchange rate selected ($A = 10$ for a 3 dB exchange rate)
  L is the noise level for that time duration

For example, an individual who worked 3 years at 90 dBA, 5 years at 88 dBA, and another 2 years at 87 dBA would have a 10-year time-weighted noise exposure level (L_{eq}, 10 year) = 10 log [(1/10)6(3610(90/10) + 5610(88/10) + 2610(87/10))] = 88.5 dBA.

Bedi[12] conducted a study on the noise level observed in the various departments of spinning mills and reported (Table 12.7).

The daily noise exposure of workers in areas like loom shed, ring frame, TFO, etc., exceeds the maximum exposure limit of 90 dBA specified by OSHA (Washington, DC).[13] The noise exposure in other work areas such as blow room, carding, combing, etc., is recorded less than 90 dBA. However, the direct application of OSHA regulations in Indian plants is also

TABLE 12.7

Noise Level in Various Departments of Spinning Mill

Department	L_{eq} dBA
Blow room	84.8
Carding	89.9
Draw frame	85.2
Combing	84.4
Simplex	87.2
Ring frame	100.1
Open-end spinning	89.9
Winding	91.4
Doubling	94.1

not valid, as most of the plants operate 8 h/day and 6 days/week, that is, exposure time is 48 h/week, which is 20% higher than the exposure time per week in the United States or in European countries.

12.4.6 Effect of Noise Pollution

Workers consistently exposed to the noise levels above 85 or even 90 dBA may reveal permanent hearing loss.[9] Prevalence of noise-induced hearing loss was highest in the workers working in weaving area, followed by the spinning workers and the workers working in TFO, doubling area. In addition to hearing loss, exposure of workers to noise levels of 90–119 dBA was also found to result in

- Cardiovascular and psycho physiologic problems
- Sleep disorders and headache
- Mental fatigue
- Annoyance, speech interference, and reduced alertness compared with those working in a relatively quiet room (60–75 dBA)
- Increased blood pressure, deep body temperature, and pulse rate
- Speed of performance was impaired significantly by noise

12.4.7 Suggestion to Eradicate Noise Pollution in Textile Industry

- Noise in spinning section can be reduced by providing elastomeric spindle mounts, elastomeric ring holders, and proper maintenance lubrication of gears, etc.[9]
- Replacement of parts with resilient materials like nylon instead of metal can provide reduction in impulse noise of looms.
- Attempts shall be made to produce complete enclosures around the loom.
- Proper maintenance by ensuring the following:
  - Reduction of imbalance through proper alignment and balancing of rotating equipment, preferably accomplished under dynamic load conditions.
  - Replacement of worn parts, such as bearing, gears, and other moving parts.

- Regular lubrication to reduce friction.
- Tightening of loose parts.
- Correct assembly of machine parts or replacements.
- Vibration isolators prevent noise from being transmitted through the base of the equipment.
- Damping—These commercially available mastic polymeric or foamed coatings reduce sound amplitude and duration and are best used in conjunction with other noise control treatment.
- Combination of foam and lead-filled vinyl provides a degree of both absorption and transmission loss where only marginal (10 dBA or less) noise reduction is necessary.
- Improving the acoustic environment by the addition of sound absorptive elements.
- A partial enclosure or sound barrier with both absorptive and sound transmission loss qualities, and correctly placed, can provide noise reduction of the order of 8–12 dB. A complete enclosure can provide a greater degree of reduction ranging from 30 dB to over 60 dB depending upon the design.

12.4.8 Preventive Measures to Control Noise Pollution

- Noise surveys to determine the degree of hazardous noise exposure by surveying any area in which workers are likely to be exposed to hazardous noise (>85 dBA). Level of hazard depends on noise intensity, duration of exposure during a typical working day, and overall exposure during working life.
- Engineering and administrative controls are undertaken to reduce exposures to <90 dBA and include design of equipment, its location and layout, selection of quieter machines, treatment of noisy rooms, administrative controls, proper maintenance, and isolation of the worker from source.
- Audiometric tests, by pre-employment and periodic follow-up testing by employers, to help determine employee effects; employee medical history and nonworkplace noise exposure should be assessed.
- Hearing protection devices such as earmuffs or helmets to reduce the amount of sound reaching the ear.

References

1. Senthil Kumar, R., Cotton dust—Impact on human health and environment in the textile industry, http://www.docstoc.com/docs/19408391/Cotton-dust-in-textile-mills, Accessed Date: November 12, 2013.
2. Klein, W., *A Practical Guide to Opening and Carding, Manual of Textile Technology*, Vol. 2, Textile Institute, Manchester, U.K., 1987.
3. Senthil Kumar, R., Cotton dust—Impact on human health and environment in the textile industry, http://www.fibre2fashion.com/industry-article/9/831/cotton-dust-impact-on-human-health-and-environment-in-the-textile-industry2.asp.

4. Cotton Dust, Occupational safety and health guideline for cotton dust, US Department of Health and Human Services, NIOSH, Washington, DC, 1988.
5. Senthil Kumar, R., Organic cotton: A route to eco-friendly textiles. *The Indian Textile Journal*, November 2007.
6. Organic Cotton Facts—Organic Trade Association, http://www.ota.com/organic/fiber/ organic-cotton-facts.html, Accessed Date: May 1, 2013.
7. Organic Cotton Production—3, The ICAC Recorder, June 1996, http://www.icac.org/cotton_ info/tis/organic_cotton/documents/1996/e_june.pdf, Accessed Date: May 1, 2013.
8. Limitations on Organic Cotton Production, The ICAC Recorder, March 2003, http://www.icac. org/cotton_info/tis/organic_cotton/documents/2003/e_march.pdf, Accessed Date: May 1, 2013.
9. Senthil Kumar, R., Noise pollution in the textile industry, http://www.scribd.com/ doc/24041262/Noise-Pollution-in-the-Textile-Industry, Accessed Date: May 1, 2013.
10. The Environment (Protection) Act, 1986, http://envfor.nic.in/legis/env/env1.html, Accessed Date: May 1, 2013.
11. Environmental Management System and ISO 14001, Federal Facilities Council Report No. 138, 1999.
12. Bedi, R., Evaluation of occupational environment in two textile plants in northern India with specific reference to noise. *Industrial Health* 44: 112–116, 2006.
13. OSHA, *A Guide for Persons Employed in Cotton Dust Environments*, US Department of Labor, Occupational Safety and Health Administration, US Department of Health and Human Services (NIOSH), and NCDOL personnel, Raleigh, NC.

13

Process Management Tools

13.1 Significance of Process Management

The term "process" means the sequence of interdependent and linked procedures that, at every stage, consume one or more resources (employee time, energy, machines, money) to convert inputs (data, material, parts, etc.) into outputs.[1] These outputs then serve as inputs for the next stage until a known goal or end result is reached. Process control is the active changing of the process based on the results of process monitoring. Once the process monitoring tools have detected an out-of-control situation, the person responsible for the process makes a change to bring the process back into control.

Process management is the ensemble of activities of planning and monitoring the performance of a process.[2] Process management is the application of knowledge, skills, tools, techniques, and systems to define, visualize, measure, control, report, and improve processes with the goal to meet customer requirements profitably.[3] Textiles consist of fibers, yarns, fabrics, and finishes. Each of these stages has a variety of processes involved to reach the next stage. Process control in textile manufacturing processes plays a crucial part in deciding the quality of the final product and productivity. Raw material is the most important factor influencing yarn quality. To a great extent, it can determine whether a product is good, and it is also responsible for the cost factor. Other factors influencing the process are machine parameters, process parameters, working procedures, and atmospheric conditions. The textile industry consists of several sectors such as spinning, weaving, knitting, and coloration. In the last two decades, spinning mills have begun to experience a quite distinct intensification of business pressures, particularly in the industrialized countries. The spinning mills are facing a tough situation and need systematic approach to survive as well as to sustain in the market. The various reasons attributed for this trend are fluctuations in raw material cost, labor shortage, and power crisis. The application of process management tools will bring cost reduction, waste elimination, and effective labor utilization, which leads to reduction in manufacturing cost and achieving good company reputation among the competitors. This chapter discusses some of the process management tools applicable to the yarn manufacturing process.

13.1.1 Process Management in Spinning Mill

With the efficient process management, the individual machine in each and every department in the spinning mill runs under optimum condition. Process management helps to tune each intermediate process with respect to final process in order to obtain optimized process-oriented compromise in terms of quality and cost. The various factors governing effective process management in spinning are listed in Table 13.1.

TABLE 13.1

Factors Governing Effective Process Management in Spinning

Testing of fiber properties before and after every important process
Bale management
Optimum machine setting in w.r.t. yarn property requirements
Effective machine maintenance
Implementation of process management tools

13.2 Process Management Tools

13.2.1 5S

5S (Table 13.2), a Japanese concept, is such a tool to enhance the efficiency, competitiveness, and survival. It is a deceptively simple system that creates an organized and productive workplace[4] (Figure 13.1).

- Sort: clearing the work area
  Any work area should only have the items needed to perform the work in the area. All other items should be cleared (sorted out) from the work area.

- Set in order: designating locations
  Everything in the work area should have a place and everything should be in its place.

- Shine: cleanliness and workplace appearance
  Cleanliness involves housekeeping efforts, improving the appearance of the work area, and, even more importantly, preventive housekeeping—keeping the work area from getting dirty, rather than just cleaning it up after it becomes dirty.

- Standardize: everyone doing things the same way
  Everyone in the work area and in the organization must be involved in the 5S effort, creating best practices and then getting everyone to "copy" those best practices the same way, everywhere, and every time. Work area layouts and storage techniques should be standardized wherever possible.

TABLE 13.2

5S—Meaning

Japanese Term	English Equivalent	Meaning in Japanese Context
Seiri	Sort	Throw away all rubbish and unrelated materials in the workplace
Seiton	Set in order	Set everything in proper place for quick retrieval and storage
Seiso	Shine	Clean the workplace; everyone should be a janitor
Seiketsu	Standardize	Standardize the way of maintaining cleanliness
Shitsuke	Sustain	Practice "Five S" daily—make it a way of life; this also means "commitment"

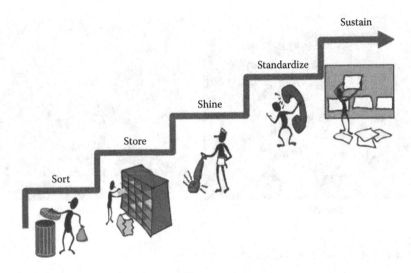

FIGURE 13.1
5S concept.

- Sustain: ingraining the 5Ss into the culture
  It is tough to keep a 5S effort, or any improvement effort for that matter, going. The 5Ss involve a culture change. And to achieve a culture change, it has to be ingrained into the organization—by everyone at all levels in the organization.

13.2.2 Application of 5S in Spinning Mill

13.2.2.1 Bale Godown

- Arrange the bales with respect to origin and lot wise.
- Put the tag with complete details such as variety, lot number, and number of bales.
- Arrange the bales in the order of using priority.
- Keep the bales transferring equipment in respective place.
- Maintain a blackboard mentioning responsible person and his duties.

13.2.2.2 Preparatory Department

- Arrange the cans and bobbins in the marked area.
- Keep a box at the center of the department and train the labors to put unwanted materials such as torn apron, nut, bolt, etc.
- Maintain a separate area with identification to keep the waste.

13.2.2.3 Spinning Department

- Keep the doff box in the allotted space.
- Keep the empties with respect to color wise and count wise.
- Maintain a common blackboard mentioning all the details of the process.

Before 5S   After 5S

FIGURE 13.2
Work place before and after implementation of 5S.

- Maintain a blackboard in each ring frames mentioning respective tenter name, count, empties to be used, roving bobbin color, spacer, etc.
- Keep the materials handling equipment in the respective places.
- Keep a box at the center of department to put unwanted materials found inside the departments.

13.2.2.4 Maintenance Department

- Keep the tool box in the respective and reachable area (Figure 13.2).
- Arrange the gears and belts and put tag.
- Keep the important spares machine wise with tag.
- Keep the traveler box with tag or sticker.
- Maintain a blackboard mentioning responsibility and responsible person.
- Keep the spare parts with respect to priority of usage.
- Keep a blackboard mentioning responsibility and responsible person for maintaining ancillary equipments such as compressor, drier, OHB, etc.

13.2.3 Advantages of 5S

The following are the benefits attained if 5S is implemented and followed diligently:

- Lower costs
- Better quality
- Improved safety
- Increased productivity
- Higher employee satisfaction
- Increased floor space
- Less wasted labor
- Better equipment reliability

13.2.4 Implementation Program of 5S

- Step 1: *start with the leadership team*[4]
  As with any improvement effort, implementation of the 5Ss must be driven from the top of the organization. Only top management can create the environment needed and give the effort the visibility and importance it needs for long-term viability.

- Step 2: *build the infrastructure*
  The 5S effort should fit within an organization's existing improvement structure. Divide and conquer by establishing 5S subcommittees for communications, training, project support, and best practices.

- Step 3: *launch communications*
  Conduct short, focused, and frequent communication sessions with all employees on the what, why, how, when, and who of the 5S initiative. Deliver the message in several formats including group meetings, using the organizations' intranet or website, bulletin board postings, and internal newsletters.

- Step 4: *train teams in 5S techniques*
  Develop a plan to train everyone in basic 5S concepts and then supplement the generic training with just-in-time training in work area–specific practices. Note that the initial teams may need to be trained in problem-solving techniques and root cause analysis.

- Step 5: *begin 5S pilots*
  Select areas that need the 5Ss (and that you project will be successful in adopting 5S practices) as pilot areas. The first pilot work areas to receive 5S treatment should be ones with high visibility.

- Step 6: *establish best practices*
  Creation (and use) of a best practices database can help multiply the impact of 5S successes by providing the means to share successes throughout the organization.

- Step 7: *develop a full roll-out plan*
  After completing the initial pilots and before involving the rest of the organization in the 5S effort, step back and evaluate how the pilots went. Get ideas from members of the pilots about how to strengthen the 5S process and use those ideas to develop a roll-out plan.

- Step 8: *continually evaluate and adjust*
  As with any process, as lessons are learned, make improvements to the 5S effort. Modify and strengthen the infrastructure, select new tools to add to the "arsenal," develop improved methods to measure and communicate progress, and challenge work areas to constantly improve.

13.2.5 5S Radar Chart

The radar chart as seen in Figure 13.3 is used to map progress using the data from the assessment checklist within the area. It should be displayed in a prominent location and updated on completion of the assessment.

This format allows progress to be monitored over a 12-month period as well as indicating clearly the current status. Today, with competition in industry at an all time high, 5S be the only thing that stands between success and total failure for some companies

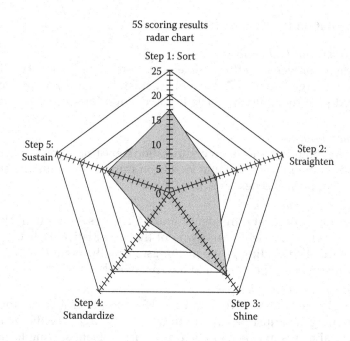

FIGURE 13.3
5S: radar chart.

especially spinning mills where lot of manpower is involved. An organized, clean working environment will help to reduce waste and errors and improve the efficiency of the operation and the working environment. One of the most important advantages of 5S is the improvement in the leadership environment. The level of satisfaction and morale among the personnel raises and a feeling of "ownership" develop toward the equipment, the product, and the company.

13.3 Total Productive Maintenance

The effective utilization of manpower and resources in the manufacturing sector leads to low manufacturing cost, which is necessary for survival in present scenario. Total productive maintenance (TPM) is one of the cost reduction techniques used in the industry, particularly in spinning mills, where a lot of maintenance activities are carried out on a daily basis.[5]

13.3.1 Conventional Maintenance System

Table 13.3 shows the maintenance activities carried out by conventional maintenance system. The major loss areas in any plant are categorized as

1. Planned downtime loss
2. Unplanned downtime loss

TABLE 13.3

Activities in the Conventional Maintenance System

Maintenance Activity	Operator (%)	Maintenance Department (%)	Project (%)	Subcontractor (%)
Routine inspection	38	62		
Periodic inspection	5	93		2
Minor repair	1	85	4	10
Shutdown	2	83	4	11

3. Idling and minor stoppages

4. Slow down

5. Process nonconformities

6. Scrap

In traditional maintenance system, the loss will be more in the aforementioned aspects due to noninvolvement of production people in basic maintenance activities that lead to frequent machine breakdown, efficiency, and utilization loss.[5] This ultimately results in production loss and increases the manufacturing cost of the product. TPM is used to overcome all these difficulties in a cost-effective manner in any manufacturing process.

13.3.2 TPM: Definition

TPM is a low-cost people-intensive system for maximizing equipment effectiveness by involving entire company in a preventive maintenance program.

- *Total*—Encompassing maintenance and production individuals working together.
- *Productive*—Production of goods and services that meet or exceed customer's expectations.
- *Maintenance*—Keeping equipment and plant in as good as or better than the original conditions at all times.

13.3.3 Objectives of TPM

- Avoiding wastage in a quickly changing economic environment
- Producing goods without reducing product quality
- Reducing cost
- Producing a low batch quantity at the earliest possible time
- Ensuring goods sent to the customers are nondefective

13.3.4 TQM versus TPM

Table 13.4 shows the comparison of total quality management (TQM) and TPM.

TABLE 13.4

Comparison of TQM and TPM

Category	TQM	TPM
Object	Quality (output and effects)	Equipment (input and cause)
Mains of attaining goal	Systematize the management	Employees' participation
Target	Quality	Elimination of losses and wastes

13.3.5 Different Modules in the Implementation of TPM in a Spinning Mill

13.3.5.1 Preparatory Module

- Step 1—*Announcement by management about TPM introduction in the organization*[5]: Proper understanding, commitment, and active involvement of the top management are needed for this module. Management must inform their employees through formal presentation on the concepts, goals, and expected benefits of TPM.

- Step 2—*Initial education and propaganda for TPM*: Educating and training of employees helps in improving the morale and also softens the resistance to change. Training is to be arranged based on the need. Take people who matters to places where TPM is already successfully implemented.

- Step 3—*Setting up departmental organization to promote TPM*: Promoting TPM organizing committees is the next step in the TPM implementation activity, which is more important for the support and successful development of TPM. TPM includes improvement, autonomous maintenance, quality maintenance, etc., as part of it. When committees are set up, it should take care of all those needs. Groups can be created by assigning leadership responsibility to section leaders and group leaders on the shop floor.

- Step 4—*Establishing the TPM policies and goals*: It is a must to establish one basic management policy committed to TPM and concrete TPM development procedures. It should specify the target of what should be achieved, how much quantity to be achieved, and when it should be achieved. For example, the utilization level of the ring frame department is 95%; it can be planned to 98% for a long-term plan, say, for 3 years.

13.3.5.2 Introduction Module

This is a public occasion, and we should invite suppliers, customers, and employers. Suppliers as they should know that we want quality supply from them. Customers will get the communication from us that we care for quality output. Employers should have awareness about the importance of TPM.

13.3.5.3 Implementation Module

In the implementation module,[5] improving equipment effectiveness is the first step in the implementation of TPM on the effectiveness of each part of equipment experiencing a loss. At the initial stage, it is better to focus team efforts on equipment suffering

TABLE 13.5

Various Causes of Losses

Losses (in terms of)	Causes
Availability	Major breakdowns, setup arrangements of the tooling, etc.
Efficiency	Minor stoppages, wrong scheduling, or expectances
Quality	Waste/rework, rejects, slow start of shift/lot

chronic losses during operation. The overall equipment effectiveness could be calculated as detailed next:

$$OEE = Availability \times Performance\ efficiency \times Rate\ of\ quality\ products$$

For example: In a draw frame machine, utilization = 95%; production efficiency = 90%; quality slivers produced = 80%

$$OEE = \left(\frac{95}{100} \times \frac{90}{100} \times \frac{80}{100} \right) \times 100 = 68.40\%$$

Though the availability of the drawing machine is 95%, the overall equipment effectiveness is 68.40% only. Similarly, calculate the OEE for other equipment also. The various causes of losses experienced are listed in Table 13.5.

13.3.6 Eight Pillars of TPM

The pillars (Figure 13.4) over which the TPM[5] implementation is developed are as follows:

1. *Autonomous maintenance*: Autonomous maintenance is one of the unique features of TPM, which believes that individuals should be responsible for their own equipment and have to perform autonomous maintenance. Autonomous maintenance consists of cleaning, lubrication, retightening, and inspection.

2. *Kaizen*: Gradual, incremental, and constant improvement in the process by involving everyone in an organization. It is a continuous program to improve quality and increase productivity. It is said that kaizen has been one of the key ingredients in Japan's competitive success in the world market.

3. *Planned maintenance*: A planned maintenance schedule should be planned for timely replacement of components, which is a must for the effective operation of equipment and long life. This has to be followed by the maintenance team. In spinning mills, components like card wires and top roller cots require timely grinding and buffing, respectively, to keep them in condition for the production of good quality slivers and yarns.

4. *Quality maintenance*: It is aimed toward customer delight through highest quality through defect-free manufacturing. Focus is on eliminating nonconformances in a systematic manner, much like focused improvement. Transition is from reactive to proactive (quality control to quality assurance).

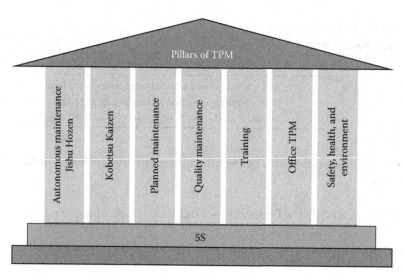

FIGURE 13.4
Pillars of TPM.

5. *Education and training*: Education and training are investments in people that yield multiple returns. Operatives in each department must be trained in such a way that they must improve the understanding about functions of their machines, early detection of abnormalities, ability to do improvements, on machines operated by them.

6. *Office TPM*: Office TPM must be followed to improve productivity, and efficiency in the administrative functions and identify and eliminate losses. This includes analyzing processes and procedures toward increased office automation.

7. *Safety, health, and environment* (*SHE*): In this area, the focus is on creating a safe workplace and a surrounding area that is not damaged by our process or procedures. This pillar will play an active role in each of the other pillars on a regular basis. The target of SHE is zero accident, zero health damage, and zero fire.

8. *5S*: It is the foundation for all the pillars. Cleaning and organizing the workplace helps the team to uncover problems. Making problems visible is the first step of improvement.

13.4 Lean Manufacturing Concepts

Lean manufacturing is one of the management techniques primarily focused on designing a robust production operation that is responsive, flexible, predictable, and consistent. Lean manufacturing is an approach that will save manufacturing costs, increase the productivity, improve the quality, and shorten the lead time. Lean manufacturing or lean production is a production practice that considers the expenditure of resources for any goal other than the creation of value for the end customer to be wasteful, and thus a target for elimination. Lean manufacturing is a management philosophy derived mostly from the Toyota Production System (TPS).

13.4.1 Lean Manufacturing: Definition

Lean manufacturing is defined as "a systematic approach to identifying and eliminating waste (non-value-added activities) through continuous improvement by flowing the product at the pull of the customer in pursuit of perfection." Lean is a set of "tools" that assist in the identification and steady elimination of waste (muda). As waste is eliminated, quality improves and production time and cost decrease.

13.4.2 Principle of Lean Manufacturing

Lean manufacturing[6] works on a simple principle. The organization must move away from departmentalized thinking and must move toward seeing the organization as one entity. Lean manufacturing is becoming lean enterprise by treating its customers and suppliers as partners. This gives the extra edge in today's cost- and time-competitive markets. Lean enterprise owners now can deliver high-quality products quickly, with low price.

13.4.3 Goals of Lean Manufacturing

Waste is defined as anything that does not add value to the final product. The goal of lean manufacturing[6] is to eliminate the eight wastes of lean:

1. Overproduction (production ahead of demand)
2. Motion (people or equipment moving or walking more than is required to perform the processing)
3. Inventory (all components, work in process and finished product not being processed)
4. Waiting (waiting for the next production step)
5. Transportation (moving products that is not actually required to perform the processing)
6. Defects (the effort involved in inspecting for and fixing defects)
7. Underutilized people
8. Inappropriate processing (due to poor tool or product design creating activity)

13.4.4 Lean Concepts in the Spinning Mills

1. *Overproduction* of feed material such as laps, sliver, roving, and ring cops leads to fly accumulation on the materials that impairs the quality of the final product.[6] Overproduction or stocking of materials is mainly due to improper balancing of machineries for process, late attending of machine breakdown in the postprocess, low productivity in the postprocess, etc.
2. *Motion*, that is, people such as tenters, siders, doffers, and winders motion, should be planned effectively. In some mills, the usage of roller skating (Figure 13.5) effectively reduces the attending time. In ring frame, it takes 15 min for one person to move around one line to check for breakages.

   Usage of roller skating facilitates only 5 min as against the earlier 15 min. This keeps the person active and avoiding unnecessary muscle movement that leads to quick tiredness and efficiency loss.

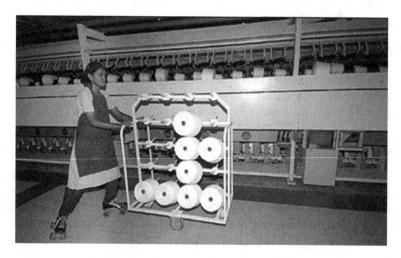

FIGURE 13.5
Labor on roller skating in spinning mill.

3. *Inventory* should be maintained as low as possible to reduce inventory cost. In the maintenance department, keep only essential spares such as belts, consumable spares, change gears, bearings, etc. It is essential to follow the JIT method to reduce the inventory cost.

4. *Waiting*—In conventional batch processing, some studies show that 90% of the time goods are waiting to be processed. This will also reduce the WIP. Waiting for the next production step occurs often due to improper production planning. This kind of waiting unnecessarily creates quality defects in yarn due to fly accumulation and occupies more space that is a constraint for material transport. Wastage of production space and increase of work in capital are the other consequences of unnecessary waiting.

5. *Transportation*—No matter how well you do transporting, it does not add value to the end product. Therefore, transportation is simply one of the wastes that have to be eliminated from the production system. In spinning mills, the transportation of laps, slivers, roving, ring cop, and cones is an unavoidable work, although we are having some automation. In the absence of automation, the huge transportation path leads to quality defects, maintenance of a higher WIP, and additional cost of transporting the goods. Inflexibility of the layout plays a big role here. This can be avoided with careful redesigning of the layouts.

6. *Defects*—Defects call for higher inspection and related costs. If you find a defect, you will have to remove it. The raw materials, time, effort, and the money put into this product will be wasted. Damage in a single dollar product can create millions of dollars of lost to your organization. Defects in final product may be due to poor raw material, mistakes from the workers, and problems in the system and machinery problems.

7. *Underutilization of human resource*—Every worker, even the people doing the most routine job in the organization, will have something to contribute to the organization than their muscle power. What lean manufacturing tries to do is to get ideas from all levels of the people in the organization and to use them for the betterment

of the organization. Therefore, not making the full use of the human resource is a waste. Wasting this without using to fight against the wastes is the biggest loss for the organization. More than that people will get motivated and will have a chain effect.

8. *Inappropriate processing*—This means using incorrect tools for the job. This does not mean that you should use complicated or expensive tools to do the job. It is about using the correct tool for the correct job.

13.4.5 Implementation of Lean Concept

There are four steps in implementing lean manufacturing:

1. Identifying the fact that there are wastes to be removed
2. Analyzing the wastes and finding the root causes for these wastes
3. Finding the solution for these root causes
4. Application of these solutions and achieving the objective

Whenever this is done go back to the stage 1 and continue this loop over and over again. When you map the process, you will start to see the

1. Value-added activities
2. Non-value-added activities

You will also have a better idea of what the avoidable non-value-added activities and unavoidable non-value-added activities are.

13.4.6 Lean Manufacturing Tools

Lean manufacturing is based on continuous finding and removal of the wastes. Value is defined from the customer's point of view. Therefore, all the tools in lean manufacturing[6] aim to identify and remove wastes from the system continuously. The tools used are as follows:

1. *Pareto analysis*: To find the root cause of the problems.
2. *Brain storming*: To understand the root causes of the problems.
3. *Ishikawa diagram or cause–effect diagram*: Representing these root causes and their relevance to the immediate problem.
4. *JIT (just in time)*: Everything is done when they are actually needed. JIT is the backbone of lean manufacturing.
5. *Work cells*: In a work cell, there will be 3–12 people depending on the job task performed by this cell. There will be many cells that will complete the total product by working together. People who are in this cell are multiskilled and can perform multiple tasks according to the requirement.
6. *Kanban tooling*: Kanban mainly focus on the reduction of overproduction.
7. *Poka-Yoke*: Mistake proof of the process.

8. Various standardization techniques like
   a. *5S*—Maintaining housekeeping and orderliness.
   b. *TPM*—Maintenance should be preventive to avoid production loss due to breakdown.
   c. *Single-minute die exchange (SMDE)*—With the aid of careful planning and coordination, it was possible to reduce the time taken to change the line into minutes from days.

13.4.7 Advantages of Lean Manufacturing

- Lead time by 50% at least (some reports say stories where lead time is being reduced up to 90%)
- Reduced work in progress (WIP) up to 80%
- Floor space savings around 30% (sometimes more than 50%)
- Increased productivity at least by 30% (even more than 100% in some cases)
- Quality improvement
- Overall cost reduction

References

1. What is process control?, http://www.itl.nist.gov/div898/handbook/pmc/section1/pmc13.htm, Accessed Date: July 15, 2013.
2. Process management, http://www.scribd.com/doc/189880389/Prg-211-Process-Management, Accessed Date: July 15, 2013.
3. Process management—Eventful solutions, http://www.eventful-solutions.co.uk/Pages/ProcessManagement.aspx, Accessed Date: July 15, 2013.
4. Senthil Kumar, R., 5S in the textile industry, http://www.docstoc.com/docs/31909437/5S-IN-THE-TEXTILE-INDUSTRY, Accessed Date: July 15, 2013.
5. Senthil Kumar, R., Application of Total Productive Maintenance (TPM) in the spinning mill. *Pakistan Textile Journal*, July 2010, http://www.ptj.com.pk/Web-2010/07-10/Practical-Hints-R.Senthil-Kumar.htm, Accessed Date: May 1, 2013.
6. Senthil Kumar, R., Lean manufacturing in spinning mills, http://www.scribd.com/doc/48784461/Lean-Manufacturing-in-the-Spinning-Mills, Accessed Date: May 1, 2013.

14

Productivity, Waste Management, and Material Handling

14.1 Productivity in the Spinning Process

In yarn manufacturing, two immediate prerequisites to success are high productivity and high quality. In the previous century, the manufacture of staple yarns was driven by the motive to achieve the desired tensile properties and esthetics at a reasonable cost. Globalization is creating a growing level of interconnectedness and competition in the world economy, and geographical clustering of activity is an important way in which firms and localities deal with these pressures. As a result, extraordinary changes are taking place in the world of textiles. Automation and online monitoring are among the three advances improving quality and productivity in short staple spinning.

Productivity in a spinning mill could be taken as an indication of the mill performance level. As the productivity increases, product cost reduces, which in turn raises the standard contribution. Productivity is simply the output expressed as the ratio of the input. In other words, productivity is defined as the efficient use of resources such as labor, capital, land, materials, and energy in the production of goods and services. Productivity is, therefore, key to the survival of a manufacturing enterprise in business. Higher productivity means accomplishing more with the same amount of resources or achieving higher output in terms of volume and quality for the same input. Productivity of spinning mill is related to total operative hours necessary to produce 100 kg of yarn. Production is affected by spindle speed, yarn twist per meter, end break rate, out of production hours, ring diameter, doffing time and type (auto-doffing or manual), and personnel organization. Productivity reduces with decrease in mill efficiency. Hence, mill performance is closely related to the total operative hours worked to produce a constant amount of production. Productivity is a comparative tool for managers, industrial engineers, and management people.

Productivity could be considered as a comprehensive measure of how organizations satisfy the following criteria:

- *Objectives*: the degree to which they are achieved
- *Efficiency*: how effectively the resources are used to generate useful output
- *Effectiveness*: what is achieved compared with what is possible
- *Comparability*: how productivity performance is recorded over time

14.1.1 Factors Influencing Productivity of a Spinning Mill

The productivity of a spinning mill is influenced by factors such as raw material, labor, machinery, maintenance, and energy consumption.

1. *Raw material*: It has an important influence on production and quality. Longer fibers require less twist per inch to give enough strength, which results in high production. Higher percentage of short fibers will result in the form of production losses due to more end breakage and higher twist per inch. Immature fibers results in production loss because the yarn made from these types of fibers has poor strength and also it produces breakage in spinning process.

2. *Labor*: Optimum number of labor employment in various departments of a spinning mill plays a crucial role in deciding the productivity. The efficiency of the labor in any department of spinning mill is of paramount importance in influencing the productivity. The right selection of labor for particular work profile requires the application of psychometric tests. The efficiency of the labor should be checked at the end of the training period. The inefficient labor allotted to any work in the spinning mill reduces the productivity. The allotment of higher work force than recommendation in a department affects the productivity of the spinning mill. Sometimes productivity is viewed as a more intensive use of labor, which should reliably indicate performance or efficiency if measured accurately. It is important to separate productivity from intensity of labor, because while labor productivity reflects the beneficial results of labor, its intensity means excess effort and is no more than work "speed-up." The essence of productivity improvement is working more intelligently, not harder.

3. *Machinery*: Productivity of any manufacturing industry relies mostly on the machines installed for manufacturing products. In the short staple spinning process, a series of machineries are employed to manufacture yarn. The speed of the machine greatly influences the productivity of the process. The selection of very high speed may affect the quality of the product delivered and running performance of the process. The right selection of machine speed and settings optimizes the productivity, quality, and running performance. The selection of machinery type such as manual or semiautomatic and automatic also has a significant effect on productivity. The significant developments in short staple spinning machinery that improved the productivity of spinning mill are listed in Table 14.1. If the process is highly automatic, then the labor allotted should be less in order to attain higher productivity.

4. *Maintenance*: The proper maintenance of the spinning machines predominantly lowers the machine downtime so that the productivity substantially improves. The important maintenance activities such as cleaning schedules and replacement of worn-out parts should be implemented effectively to enhance the productivity by reducing the unnecessary machine break downs. The deviation in the replacement of worn-out machine parts such as carding wire points, top roller cot, aprons, spindle tapes, roller bearing, and spindle may significantly affect the quality and process performance so that productivity gets dropped.

5. *Energy consumption*: The energy consumption in units by the spinning machines for a particular process should be closer to or less than the norms. The energy consumed for producing 1 kg of yarn is normally taken and compared with industry

TABLE 14.1

Significant Machine Developments in Short Staple Spinning Process That
Improved Productivity of Spinning Mill

Bale plucker instead of manual mixing operation
Chute feed system instead of lap feed system
Automatic waste evacuation system instead of manual waste collection
Automatic doffing of cans in carding and draw frame machines
Usage of bigger diameter cans in carding and draw frame departments
Automatic lap changing system in comber machine
Automatic roving bobbin transport system
Automatic spinning bobbin transport system
Automatic doffing in ring frame department
Automatic winding machine
Improved machine design that facilitates higher machine speed, user-friendly operations

norms. The various factors influencing the energy consumption of the machines
are machine load, process type, motor specifications, and motor maintenance.
Units per kilogram (UKG) is a parameter used in the spinning mill to measure the
energy consumption level.

14.1.2 Productivity Measurement

14.1.2.1 Production per Spindle

Production achieved by a spindle for 8 hour running time in a ring frame is termed as production per spindle. Production per spindle is normally expressed in grams or kilograms.
Production per spindle is one of the vital productivity parameters in the spinning mill.
A higher level of production per spindle offers a number of benefits to a mill such as reduction in overheads and administration and wages cost per unit production as well as greater
marginal profits accruing from the increased sales. A report by *SITRA Focus*[1] indicates that
a 10% increase in production per spindle would lead to a saving of about Rs. 10 lakhs a
year for a 25,000 spindle mill. Production per spindle in a spinning mill is highly influenced by speed, raw material, process parameters, maintenance, efficiency, end breakage
rate, and modernized machines. In general, the higher the count, the lower the production
per spindle. Similarly, the higher the twist per inch, the lower the production per spindle.
Operational deficiencies such as poor machinery condition, bad housekeeping, improper
material handling, and inefficient labor may significantly affect the production per spindle.[3]

The various factors to be controlled in order to achieve required production per spindle
in the ring frame department are

- Pneumafil and lapping waste
- Twist contraction
- Idle spindles
- Time loss during doffing and maintenance

The level of pneumafil and lapping waste for a given count is mostly decided by the end
breakage rate and the work assignment for the ring frame tenter. The production per spindle of a spinning mill can be improved by reducing end breakage rate. This can be possible

by imposing better supervision and control on the tenters. The other causes of efficiency loss[2] occurred in the ring spinning department are summarized as

- Maintenance—1.5%
- Idle spindles—0.1%
- Doffing—0.8% per doff per shift

14.1.2.2 Labor Productivity

Labor productivity is measured by the term "HOK." HOK is defined as operating hours required to produce 100 kg of yarn. HOK enables comparison between mills running in various environments. Rapid developments in the technology enable mills to run with reduced work force.

14.1.2.3 Operatives per 1000 Spindles (OHSAM)

The average number of operatives engaged per 1000 spindles.

14.1.3 Measures to Improve Productivity in the Spinning Mill

1. Selecting appropriate raw material
2. Correct balancing of machineries
3. Selection of optimum spindle speed
4. Selection of optimum twist
5. Reducing end breakage rate in ring spinning
6. Reducing soft waste generation
7. Reduce the machine efficiency loss during doffing and maintenance
8. Correct follow-up of HOK and OHS as per recommendations
9. Energy-efficient motor and spindles usage
10. Correct maintenance of R.H%
11. Minimize the dust accumulation and fly liberation
12. Conduct energy audit to avoid unnecessary energy losses
13. Use of automated machines
14. Linking systems for transporting roving bobbin and ring bobbin

14.2 Yarn Realization

Yarn realization (YR)[3] is a term used to represent the quantity of yarn produced from a given weight of raw material expressed as a percentage. A high yarn realization is a matter of paramount importance in the production economics of a spinning mill. A 1% reduction in yarn realization has a significant economic impact on mill's profit. The resale value of wastes is much less than the actual price of cotton till it reaches the yarn stage. The control of raw material cost is mainly influenced by yarn realization. The yarn realization normally ranges from 85% to 90% for carded cotton yarn and from 67% to 75% for combed cotton yarn, depending on the trash% in cotton and quantity of waste extracted during fiber to yarn conversion.[3]

$$\text{For cotton carded yarns, YR\%} = (100 - (W_{br} + W_k + W_h + W_s + W_g) - I)$$

$$\text{For cotton combed yarns, YR\%} = (100 - (W_{br} + W_k + W_c + W_h + W_s + W_g) - I)$$

where
W_{br} is the blow room waste%
W_s is the sweep waste%
W_k is the card waste%
W_g is the gutter/filter waste%
W_c is the comber noil%
I is the invisible loss%
W_h is the yarn waste%

Yarn realization plays a significant role in production economics of a spinning mill. A 1% improvement in yarn realization would lead to a saving of Rs. 20 lakhs per year for a 30,000 spindle mill manufacturing 40^s carded yarn.[3]

14.2.1 Measures to Improve Yarn Realization

- It should be ensured that the overall lint in waste is no more than 40% in cottons with high amount of trash and 50% for cottons with a low level of trash.
- Bale weight variation due to moisture loss should be prevented. Variation between invoice weight and actual weight of cotton bale should be controlled.
- Contamination content in the cotton bale should be at a very low level.
- Addition of soft waste in mixing should be at a minimum level.
- Control fly liberation in blow room department by maintaining recommended R.H%.
- Sand liberated from the trash removed also should be taken into account for realization calculation.
- Fitting of automatic waste evacuation system in blow room and carding to control effective waste management.
- Addition of right amount of water and mixing oil during stack mixing in order to control fly liberation and invisible loss.
- Weighing of waste during every shift wise should be proper, and there should not be any manipulation under any circumstances.
- Weighing balance used for bale weighment, waste weighment, and cone bag weighment must be calibrated as per recommendation.
- Standard procedure must be adopted while taking process stock and the person responsible for process stock should be trained properly.
- SITRA reported that there was a good scope for maintaining the required yarn quality[3] without increasing the comber waste (i.e., by maintaining mill standard comber waste) but improving the process proficiency (i.e., nep removal efficiency).

- The incidence of yarn waste should be below 0.4%. Cop rejection in the automatic cone winding should be checked regularly, ring frame wise, and it should be ensured that it does not exceed 6%.

- Appropriate roving empties and ring tubes with microgrooves should be used to avoid frequent slough-off of roving and yarn, respectively.

- Ring frame doffers should properly follow simultaneous doffing and donning method with which the starting breaks could be reduced.

- R.H% to be maintained in each and every department to control fly liberation that increases the invisible loss.

- Yarn conditioning plant improves the yarn realization to some extent by increasing the moisture in final package.

- R.H% should be maintained properly in cone packing area to prevent any weight loss.

14.2.2 Waste Management in Spinning Mill

With the ever-increasing price of cotton, huge investments on sophisticated machines, and increasing labor wages, it is a highly challenging task for any spinning mill to enhance the productivity.[4] In order to survive the huge competition, it is absolutely essential that waste incurred during the yarn manufacturing processes should be kept under control. In buying cotton, the price includes the nonlint material and has to be considered in valuing the material. The waste generated in the spinning mill can be categorized as reusable waste or soft waste, cleaning waste, and microdust. The wastes such as lap bits, sliver bits, roving ends, pneumafil waste, and lapping waste obtained from each department of spinning process are termed as reusable waste or soft waste. These wastes can be collected from each department and can be fed back into the mixing or blow room process in recommended proportion. The various causes attributed for excess soft waste generation are listed in Table 14.2.

The cleaning waste is that which is extracted from cotton deliberately to obtain the desired yarn quality. The cleaning waste such as blow room droppings, licker-in dropping, flat strips, and cylinder flies is mainly obtained from blow room and carding departments. The cleaning waste mainly consists of vegetable matters such as broken seeds, leaf bits, and so on. Cleaning machines are unable to remove the nonlint material without removing some usable fiber; at times the waste may contain up to 50% usable fiber. Indeed, the removal

TABLE 14.2

Causes of Excess Reusable or Soft Waste

Errors and mistakes during weighing
Loose cotton falling from bales while transportation
Rejection from faulty work
Poor sliver can topping and replenishing procedure
Poor housekeeping
Quality standards not met
Poor work practices
Improper material handling
Poor condition and performance of machines
Higher end breakage rate and lapping tendency

rates vary from 40% to 70%, depending on the type of waste, the type of machine, and the running conditions. The increase in fly generation in the department leads to higher amount of sweep waste and invisible loss. The sweep waste in the department increases with poor working conditions such as higher end breakage rate and higher roller lapping tendency. The poor work practices also tend to generate higher amount of sweep waste.

14.2.2.1 Waste Investigation

The waste investigation[4] should be done in order to take preventive measure as well as to take action where abnormality is noted in processes. The following steps to be implemented for effective waste control in the process:

- Investigate the existing system of waste collection for different types of waste.
- Follow up collection procedure for routing by
  - Checking the quantity and quality of waste in a set of machines in each department shift wise.
  - Work-out norms for comparing against actual waste made for corrective action.
- Conducting a 1000 kg waste test for effective departmental-level waste control.
- More effective and reliable method of collecting and reporting of wastes could be considered where there is malfunctioning.

14.2.2.2 Recording of Waste

The waste generated from each machine of every department should be recorded in proper manner. The following points should be considered while recording the waste:

- Shift wise, waste should be recorded type wise along with actual machine production for working out waste%.
- The tendency for wrongful disposal of excess waste en route when the department and waste shed are distantly placed.
- Incorrect collection and recording procedure of waste reflect high invisible loss and incorrect waste stocks.
- The variations in tare weight of waste bags and incorrect weighing balance lead to wrong recording of waste.
- The maintenance of correct waste recording will provide any fluctuations in waste extraction that may help to identify any deterioration in mixing or process.
- Incorrect stocks will mislead decision on reuse or sale of wastes.
- Periodic surprise checks will bring out any malpractices.

14.2.2.3 Waste Reduction and Control

Step 1: Investigation of waste problems[4]:

- Determine the amount of waste made
- Effect of waste on production, quality, unit cost, emplyoyees' pay, and job security

- Cause of waste due to material, machine, method, employee, and poor supervision
- Loss in rupees on waste at each stage
- Present schedule of collecting waste, weighing, and reporting

Step 2: Suggestions needed:

Use brainstorming technique with raw materials purchase, standards, costing, marketing, and "line" personnel.

- What to do to improve the material?
- What improvement in machine settings, speeds, and maintenance is possible?
- Changes required to improve the method of handling and processes.
- Skills needed to improve labor.
- What supervisory controls are required?
- Changes needed in collecting, weighing, and reporting system.

Step 3: Work out details

- How much waste can be reduced? Set target.
- List details to improve materials, machines, workers, and processes.
- Outline training action at all levels to create waste consciousness, through use of posters, graphs, handouts, etc.
- List details of supervisory control required to be adopted.
- Recommendations and suggestions to other departments.

Step 4: Set-up control

- Set up waste standards and target for 3 months, 6 months, and 1 year.
- Fix variance to be allowed.
- Fix persons responsible for controlling and coordinating wastes.
- Decide frequency of checking.
- Structure system of communicating to workers and management about the progress and setbacks.

14.2.3 Invisible Loss

Invisible loss is a part in the estimation of yarn realization. Invisible loss in a spinning mill arises mainly due to the following reasons:

- Higher liberation of short fibers and fluff from the departments
- Error in weighment of cotton bale and waste during purchase and sale, respectively
- Weighment errors in the yarn sold
- Higher differences in moisture regain between incoming cotton and outgoing yarn

TABLE 14.3

Invisible Loss: Norms

Rating	Invisible Loss (%)
Good	0.5
Average	1.0
Poor	1.5

Source: Courtesy of SITRA, Coimbatore, India.

- Wrong method of estimating process stock
- Wrong method of waste accounting

Table 14.3 shows the norms of invisible loss as suggested by SITRA.

14.3 Material Handling

Material handling[5] can be defined as "art and science of conveying, elevating, positioning, transporting, packaging and storing of materials." Material handling plays a vital role in any manufacturing industry by easily transporting the material from one place to another place. A material may be handled even 50 times or more before it converts to finished product. It has been estimated that the average material handling cost is roughly 10%–30% of the total production cost depending on the product to process. The cost of the production can be reduced considerably by saving the material handling cost. Material handling involves the movement of materials, manually or mechanically in batches or one item at a time within the plant.[5] The movement may be horizontal, vertical, or the combination of these two. Material movement adds to the cost but not to the product value. The ideal manufacturing plant would have an absolute minimum of materials handling and more use of mechanical material handling equipment. The multiple benefits associated with effective material handling in a manufacturing industry are listed in Table 14.4.

TABLE 14.4

Benefits Associated with Effective Material Handling

Improving productivity
Increasing the handling capacity
Reducing manpower
Increasing the speed of material movement
Reducing materials wastage
Promoting easier and cleaner handling
Eliminating idle time of machines, equipment, and workers
Reducing fatigue incurred by the workers
Increasing safety and minimizing accidents
Locating and stocking material better and in less space
Minimizing production cost, etc.

14.3.1 Principles of Material Handling

- Reduce the unnecessary movements (manual and mechanical) involved in a production process.
- Adopt possible shortest routes for transporting materials.
- Employ mechanical trolleys and forklifts wherever essential instead of manual labor for material handling in order to accelerate the material movements.
- Adopt principle of containerization or palletization to transport bulk number of materials in single unit (bobbins and cones).
- Overloading of material in the material handling equipment should be avoided in order to prevent any material quality deterioration.
- Appropriate, standard, efficient, effective, flexible, safe, and proper-sized material handling equipment should be selected.
- Utilize gravity for assisting material movements wherever possible.
- Design trolleys and containers in such a way that the material transported should not get damage during handling.
- Design material handling equipment in such a way that it economizes the material handling process.
- The movement of material handling equipment should not interfere with neighboring machine operations.
- The maintenance of material handling equipment is vital in preventing any interruption in handling.

14.3.2 Factors Governing Selection of Material Handling Equipment

1. *Dimensions and parameters of the materials being transported*: The dimensions of the material being transported such as length, width diameter, and material parameters such as weight, surface characteristics, delicacy, and its chances of getting damaged during handling should be considered.[6]

2. *Layout of department*: The path of material transportation, door dimensions, equality in floor levels between departments, height of the ceiling, strength of floor and walls, columns and pillars influence to a great extent the choice of a material handling equipment.

3. *Machine production*: The machines may have different production rates per unit time. The material handling equipment should be able to handle the maximum output.

4. *Type of material flow pattern*: A horizontal flow pattern may need trucks, overhead bridge cranes, conveyors, etc., whereas a vertical flow pattern will require elevators, conveyors, pipes, etc.

5. *Production type*: The selection of the material handling equipment depends on the type of production such as mass production and batch production. Conveyors are more suitable for mass production on fixed routes and powered trucks and trolleys for batch production.

6. *Other factors*: Some other factors also considered during the selection of material handling cost such as cost of material handling equipment, handling costs, life of the equipment, and amount of care and maintenance are required for the equipment.

TABLE 14.5

Material Handling Equipment in Different Departments of Spinning Mill

Department	Material Type	Material Handling Equipment
Bale godown	Bale	Bale trolley or four-wheeler truck
Mixing	Bale	Bale trolley or four-wheeler truck
Blow room	Lap	Lap trolley
	Loose cotton	Chute feed by pneumatic transport duct
Carding	Sliver cans	Can trolley or can fitted with castor wheels
Comber	Lap; sliver can	Lap trolley; sliver can with castor wheels
Draw frame	Sliver cans	Can trolley or can fitted with castor wheels
Speed frame	Roving bobbins	Bobbin trolley (manual); Bobbin transport system (automatic)
Ring frame	Ring bobbins	Doffing trolleys or link system
Winding	Cones	Cone trolley
Yarn conditioning	Cones	Specially designed cone trolley
Packing	Cone bag	Pallet truck or conveyor system
Waste godown	Waste	Waste trolley

14.3.3 Material Handling Equipment in Spinning Mills

In spinning mill, there are many departments that involve handling of raw material, intermediate products, wastes, finished goods, stores, and maintenance tool equipment. During fiber to yarn conversion, materials (raw material, laps, sliver, roving, yarn, finished goods, and wastes) are stored at different places and transported between departments. The selection of appropriate production machinery and proper layout of spinning mill eliminate as far as possible the need of material handling. In a spinning mill, the chute feed system and automatic material transportation (ribbon lap, roving bobbin, ring bobbin and cone) reduce the material handling activities to a greater extent that significantly contribute to productivity enhancement. The selection of appropriate material handling equipment for performing a particular task is crucial in terms of safety and cost minimization. The various material handling equipments used in different departments of spinning mill are summarized in Table 14.5.

14.3.3.1 Bale Godown

Spinning mills receive cotton fibers in the form of bales in lorries or trucks.[2] Cotton bales usually weigh around 170 kg. In conventional mills, bales are transported with the help of two-, three-, or four-wheeled industrial trolleys[7] (Figure 14.1) for storing in bale godown one by one which consumes time and requires more workers. Modern mills normally handling a large number of bales can use forklifts[7] (Figure 14.2) exclusively for unloading bales from lorry and transporting them to the godown and stackers can be used to stack the bales.

14.3.3.2 Mixing and Blow Room Department

In a conventional blow room line, cotton from the mixing to the mixing bale opener is transported by means of a mixing trolley. Use of liftable spring–type pedal–operated mixing trolley eliminates spilling of the fiber tufts on the floor leading to poor housekeeping

FIGURE 14.1
Bale trolley.

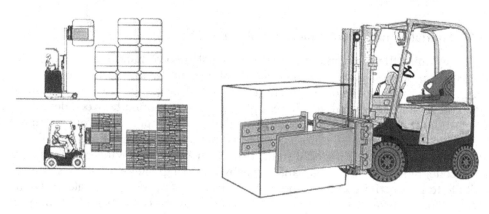

FIGURE 14.2
Forklift truck.

and more waste. This type of trolley can transport up to 30 kg of material at a time. In a conventional blow room line, blow room laps are transported manually by the workers keeping one lap at a time on the shoulder. Some modern mills having lap feed system are using lap trolley that can transport four to six laps at a time.[7] The vertical stacking of laps closer together in the lap trolley may lead to the damage on the lap surface. Lap produced from blow room must be covered with synthetic cover cloth in order to correct this problem.

14.3.3.3 Trolleys Used in Carding, Drawing, and Comber

Sliver cans (full and empty) are to be transported between cards, draw frames, comber preparatory machines, combers, and fly frames. The cans in which sliver get deposited are of various diameter such as 14″, 16″, 42″, and 48″ that are provided with or without castor wheels. Cans that are transported manually by dragging them on the floor would not only spoil the floor but also damage the can and result in wastage of sliver.

To overcome the aforementioned problem, the sliver can is normally transported by means of can trolley.

14.3.3.4 Roving Bobbin Trolley

The doffed roving bobbins are normally placed on the top arm frames and the doffers then carry 8–10 bobbins by hand to the storage place. This practice affects the package quality and sometime may also cause injury to the worker. Roving bobbin trolley as shown in Figure 14.3 can be moved in between fly frames and up to 60 bobbins (30 on each side) can be stacked easily.[7] After doffing, roving bobbins are usually kept in racks and transported to the ring frame by open-type trolley or automatic roving bobbin transporting system.[8]

In another model of roving bobbin trolley, the spring-loaded bottom (Figure 14.4) keeps the bobbins at an almost constant height and thus avoids physical strain on the workers and protects the yarn when loading and unloading the trolley.[8]

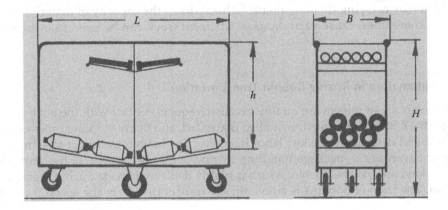

FIGURE 14.3
Roving bobbin trolley.

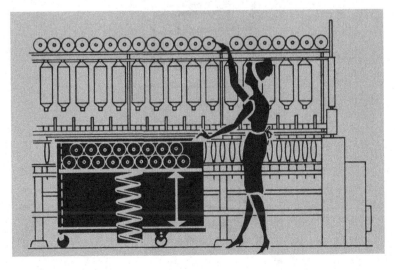

FIGURE 14.4
Spring-loaded roving bobbin trolley.

14.3.3.5 Ring Bobbin Trolley

In conventional mills, doffed cops are transferred to baskets or big containers and they are transported to postspinning departments in trolleys.[7] This practice results in damage to some cops and entanglements, leading to yarn waste. Today, the ring bobbin trolley is used, which has two compartments such as one for carrying empty cops and another for keeping doffed cops. Plastic crates can be fitted in the compartments. Each doffer must be given one trolley. Modern mills are equipped with automatic doffing and ring bobbin transportation that drastically reduces the yarn waste and labor requirements and improves yarn and package quality.

14.3.3.6 Winding and Packing

Cone/cheese trolleys as shown in Figure 14.5 are used to transport the cone from winding machine to packing section. Packed cone bags or cartons are transported to cone bag godown by carrying them manually.[6] This reduces the capacity of material handling and may, sometimes, cause yarn damage. Platform truck can be used to overcome this problem.

14.3.4 Automation in Roving Bobbin Transportation

Automation is about improving quality because frequent contact with the bobbins, which damages the material and impairs quality, is avoided, and many working processes can be simplified and designed to be less labor intensive. The bobbin transport system provides economic advantage to package handling in ring spinning. It ensures the protection of roving wound around the bobbin, which is highly liable to damage during manual transportation. The train of bobbins is automatically transported from the speed frame to the

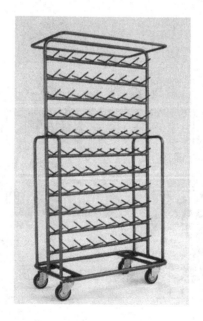

FIGURE 14.5
Cone trolley.

storage area and respective ring frame by selecting the appropriate program in the PLC.[9] Empty bobbins in the ring frame are manually interchanged with full bobbins.

14.3.5 Automation in Ring Bobbin Transportation

The maximum level of automation on the winding process is achieved by a direct link of the ring spinning frames to the winder.[10] The advantage is continuous automatic transfer of the ring bobbins from one machine to the other. The main advantages are

- Yarn quality of the wound package is preserved as bobbins are prepared and automatically doffed on the ring frames, without any operator handling or loading into separate boxes
- Higher efficiency on winder
- Production output flow consistency and timed as a single machine
- Minimization of winding room personnel

References

1. SITRA, How to control invisible loss in spinning mills?—A case study. *SITRA FOCUS* 23(4), November 2005, pp. 1–5.
2. Ratnam, T.V. and Chellamani, K.P., *Maintenance Management in Spinning*, SITRA Monograph, Coimbatore, India, 2004, pp. 2–5.
3. Mariappan, S. and Shanmuganandam, D., How to improve yarn realization in a spinning mill?—A case study. *SITRA Focus* 28(6), March 2011, pp. 2–6.
4. Nithyanand, A.R., Waste investigation and control for a spinning mill. *The Indian Textile Journal* 79–86, September 1984, pp. 79–86.
5. Khanna, O.P., *Material Handling, Industrial Engineering and Management*, Dhanpat Rai Publications (P) Ltd., New Delhi, India, 2009, pp. 25.1–25.14.
6. Uttam, D., Material handling in textile industries. *International Journal of Advanced Research in Engineering and Applied Sciences* 2(6), June 2013, pp. 52–56.
7. A study on materials handling in spinning mills. *The Indian Textile Journal*, http://www.indian-textilejournal.com/articles/FAdetails.asp?id = 1308, Accessed Date: November 12, 2013.
8. Material handling equipment for the textile industry, GMohling Catalogue, 2005.
9. Techno front, Automation from one source: From roving frame to package. *The Indian Textile Journal*, September 2012, http://www.indiantextilejournal.com/articles/FAdetails.asp?id=4720, Accessed Date: November 12, 1979.
10. Bobbin transport system, http://www.elgielectric.com/bobbin.html, Accessed Date: November 12, 2013.
11. Nair, A.U., Nachane, R.P., and Patwardhan, P.A., Comparative study of different test methods used for the measurement of physical properties of cotton. *Indian Journal of Fibre & Textile Research* 34: 352–358, December 2009.

15

Case Studies

15.1 Mixing-Related Problems

15.1.1 Higher Needle Breakage in Knitting

A mill supplying yarn of 40ˢNe carded hosiery got a customer complaint of higher needle breakages in knitting.

Analysis and interpretation from the process:

- The cotton used in the mixing is found to have higher short-fiber content and poor length uniformity.
- The yarn hairiness, especially S3 value, is found to be higher, and this led to higher needle breakage in knitting. One of the reasons for higher hairiness is using cotton of higher short-fiber content and poor uniformity index.

Suggestions:

- Use cotton having lower short fiber content and good uniformity index in mixing.
- Hairiness of the yarn, especially S3 value, to be kept under control in ring spinning by reducing spindle speed or using appropriate traveler.

15.1.2 Barre Effect in the Woven Fabric

A mill producing 60ˢNe carded warp received a market complaint of barre effect in the woven fabric produced from the aforementioned yarn.

Analysis and interpretation from the process:

- The mill mixes two varieties of cotton (50:50) with micronaire value of 2.8 and 3.5, which is not recommendable.
- Proper bale lay-down planning is not implemented consistently.

Suggestions:

- The difference in micronaire value between two cottons used for mixing should not be more than 0.3.
- The difference in reflectance value Rd (from HVI) of two cottons to be mixed as per recommendations.
- Ensure proper blending homogeneity by proper implementation of bale lay-down planning.

15.1.3 Poor Fabric Appearance due to Black Spots in the Knitted Fabric

A mill producing 30ˢNe carded hosiery got a customer complaint of poor fabric appearance with lots of small black spots.
Analysis and interpretation:

- The mill using cotton with higher trash% of around 4.5% in mixing is one of the reasons for this problem.
- The combined cleaning efficiency achieved in the process is only 90%, which is lower as against 95%–98%.
- Waste% removed in the carding is less for the trashy cotton used.

Suggestions:

- Use cotton with less trash%.
- Waste removal in blow room and carding to be improved.
- Frequent checkup of cleaning efficiency beater wise and department wise.

15.1.4 Higher Sliver Breakage in Carding

A mill experienced higher nep generation in blow room and carding process of 60ˢNe carded warp.
Analysis and interpretation:

- The number of openers and cleaners is found correct.
- The grid bar settings are also found optimum.
- The air velocity in the pneumatic duct is also found optimum.
- The soft waste addition in mixing is found higher, which increases the tendency of nep generation.

Suggestions:

- Soft waste addition should be less than 4% for fine count yarns.
- Usage of twisted roving soft waste should be avoided.

15.1.5 Higher Roller Lapping in Spinning Preparatory Process

A mill processing imported cotton experienced higher roller lapping tendency in spinning preparatory process especially drawing and speed frame.
Analysis and interpretation:

- R.H% is maintained correctly in every department as per recommendations.
- Honeydew content in the cotton is found higher than recommendation.
- Antistatic oil is not used in mixing, which controls lapping tendency.

Suggestions:

- Check the honeydew content of cotton before mixing especially in imported cottons where honeydew is found higher.
- Use proper antistatic oil in mixing if cotton is having higher honeydew content.

15.1.6 Higher Polypropylene Contamination in Yarn

A mill producing 40^{s}Ne carded warp had higher contamination-based breaks in autoconer.
Analysis and interpretation:

- Cotton bales wrapped with polypropylene (PP) cloth removed during bale lay down got torn out and small amount of PP got mixed with cotton bales.
- Contamination sorting methods not adopted in mixing.
- Contamination sorter in blow room is less efficient in ejecting white PP.

Suggestions:

- Remove the bale cover in separate place to avoid mix-up.
- Proper usage of material handling equipment.
- Adopt contamination sorting methods in mixing area.
- Replace optical principle–based contamination sorter with optosonic (optical + acoustic) principle–based contamination sorter that ejects white PP effectively.
- Check the efficiency of contamination sorter in blow room frequently.

15.2 Blow Room–Related Problems

15.2.1 Higher Sliver Breakages in Card Sliver

The carding department of a cotton spinning mill experienced higher sliver breakages at the delivery zone.
Analysis and interpretation:

- Delivery speed of carding is found optimum.
- R.H% in carding department is found satisfactory.
- Within lap, C.V% was checked and found to be 2.0%, which is very high.

Suggestions:

- Within lap, C.V% should be under 1%.
- Piano feed regulating motion should be checked periodically and ensure all the links are in position.
- Ensure proper opening of tuft and tuft size should be as per recommendations.

15.2.2 Higher End Breakage in Rotor Groove

A rotor spinning mill producing 20ˢNe cotton warp experienced higher production loss due to abnormal end breakage rate especially in rotor groove.
Analysis and interpretation:

- Rotor speed is found optimum.
- R.H% is found as per recommendations.
- Trash% in the feed sliver was checked and found higher than recommended.
- Trash removal and cleaning efficiency are found lower in blow room.

Suggestions:

- Trash% in feed sliver should be less than 0.3%.
- The combined cleaning efficiency should be 95%–98%.

15.2.3 Holes in the Blow Room Lap

A conventional blow room line produced laps with horizontal holes at a discrete interval.
Analysis and interpretation:

- The wire point of last beater was checked and some points were found hooked inwards that are responsible for uneven deposition of tufts on cage.
- Opened material from the last beater was deposited on the cage condenser in uneven manner due to the obstruction in the movements of dampers inside the cage.

Suggestions:

- Wire point of last beater to be checked and corrected.
- Obstruction in cage condenser to be corrected.

15.2.4 High Short Thick Places in Yarn

A mill producing 40ˢNe carded warp found higher incidence of short thick places in the yarn.
Analysis and interpretation:

- The process parameters in drawing and speed frame are found satisfactory.
- The flat setting and licker-in speed are also found optimum.
- The opening intensity in blow room was checked and found very low.
- The tuft size delivered from the last beater of blow room is found higher than recommendations.

Suggestions:

- Employ one more opener or increase opener speed.
- Optimum setting of air velocity in ducts.

15.3 Carding-Related Problems

15.3.1 Higher Yarn Imperfections

A mill producing 30^sNe carded hosiery yarn experienced higher level of imperfections, especially neps and thick places.

Analysis and interpretation:

- The speed and roller setting in ring frame, speed frame, and draw frame are found optimum.
- The carding delivery rate is found higher where low delivery rate is preferred for hosiery applications.
- The cylinder flat setting is also found wider than recommended.

Suggestions:

- Carding delivery rate to be kept around 150 m/min.
- Cylinder–flat setting to be optimum in order to improve better fiber to fiber separation.

15.3.2 Higher Creel Breakages in Drawing

A mill experienced higher creel breakages in draw frame department in a particular process.

Analysis and interpretation:

- Level of sliver breakages in the carding is found satisfactory.
- Creel parts of draw frame are checked for any surface damages and found in good condition.
- While observing, it is found that the licking between the slivers in the can was the reason for creel breakages.
- Can content in the carding was kept more than recommendation.
- Can plate setting is not concentric in many cans, which increases the sliver abrasion with the can sides that can create sliver breakage.

Suggestions:

- Can content to be kept as per recommendation.
- Can plate setting to be maintained in concentric condition.

15.4 Draw Frame–Related Problems

15.4.1 Poor Fabric Appearance

The appearance of fabric produced from 90^sNe warp yarn supplied by mill "A" found poor.

Analysis and interpretation:

- The yarn unevenness and roving $U\%$ was checked and found higher.
- The process parameters in ring frame and speed frame were found optimum.

- The draw frame sliver U% was checked and found high.
- Roller setting in draw frame was checked and found wider than recommended.
- Delivery rate of finisher draw frame was found higher.

Suggestions:

- Maintain roller setting w.r.t. 2.5% span length of fiber used.
- Check the exactness of roller setting frequently using carbon paper.
- Check the sliver U% at frequent interval.
- Delivery rate of draw frame to be kept as per norms.

15.4.2 High Yarn Count C.V%

The end breakage at autoconer processing 40^{s}Ne carded warp found higher especially count channel–based breaks.

Analysis and interpretation:

- Delivery speed at autoconer and ring frame was found optimum.
- Roving bobbin count C.V% was checked and found at a higher rate.
- Delivery rate of speed frame was checked and found optimum.
- Draw frame sliver count C.V% was checked and found higher.
- A% of autoleveler was checked and found poor.
- There was a malfunction at one head of creel, so there was a possibility of occurrence of singles.

Suggestions:

- Check the A% of autoleveler at frequent intervals.
- Check the sliver count C.V% thrice per shift for effective control of variations.

15.5 Comber

15.5.1 High Yarn Unevenness

A mill produced 40^{s}Ne combed yarn found higher unevenness.

Analysis and interpretation:

- Ring frame and speed frame process parameters were checked and found satisfactory.
- Lap licking found more at many heads due to improper R.H%.
- Some stop motions at web delivery side were not functioned properly.

Suggestions:

- R.H% should be kept under control.
- Precomber draft to be optimum to avoid lap licking.
- Ensure proper functioning of all stop motions.

15.6 Speed Frame

15.6.1 High Level of Thin Places in the Yarn

Mill "A" produced 30^{s}Ne combed hosiery yarn in which high level of thin places found in the yarn.

Analysis and interpretation:

- The functioning of bobbin holders was checked for any false drafts and found satisfactory.
- Roving bobbin unevenness was checked and found on the higher side.
- The functioning of several rows of creel rollers was checked and found one of the creel rollers not in rotating condition and developed false draft in sliver.
- The path of sliver in the speed frame creel was also not found in straight condition w.r.t. respective drafting head that developed stretch in combed sliver having less cohesion.

Suggestions:

- Proper functioning of the creel roller and creel drive to be checked frequently.
- The surface of the creel guides to be maintained in a smooth manner.
- Sliver path in the creel should be maintained properly.

15.7 Ring Frame

15.7.1 Higher Hard Waste in Winding

A mill generated higher hard waste in winding department especially from the spinning bobbins.

Analysis and interpretation:

- Ring yarn quality is found satisfactory.
- Ring frame end breakage rate is less.
- End-breakage rate immediately after doffing is found higher where more number of over-end piecings are carried out.
- The unwinding of last few layers got disturbed due to the above over-end piecing that generates higher level of breakage in autoconer, leading to higher cop rejection rate.
- Usage of knife to clear the last few layers on spinning tubes damaged the empties.
- Unwinding was poor with the damaged spinning tubes.

Suggestions:

- Reduce the end breakage rate after doffing to prevent over-end piecing.
- Strict instruction not to use knife or any sharp object that may damage empties.
- Avoid using damaged empties.

15.7.2 Shade Variation in Cone

Some cones produced from a cone winding machine were found to have shade variation.
Analysis and interpretation:

- Yarn belonging to different shade in a cone was taken for count checkup.
- Count of the yarn was found the same, so there was no chance of count mix-up.
- Roving bobbins in the respective frames were checked and found mixed up in some places.
- Roving hank of normal and mixed bobbins was the same, but the cotton variety of those two bobbins was different. The reason for shade variation was roving bobbin mix-up.

Suggestions:

- Proper care has to be taken while feeding roving bobbins to ring frame.
- Ensure every cone should undergo shade variation test in the winding department.
- Follow proper color codes in every department to avoid mix-up.

15.7.3 Higher Yarn Breakages in Weaving

A customer complaint of higher breakages in weaving preparatory and subsequently in weaving process got from a weaving mill processing the 60^{s}Ne carded warp yarn.
Analysis and interpretation:

- Yarn tenacity at ring frame and winding were checked and found satisfactory.
- Yarn tenacity was checked at the customer place (weaving mill) and found lower due to improper humidification in the weaving plant. This was the reason for higher breakages, and yarn tenacity was found satisfactory at recommended humidity conditions.

Suggestions:

- Maintain proper humidity level at every department to attain the same yarn tenacity and avoid breakages.

15.7.4 Barre in Fabric

A weaving mill produced fabric from 60^{s}Ne high twist yarn found that the fabric had barre effect after dyeing.
Analysis and interpretation:

- Count C.V% is found satisfactory between frames.
- While checking twist C.V%, some spindles showed higher twist variation.
- Spindle tape tension was checked and found not in optimum condition in many spindles.
- TPI C.V% was the reason for the barre effect.

Suggestions:

- Spindle tape to be replaced after its life cycle.
- Worn-out spindle tapes to be replaced.
- Spindle tape tension to be checked at the recommended interval.

Bibliography

Actual draft and mechanical draft, http://www.most.gov.mm/techuni/media/TE_03022_7.pdf, Accessed Date: May 1, 2013.

Adanur, S., Weaving speeds up. *Textile World*, http://textileworld.com/Articles/2000/May/Features/Weaving_Speeds_Up.html, Accessed Date: May 1, 2013.

Addisu, F. and Abdul Hameed, P.M., Investigation into the periodicity of mass variation of yarn and its effect on fabric appearance. *AUTEX Research Journal* 7(2): 89–94, June 2007.

Afzal, M.I. 2001. Cotton stickiness—a marketing and processing problem. In J.-P. Gourlot and R. Frydrych, eds., Improvement of the Marketability of Cotton Produced in Zones Affected by Stickiness. Proceedings of the Final Seminar, Lille, France, July 4–7, 2001, pp. 105–111. CIRAD, Montpellier, France.

Alon, G. and Alexander, E., Mechanism of nep formation. *Melliand Textilberichte* 59: 792–795, 1978a.

Alon, G. and Alexander, E., Mechanism of nep formation. *Melliand Textilberichte* (English edn.) 59: 753–756/792–795, October 1978b.

American Society for Testing and Materials (ASTM), Standard test method for measurement of cotton fibers by high volume instruments (HVI), Annual Book of ASTM standards, 07.02, ASTM standard D4604-86, 475–485, 1994.

Anne Claflin, *A Guide to Noise Control in Minnesota*, Minnesota Pollution Control Agency, Saint Paul, MN, 2008.

Artzt, P., Short staple spinning: Quality assurance and increased productivity. *ITB—International Textile Bulletin* 49(6): 10, 2003.

Artzt, P. and Jehle, V., The high-production carding process—Challenge for the new millennium. *International Textile Bulletin* 1: 37–43, 2000.

Artzt, P., Maidel, H., and Heimpel, F., Effect of preliminary opening at the high-performance card on the carding result. *Textil Praxis International* 37(5): 465–475, 1982.

Artzt, P., Maidel, H., and Messmer, R., Optimization of the card clothing for processing cotton. *Melliand Textilberichte* (with English translation) 69(3): 167–170, 1988.

Artzt, P., Maidel, H., and Rost, G., Optimization of card clothing and carding conditions in cotton processing, *Proceedings in International Cotton Conference*, Bremen, Germany, 1992, p. 173.

Artzt, P. and Schreiber, O., Fibre strain in high efficiency cards due to the licker-in at production rates above 30 kg/hr. *Melliand Textilberichte* (English edn.) (2): 107–115, 1973.

Ashdown, T.W.G. and Townend, P.P., The effect of the density of card clothing on the carding process. *Journal of the Textile Institute* 52: T171, 1961.

Austrian Energy Agency, *Step by Step Guidance for the Implementation of Energy Management*, Benchmarking and Energy Management Schemes in SMEs Project of Intelligent Energy—Europe, Austrian Energy Agency, Vienna, Austria, http://www.iee-library.eu/index.php?option, 2007, Accessed Date: May 1, 2013.

Autoconer 338 D/V, The Caddy-System—Innovation, Efficient, Flexible, Schlafhorst Saurer Group Information Brochure.

Autoconer 338 RM/K/E, Sensor-controlled package winding, Schlafhorst Saurer Group Information Brochure.

Azzam, H.A. and Mohamed, S.T., Adapting and tuning quality management in spinning industry. *AUTEX Research Journal* 5(4): 246–258, December 2005.

Azzouz, B., Adjustment of cotton fiber length by the statistical normal distribution: Application to binary blends. *Journal of Engineered Fibers and Fabrics* 3(3): 35–46, 2008.

Backer, S., Itani, W., and Park, J., On the mechanics of nep formation. Part 2: Nep formation in mill processes, Final Report, USDA Contract No. 12-14-100-5757 (72).

Badurdeen, A., History of lean manufacturing, http://www.lean6sigma.vn/Download…/26-Lean-Manufacturing-Basics.htm, Accessed Date: May 1, 2013.

Balasubramanian, N., The effect of top-roller weighting, apron spacing, and top-roller setting upon yarn quality. *Textile Research Journal* 45: 322, 1975.

Balasubramanian, N., Rotor spinning: Influence of opening roller & transport tube parameters. *The Indian Textile Journal*, May 2013, http://www.indiantextilejournal.com/articles/FAdetails.asp?id=5233, Accessed Date: June 15, 2013.

Balasubramanian, N., Rotor spinning—Influence of fibre properties, yarn quality compared to ring spinning, preparatory and post spinning, http://balajamuna.hpage.co.in/rotor-spinning-influence-of-fibre-properties-yarn-quality-compared-to-ring-spinning-preparatory-an_48667222.html, Accessed Date: May 1, 2013.

Bale cover materials, http://www.cottonguide.org/cotton-guide/cotton-value-addition-bale-cover-materials/, Accessed Date: August 29, 2013.

Bale management in spinning, http://texpedia.org/index.php/tex-shelf/bale-management-system-in-spinning, Accessed Date: August 29, 2013.

Bale strapping and banding materials, http://www.cottonguide.org/cotton-guide/cotton-value-addition-bale-strapping-or-banding-materials/, Accessed Date: 29 August 2013.

Banerjee, P.K. and Alagirusamy, R., *Yarn Winding*, NCUTE Publication, Delhi, India, March 1999.

Barclay, S. and Buckley, C., *Waste Minimization Guide for the Textile Industry*, available at: http://www.c2p2online.com/documents/Wasteminimization-textiles.pdf, 2000, Accessed Date: May 1, 2013.

Barella, A., Manich, A.M., Marino, P.N., and Ganofalo, J., Factorial studies in rotor spinning. Part I: Cotton yarns. *Journal of the Textile Institute* 74(6): 329, 1983.

Basu, A., Neps—Sources and controls. *Asian Textile Journal* 22(33): 119–124, February 1994.

Basu, A. and Gotipamul, R., Effect of some ring spinning and winding parameters on extra sensitive yarn. *Indian Journal of Fibre & Textile Research* 30: 211–214, June 2005.

Berkolizing–Bracker AG, http://www.bracker.ch/, Accessed Date: August 29, 2013.

Booth, W.H., The modern cotton spinning industry, http://www.cs.arizona.edu/patterns/weaving/articles/csr_spn2.pdf, Accessed Date: May 1, 2013.

Bornet, G.M., The rating of yarns for short-term unevenness. *Textile Research Journal* 34(5): 385–390, 1964.

Borzunov, I.G., Conditions for the efficient opening of the fibrous material by the licker-in of the carding engine. *Technology of the Textile Industry, U.S.S.R.* 4: 47–51, 1965.

Bottom Rollers, Oerlikon Textile components, Texparts Information Brochure.

Broughton, M., Mogahzy, Y.E., and Hall, D.M., Mechanism of yarn failure. *Textile Research Journal* 62: 131–134, 1992.

Brydon, A.G., *Flexible Card Clothing*, The Textile Institute, Manchester, U.K., 1988.

Bureau of Energy Efficiency (BEE), A case study by Kesoram Rayon: Dryers, available at: http://www.bee-india.nic.in/index.php?module=intro&id=10, 2003, Accessed Date: August 29, 2013.

Buvanesh Kumar, K., Vasantha Kumar, R., and Thilagavathi, G., Effect of spacers & shore hardness on yarn quality. *The Indian Textile Journal*, December 2006, http://www.indiantextilejournal.com/articles/FAdetails.asp?id=195, Accessed Date: May 1, 2013.

C701 Carding, Marzoli Spinning Solutions, Marzoli, Italy, http://www.marzoli.com/sites/default/files/marzoli_product/c701/esecuitvo_carding_section.pdf, Accessed Date: May 1, 2013.

Caffal, C., Energy management in industry. Centre for the Analysis and Dissemination of Demonstrated Energy Technologies (CADDET), Sittard, the Netherlands, 1995.

Canadian Industry Program for Energy Conservation (CIPEC), Team up for energy savings—Fans and pumps, available at: http://faq.rncan.gc.ca/publications/infosource/, 2007, Accessed Date: May 1, 2013.

Canoğlu, S., Effects of ring machine apron and cot components on yarn quality. *Tekstil ve Konfeksiyon* 23(3): 213–219, 2013.

Cantu, J., Krifa, M., and Beruvides, M., Fiber neps generation in cotton processing, https://www.icac.org/meetings/wcrc/wcrc4/presentations/data/papers/Paper1864.pdf, Accessed Date: May 1, 2012.

Carding Machine-Operating Principle, http://www.cottonyarnmarket.net/OASMTP/Carding%20 Machine%20-%20Operating%20Principle.pdf, Accessed Date: May 1, 2012.

Carding machine—Operating principle, http://www.cottonyarnmarket.net/OASMTP/Carding%20 Machine%20-%20Operating%20Principle.pdf. Accessed Date: May 1, 2013.

Carreira, B., *Lean Manufacturing That Works: Powerful Tools for Dramatically Reducing Waste and Maximizing Profits*, AMACOM Division American Management Association, New York, 2005.

Cergel, Y.A., Shiva Prasad, B.G., Turner, R.H., and Cerci, Y., Reduce compressed air costs. *Hydrocarbon Processing* 57–64, December 2000.

Chandran, K.R. and Muthukumaraswamy, P., SITRA Energy Audit—Implementation strategy in textile mills, available at: http://www.emt-india.net/process/textiles/pdf/SITRA%20Energy%20Audit.pdf, 2002, Accessed Date: May 1, 2013.

Chattopadhyay, R., Quality consideration in blow room, NCUTE Pilot Program on Spinning—Blow room and carding, IIT, Delhi, India, October 9–11, 1998.

Chauhan, R.S., Yarn hairiness: Measurement, effect & consequences. *Indian Textile Journal*, February 2009, http://www.indiantextilejournal.com/articles/fadetails.asp?id=1927, Accessed Date: November 12, 2013.

Chellamani, K.P., Chattopadhyay, D., and Thanabal, V., Influence of wire point density in cards and combers on neps in sliver and yarn quality. *Indian Journal of Fibre & Textile Research* 28: 9–15, March 2003.

Comber machine—Technical Brochure, Marzoli machinery.

Contamination issues in cotton—Agrocel, *OE Regional Conference*, January 2008.

Çoruh, E. and Çelik, N., Influence of nozzle type on yarn quality in open-end rotor spinning. *FIBRES & TEXTILES in Eastern Europe* 21(2(98)): 38–42, 2013.

Cotton fibre growth, http://textlnfo.wordpress.com/2011/10/27/cotton-fibre-1/, Accessed Date: May 1, 2013.

Cotton fibre, http://www.cottonyarnmarket.net/books/cottonfibre.htm, Accessed Date: May 1, 2012.

Cotton fibre selection for optimum mixing, http://www.scribd.com/doc/13393448/Practice-of-Cotton-Fibre-Selection-for-optimum-mixing, Accessed Date: May 1, 2012.

Cotton fibre testing, http://www.cottonyarnmarket.net/OASMTP/cotton_fibre_testing.htm, Accessed Date: May 1, 2012.

Cotton mixing, http://www.cottonyarnmarket.net/OASMTP/cotton_mixing.htm, Accessed Date: May 1, 2013.

Cotton yarn manufacturing, http://www.ilo.org/oshenc/part-xiv/.../880-cotton-yarn-manufacturing, Accessed Date: May 1, 2013.

Daniell, W.E., Swan, S.S., Mcdaniel, M.M., Camp, J.E., Cohen, M.A., and Stebbins, J.G., Noise exposure and hearing loss prevention programmes after 20 years of regulations in the United States. *Occupational and Environmental Medicine* 63: 343–351, 2006.

Das, A., Ishtiaque, S.M., and Rajesh, K., *Indian Journal of Fibre & Textile Research* 29: 173, 2004.

Das, A., Yadav, P., Sharma, V., Ishtiaque S. M., and Kumar, R., Design and Development of Draftometer and a Critical Study on Drafting Force of Roving—44th JTC At SITRA, Coimbatore, March 2003, pp. 114–120.

Das, A., Yadav, P., and Ishtiaque, S.M., Apron slippage in ring frame: Part II—Factors affecting apron slippage and their effect on yarn quality. *Indian Journal of Fibre & Textile Research* 27: 135–141, June 2002.

Datta, B., Kanjilal, S.K., Anil, B., and Mehta, N.C., Optimum package size in ring spinning frame-I. *The Indian Textile Journal*, August 1983.

De Swaan, A., The function of the doffer in carding. *Journal of the Textile Institute* 42: 209–212, 1951.

Debarr, A.E. and Catling, H., *The Principles and Theory of Ring Spinning*, The Textile Institute, Manchester, U.K., 1965, p. 122.

Dehghani, A., Lawrence, C.A., Mahmoudi, M., Greenwood, B., and Iype, C., Fibre dynamics in a revolving-flats card: An assessment of changes in the state of fibre mass during the early stages of the carding process. *Journal of the Textile Institute* 91: 359, 2000.

Deluca, B. and Thibodeaux, D.P., The relative importance of fibre friction and torsional and bending rigidities in cotton sliver, roving and yarn. *Textile Research Journal* 62(4): 192–196, 1992.

Devadasan, M., Evaluation and control of noise hazards in textile mills. *Journal of Information, Knowledge and Research in Civil Engineering* 1(2): 47–51, November 2010.

Dimensions and Density, http://www.cottonguide.org/cotton-guide/cotton-value-addition-dimensions-and-density/, Accessed Date: May 1, 2013.

Douglas, K. (ed.), Measurement of the quality characteristics of cotton fibers. *Uster News Bulletin* 38: 23–31, July 1991.

Drafting aprons—Torque value—Ring frame—Cotton Yarn Market, http://www.cottonyarnmarket.net/OASMTP/Drafting%20aprons.pdf, Accessed Date: May 1, 2013.

Dust filtering, http://www.rieter.com/en/rikipedia/navelements/mainpage/, Accessed Date: May 1, 2013.

Dutta, B., Salhotra, K.R., and Qureshi, A.W., Variability studies on drafting force in the roller drafting of polyester-viscose blends, Paper presented at the *38th All India Textile Conference*, Mumbai, India, November 1981.

Editorial Team, UTUS, Uster AFIS PRO 2, Process control in spinning mills by single fibre testing—A field report, March 2010.

El-Moghazy, Y.E. and Krifa, M., Fiber length utilization efficiency—A new approach to utilizing fibre length, *Proceedings of the Beltwide Cotton Conference*, New Orleans, LA, 2005, pp. 3056–3066.

Elongation, http://www.cottonguide.org/cotton-guide/cotton-value-addition/elongation/, Accessed Date: May 1, 2013.

Energy control system for humidification plants. *ITJ* (*Indian Textile Journal*), http://www.indiantextilejournal.com/products/PRdetails.asp?id=530, 2008, Accessed Date: May 1, 2013.

Energy Manager Training (EMT), Best practices/case studies—Indian industries, Energy-efficiency measures in Nahar Industrial Enterprises Ltd Punjab, available at: http://www.emtindia.net/eca2005/Award2005CD/32Textile/NaharIndustrialEnterprisesLtdPunjab.pdf, 2005a, Accessed Date: May 1, 2013.

Energy Manager Training (EMT), Best practices/case studies—Indian industries, Energy-efficiency measures in Raymond Limited Madhya Pradesh, available at: http://www.emtindia.net/eca2005/Award2005CD/32Textile/RaymondLimitedMadhyaPradesh.pdf, 2005b, Accessed Date: May 1, 2013.

Energy Manager Training (EMT), Best practices/case studies—Indian industries, Energy-efficiency measures in Grasim Industries Ltd Staple Fibre Division Nagda, available at: http://www.emtindia.net/eca2006/Award2006_CD/32Textile/GrasimIndustriesLtdStapleFibreDivisionNagda.pdf, 2006, Accessed Date: May 1, 2013.

Energy Manager Training (EMT), Best practices/case studies—Indian Industries, Energy-efficiency measures in DCM Textiles, Haryana, available at: http://www.emt-india.net/eca2008/Award2008CD/31Textile/DCMTextilesHissar-Projects.pdf, 2008a, Accessed Date: May 1, 2013.

Energy Manager Training (EMT), Best practices/case studies—Indian industries, Energy-efficiency measures in Rishab Spinning Mills, Jodhan, available at: http://www.emt-india.net/eca2008/Award2008CD/31Textile/RishabSpinningMillsJodhan-Projects.pdf, 2008b, Accessed Date: May 1, 2013.

Energy Manager Training (EMT), Best practices/case studies—Indian industries, Energy-efficiency measures in Vardhman Yarns & Threads Ltd, Hoshiarpur, available at: http://www.emtindia.net/eca2008/Award2008CD/31Textile/VardhmanYarns&ThreadsLtdHoshiarpur-Projects.pdf, 2008c, Accessed Date: May 1, 2013.

Energy Saving Audit in Textile Industry, http://www.energymanagertraining.com/textiles/pdf/Energy%20savings%20audit%20in%20textile%20industry.pdf, Accessed Date: May 1, 2012.

Erbil, Y., Babaarslan, O., and Baykal, P.D., Influence of navel type on hairiness of rotor spun blend yarns. *FIBRES & TEXTILES in Eastern Europe* 16(2(67)): 31, April/June 2008.

EsperoVolufil, Savio—ESPERO, http://www.saviotechnologies.com/savio/en/Products/Automatic-Winders/Documents/EsperoVolufil_en_es_062012.pdf, Accessed Date: May 1, 2013.

European Commission, Reference document on best available techniques for the textiles industry, http://eippcb.jrc.ec.europa.eu/reference/brefdownload/download_TXT.cfm, 2003.

Faufmann, D., Neps and carding. *Textil Praxis* (English edn.) 4: 151–156, November 1957.

Feil, R.W., Fibre separation and cleaning on the cotton card. *Melliand Textilberichte* 61(9): 765–768, 1980a.

Feil, R.W., Fibre separation and cleaning on cotton card. *Melliand Textilberichte* (English edn.) 61(9): 1207–1213, September 1980b.

Feld, W.M., *Lean Manufacturing—Tools, Techniques and How to Use Them*, CRC Press, Boca Raton, FL, 2001.

Fibre dynamics in the revolving flat card, http://www.cottonyarnmarket.net/OASMTP/Fibre%20 Dynamics%20in%20the%20Revolving-Flats%20Card.pdf, Accessed Date: May 1, 2013.

Fibre neps generation in cotton processing, https://www.icac.org/meetings/wcrc/wcrc4/presentations/data/papers/Paper1864.pdf, Accessed Date: May 1, 2013.

Fibre2fashion, Pollution by textile industry—Pollutants of water, air, land, http://www.fibre2fashion.com/industry-article/41/4052/various-pollutants-released1.asp Formation in Mill Processes, Final Report, USDA Contract No. 12-14-100-5757 (72), May 2012, Accessed Date: May 1, 2013.

Fibres, http://textilesworldwide.blogspot.com/, Accessed Date: May 1, 2013.

Furter, R., *Evenness Testing in Yarn Production: Part I & II*, The Textile Institute, Manchester, U.K., 1982, pp. 15–39, 53–73.

Furter, R., Physical properties of spun yarns—Application report, June 2004, 3rd edn., June 2009.

Furter, R. and Frey, M., Analysis of the spinning process by counting and sizing neps, Zellweger Uster, Uster, Switzerland, 1990, SE476.

Garde, A.R. and Subramanian, T.A., *Process Control in Spinning*, ATIRA, Ahmedabad, India, 1978.

Gebald, G., Quality in splicing. *Textile Asia* 5: 65–72, 1984.

Ghatage, R., Modern concept in blow room and carding, NCUTE—Program on Latest Textile Machinery Used Globally, DKTE Textile and Engineering Institute, Ichalkaranji, India.

Ghorashi, H., The universal transition from manual to instrument cotton classing, Report to ITMF HVI Working Group Meeting, Bremen, Germany, 2006.

Gordon, S. and Hsieh, Y.L., *Cotton: Science and Technology*, Woodhead Publications, Cambridge, U.K., 2007.

Graf metallic card clothing, http://www.graf.ch/products/metallic.pdf, 2003, Accessed Date: May 1, 2013.

Graham, J.S. and Bragg, C.K., Drafting force measurement as an aid to cotton spinning, *Textile Research Journal* 42(3): 180, 1972.

Grishin, P.F., A theory of drafting and its practical applications. *Journal of the Textile Institute* 45: T167–T271, 1954.

Grosberg, P., Open end spinning, *Proceedings of International Conference*, New Delhi, India, 1973, p. 112.

Grosberg, P. and Iype, C., *Yarn Production: Theoretical Aspects*, The Textile Institute, Manchester, U.K., 1999.

Grover, G. and Lord, P.R., The measurement of sliver properties on the draw frame. *Journal of the Textile Institute* 83(4): 560–572, 1992.

Gruarin, R., Inter-linking of lap preparation, combing and drawing—A modern logistical solution. *International Textile Bulletin* 3: 28–34, 1994.

Gurumurthy, G., Contamination of cotton falling, says survey, *The Hindu—Business Line*, December 2005.

Handley, J.M., Noise control for industry. *Water, Air, and Soil Pollution* 2: 331–353, 1973.

Hannah, M., The theory of high drafting. *Journal of the Textile Institute* 41(3): T57–T123, 1950.

Hannak, R., Steinberg, J., and Balakrishnan, K., Job hazards profiling and workplace improvements in SMES—Experiences from India, *Safety Science Monitor* 9(1): 1–4, 2005; Short communication 3.

Hasler, F., Comb, C., and Comb, T., *Spinnovation—The Magazine for Spinning Mills* 27, April 2013, pp. 13–15, Spindelfabrik Suessen GmbH.

Hasler, F., Circular comb and top comb, graf. *Spinnovation—The Magazine for Spinning Mills* 27: 13–15, April 2013.

Hebert, J.J., Boyleston, E.K., and Thibodeaux, D.P., Anatomy of a nep. *Textile Research Journal* 58: 380–382, 1988.

Hequet, E.F., Abidi, N., and Ethridge, D., Processing sticky cotton: Effect of stickiness on yarn quality. *Textile Research Journal* 75(5): 402–410, 2005.

Hequet, E. and Abidi, N., Processing sticky cotton: Implication of trehalulose in residue build-up. *The Journal of Cotton Science* 6: 77–90, 2002.

Hequet, E. and Ethridge, D.A., Effect of cotton fibre length distribution on yarn quality, *Proceedings of the Beltwide Cotton Conference*, San Antonio, TX, 2000, pp. 1507–1514.

Hergeth, H., Copeland, A., and Smith, G., Effect of the distance between the rotor and navel (doffing tube) in OE spinning on yarn characteristics. *Meliand Textilberischte* 78(3): 134, 1997.

Heyn, A.N.J., Causes and detection of damage in raw cotton. *Textile India* 120: 137–145, 1956.

Hori, Y., A statistical analysis of yarn breakage rates on a spinning frame. *Journal of the Textile Machinery Society of Japan* 8(2): 54–59, 1955.

Importance of cotton mixing, http://www.textiletoday.com.bd/magazine/2008–09_issues/2009jul/technical_article/importance_mixing_cotton_fiber.html, Accessed Date: May 1, 2012.

Improvement in cotton fibre properties, http://www.cottonyarnmarket.net/OASMTP/improvements_in_cotton_fiber_pro.htm, Accessed Date: May 1, 2012.

Ingersoll-Rand, Air solutions group—Compressed air systems energy reduction basics, http://www.air.ingersoll-rand.com/NEW/pedwards.htm (June 2001), 2001, Accessed Date: May 1, 2012.

Ishtiaque, S.M. and Das, A., Fibre friction and process performance, Paper presented at the *Seminar on Process Control in Spinning*, IIT, Delhi, India, March 2001, Accessed Date: May 1, 2012.

Jannet, J.V. and Jeyanthi, G.P., Biochemical profile of gin women laborers in Tirupur. *Indian Journal of Occupational and Environmental Medicine* 11(2): 65–70, May–August 2007.

Jhatial, A.K., Gianchandan, P.K., Syed, U., and Sahito, I.A., Influence of traveler weight on quality and production of cotton spun yarn. *Science International (Lahore)* 24(3): 299–301, 2012.

Jones, P.C. and Baldwin, J.C., The influence of seed-coat neps in yarn manufacturing. *Uster Solutions* 3, September 1995.

Kadoglu, H., Influence of different rotor types on rotor yarn properties. *Melliand Textilberichte* (English edn.) 80(3): E36, 130, 1999.

Kaplan, S., Araz, G., and Goktepe, O., A multi criteria decision and approach on navel selection problems in rotor spinning. *Textile Research Journal* 76: 896, 2006.

Karasev, G.I., On the efficient utilization of the cylinder clothing of the card. *Technology of the Textile Industry U.S.S.R.* 3: 159–164, 1964.

Kawakami, J. and Hashimoto, M., Theoretical estimation of end breakage rate in spinning. *Journal of the Textile Machinery Society of Japan* 27(1): 1–7, 1955.

Kirecci, A., http://textile2technology.com/2009/08/what-is-drawframe.html, Accessed Date: May 1, 2012.

Klein, W., *A Practical Guide to Opening and Carding, Manual of Textile Technology*, Vol. 2, Textile Institute, Manchester, U.K., 1987.

Koç, E. and Kaplan, E., An investigation on energy consumption in yarn production with special reference to ring spinning. *FIBRES & TEXTILES in Eastern Europe* 15(4): 63, 2007.

Krause, H.W. and Soliman, H.A., Open end spinning—The problem of fibre forming and yarn formation, *Textile Research Journal*, 41: 101, 1971.

Křemenáková, D. and Militký, J., Identification of bundles in blended yarns, centrum.tul.cz/centrum/publikace/projekt/[25] KremMil.doc, Accessed Date: May 1, 2012.

Krenz, R., The quality and grading of Egyptian cotton, Study No. 61 February 1999.

Krishnan, K.B., Pillay, K.R.P., and Balasundaram, D., The effect of trash content in draw frame slivers on open end spinning performance and yarn quality, Paper presented at the *28th Joint Technological Conference of ATIRA, BTRA, SITRA and NITRA*, SITRA, Coimbatore, India, February 1987, p. 6.1.

Krishnaswamy, R., Paradkar, T.L., and Balasubramanian, N., Influence of winding on hairiness: Some interesting findings. *BTRA Scan* (6): 8–10, 1990.

Kumar, A., Ishtiaque, S.M., and Salhotra, K.R., Impact of different stages of spinning process on fibre orientation and properties of ring, rotor and air-jet yarns: Part 1—Measurements of fibre orientation parameters and effect of preparatory processes on fibre orientation and properties. *Indian Journal of Fibre & Textile Research* 33: 451–467, December 2008.

Kumar, S.R., Rotor spinning hand book, http://wwwsen29iitcom.blogspot.in/2010/06/rotor-spinning-hand-book.html, Accessed Date: May 1, 2013.

Lang, T., *HP-GX4010–New Top Weighting Arm for Roving Frames, Spinnovation—The Magazine for Spinning Mills* 23: 14–15, August 2007.

Lamb, P.R., The effect of spinning draft on irregularity and faults Part I: Theory and simulation, *Journal of the Textile Institute* 78: 101, 1987a.

Lamb, P.R., The effect of spinning draft on irregularity and faults Part II: Experimental studies, *Journal of the Textile Institute* 78: 88, 1987b.

Lawrence, C.A., Fibre Dynamics in the Revolving Card, http://www.cottonyarnmarket.net/OASMTP/Fibre%20Dynamics%20in%20the%20Revolving-Flats%20Card.pdf, Accessed Date: May 1, 2012.

Lawrence, C.A., *Fundamentals of Spun Yarn Technology*, CRC Press, Boca Raton, FL, 2003.

Lawrence, C.A., Dehghani, A., Mahmoudi, M., Greenwood, B., and Iype, C., Carding—A critical review, https://sites.google.com/site/spinningtextile/theoryofcarding, Accessed Date: May 1, 2012.

Length and length uniformity, http://www.cottonguide.org/cotton-guide/cotton-value-addition/length-and-length-uniformity/, Accessed Date: May 1, 2012.

Length and length uniformity, http://www.cottonguide.org/cotton-guide/cotton-value-addition/length-and-length-uniformity/, Accessed Date: May 1, 2013.

Liu, X., Su, X., and Wu, T., Effects of the horizontal offset of the ring spinning triangle on yarn. *FIBRES & TEXTILES in Eastern Europe* 21(1(97)): 35–40, 2013.

Lord, P.R., *Hand Book of Yarn Technology: Technology, Science and Economics*, Woodhead Publishing Ltd, Cambridge, U.K., 2003.

Lord, P.R. and Grover, G., Roller drafting. *Textile Progress* 23(4), 1993 (Textile Institute).

Lotka, M. and Jackowski, T., Yarn tension in the process of rotor spinning. *AUTEX Research Journal* 3(1): 23–27, March 2003.

Manohar, J.S., Rakshit A.K., and Balasubramanian, N., Influence of Rotor Speed, Rotor Diameter, and Carding Conditions on Yarn Quality in Open-End Spinning, *Textile Research Journal*, 53(8): 497–503, August 1983.

Material handling storage equipment, http://www.indiamart.com/rabatex-industries/products.html#material-handling-storage equipment, Accessed Date: May 1, 2012.

Material Handling & Storage Equipment, Rabatex, Industries, Gujarat, India, http://www.rabatex.com/material%20handling%20&%20storage%20equipment.html, Accessed Date: November 12, 2013.

May, O.L., Quality improvement of upland cotton, in: Basra, S.A. and Randhawa, L.S. (eds.), *Quality Improvements in Field Crops*, The Haworth Press Inc., New York, 2002, pp. 371–394.

Mesdan System for knot-free yarns, Information Brochure of MESDAN® SpA.

Morvay, Z.K. and Gvozdenac, D.D., *Applied Industrial Energy and Environmental Management*, John Wiley & Sons Ltd., Chichester, U.K., 2008.

Nag, R.K., Effect of fibre hook on comber performance and yarn quality. *Textile Today*, January 2008, https://www.yumpu.com/en/document/view/11743671/effect-of-fibre-hook-on-comber-performance-and-textile-today, Accessed Date: May 1, 2013.

Narkhedkar, R.N. and Lavate, S.S., Cotton contamination removal systems in blow room. *Indian Textile Journal*, June 2011, http://www.indiantextilejournal.com/articles/FAdetails.asp?id=3622, Accessed Date: November 12, 2013.

Nawaz, M., Effect of some splicing variables upon strength characteristics of polyester/cotton blended yarn. *Journal of Agriculture and Social Sciences* 01(1): 35–37, 1813–2235, 2005.

Nawaz, S.M., Shahbaz, B., Tusief, M.Q., and Ismail, Effect of wire point density of cylinder and stationary flats and flat speed on fibre growth. *Pakistan Journal of Science* 60(3–4): 82–84, December 2008.

Neps Devaluate Cotton, http://www.fibre2fashion.com/industry-article/16/1550/neps-devalue-cotton1.asp, Accessed Date: May 1, 2012.

Nemailal, T., Effect of speed, twist, draft on ring spun yarn. *Indian Textile Journal* 11: 10, 2002.

Nikolić, M., Stjepanovič, Z., Lesjak, F., and Štritof, A., Compact spinning for improved quality of ring-spun yarns. *FIBRES & TEXTILES in Eastern Europe* 11(4): 43, October/December 2003.

Novibra: Bottom rollers, http://www.novibra.com/index.php?id=119, Accessed Date: May 1, 2012.

Noweir, M.H. and Jamil, A.T.M., Noise pollution in textile, printing and publishing industries in Saudi Arabia. *Environmental Monitoring and Assessment* 83: 103–111, 2003.

Oerlikon Textile Components, PK 5000 Series Weighting Arms-For short staple roving frames, http://texparts.saurer.com/fileadmin/Texparts/Dokumente/Texparts_PK_5000_Series_en.pdf, Accessed Date: May 5, 2013.

Optimization of draw frame roller settings, http://www.fibre2fashion.com/industry-article/ technology-industry-article/optimisation-of-draw-frame-roller-settings/optimisation-of-draw-frame-roller-settings1.asp.

Organic Cotton Production—4, The ICAC Recorder, December 1998, pp. 5–6, https://www.icac. org/cotton_info/tis/organic_cotton/documents/1998/e_december.pdf,   Accessed   Date: November 12, 2013.

Orion M/L, Savio Macchine Tessili, Information Brochure.

OSHA, *A Guide for Persons Employed in Cotton Dust Environments*, Occupational Safety and Health Division, Department of Labor, Occupational Safety and Health Administration, Washington, DC, 2007, http://www.nclabor.com/osha/etta/indguide/ig5.pdf, Accessed Date: 29 August 2013.

Oxenham, W., Developments in spinning. *Textile World*, May 2003, http://www.textileworld.com/ Issues/2003/May/Textile_News/Developments_In_Spinning, Accessed Date: November 12, 2012.

Oxtoby, E., *Spun Yarn Technology*, Butterworth-Heinemann, Boston, MA, 1987, pp. 58–61, Chapter 5— Roller Drafting, Doubling and Fiber Control.

Palm, G., Card clothing for high-production carding of cotton. *Textil Praxis International* (Foreign Edition with English Supplement) 39(8): 744–747, 1984.

Palm, G., The card and its metallic card clothing. *Melliand Textilberichte* (with English translation) 69(6): 388–391, 1988.

Patil, U.G., Raichurkar, P.P., and Mukherjee, S., Effect of cleaning point of uniclean machine in blow room on cleaning efficiency and yarn quality, http://www.fibre2fashion.com/industry-article/38/3737/effect-of-cleaning-point-of-uniclean-machine1.asp, Accessed Date: May 1, 2012.

Peer Mohamed, A. and Veerasubramanian, D., Roving twist & its significance. *The Indian Textile Journal*, June 2009, http://www.indiantextilejournal.com/articles/FAdetails.asp?id=2128, Accessed Date: November 12, 2013.

Pierce, F.T. and Lord, E., The fineness and maturity of cotton. *Journal of the Textile Institute* 30: T173–T210, 1939.

Pillay, K.P.R., A study of the hairiness of cotton yarns—Part I: Effect of fiber and yarn factors, *Textile Research Journal*, 34(8): 663–674, 01/1964.

Prevention   of   Barre,   http://www.cottonyarnmarket.net/OASMTP/PREVENTION%20OF%20 BARRE.pdf, Accessed Date: May 1, 2012.

Prevention of noise-induced hearing loss, Report of a WHO-PDH Informal Consultation, Geneva, Switzerland, October 28–30, 1997.

Properties of Textile Fibres, http://textilelearner.blogspot.in/2012/01/fiber-properties-properties-of-textile.html, Accessed Date: May 1, 2013.

Purushothama, B., *A Practical Guide for Quality Management in Spinning*, Woodhead Publishing India Pvt. Ltd., New Delhi, India, 2012a.

Purushothama, B., *Training and Development of Technical Staff in the Textile Industry*, Woodhead Publishing India Pvt. Ltd., New Delhi, India, 2012b.

Rabinowitz, P.M., Galusha, D., Dixon-Ernst, C., Slade, M.D., and Cullen, M.R., Do ambient noise exposure levels predict hearing loss in a modern industrial cohort? *Occupational and Environmental Medicine* 64: 53–59, 2007.

Rafiq Chaudhry, M. (Head, Technical Information Section, ICAC), Suitable varieties for organic cotton production, *International Conference on Organic Cotton*, Cairo, Egypt, September 23–25,   1993,   pp.   10–12,   https://www.icac.org/cotton_info/tis/organic_cotton/ documents/1993/e_suitablevarieties.pdf, Accessed Date: November 12, 2013.

Ramey Jr., H.H., *The Meaning and Assessment of Cotton Fibre Fineness*, International Institute for Cotton, Manchester, U.K., 1982, 30pp.

Rane, V., Effect of process parameters & machine design on winding. *The Indian Textile Journal*, March 2012, http://www.indiantextilejournal.com/articles/FAdetails.asp?id=4361, Accessed Date: November 12, 2013.

Ratnam, T.V. and Chellamani, K.P., *Quality Control in Spinning*, 3rd revised edn., The South India Textile Research Association, Coimbatore, India, 1999.

Rengasamy, R.S., Generation and control of hairiness of spun yarns, NCUTE Programme Series—Ring Spinning, Doubling and Twisting, NCUTE, IIT, Delhi, India, March 2000.

Rikipedia, http://www.rieter.com.

Ring spinning, Spinning triangle, Spinning process, http://www.fibre2fashion.com/.../origin-of-ends-down-in-ring-spinning1.asp.

Ripka, J., Present status of OE rotor spinning and influence of some raw material parameters on yarn quality. *Indian Journal of Fibre & Textile Research* 17: 231–237, December 1992.

Rodionov, V.A. and Illarionova, E.V., Equations for calculating the optimum rotation rate of spindles on ring twisting frames. *Fibre Chemistry* 32(6): 456–460, 2000.

Saha, S.K. and Hossen, B.J., Optimization of doubling at draw frame for quality of carded ring yarn. *International Journal of Engineering & Technology IJET-IJENS* 11(06): 92–97, 2011.

Salhotra, K.R., Significance of modern developments in blow room, NCUTE Pilot Program on Spinning—Blow room and carding, IIT, Delhi, India, October 9–11, 1998.

Salhotra, K.R., Mechanism of end breakage in ring spinning, in: Chattopadhay, R. (ed.), *Advances in Technology of Yarn Production*, NCUTE, IIT, Delhi, India, 294pp.

Salhotra, K.R. and Balasubramanian, P., An approach to optimisation of rotor spinning machine parameters. *Journal of the Textile Institute* 77(2): 128, 1986.

Sathaye, J., Price, L., de la Rue du Can, S., and Fridley, D., Assessment of energy use and energy savings potential in selected industrial sectors in India, Report No. 57293, Lawrence Berkeley National Laboratory, Berkeley, CA, available at: http://industrial-energy.lbl.gov/node/130, 2005, Accessed Date: May 1, 2012.

Savio to display P Polar/I automatic link winder olar/I automatic link winder, *ITM Expo Eurasia 2013*, Istanbul, Turkey, http://www.ptj.com.pk/Web-2013/05-2013/PDF-May-2013/ITM-Texpo-Eurasia-Savio.pdf, 2013, Accessed Date: May 1, 2012.

Schlafhorst Autoconer 338: Technical Information Manual.

Schwippl, H., *Rieter Technology Handbook for Service Engineers*, Rieter, Winterthur, Switzerland, 2006 edn.

Sehrndt, G.A., Parthey, W., and Gerges, S.N.Y., Noise sources by WHO report.

Senthil Kumar, R., How to control end-breaks in ring frames?, http://www.scribd.com/doc/.../How-to-Control-End-Breaks-in-Ring-Frames, Accessed Date: May 1, 2012.

Senthil Kumar, R., Process control in spinning and weaving—Class notes, KCT, Coimbatore, India, 2013, Accessed Date: May 1, 2012.

Senthil Kumar, R., Process control in spinning and weaving-class notes, http://www.scribd.com/doc/92597372/Process-Control-in-Spinning-and-Weaving-Class-Notes, Accessed Date: May 1, 2012.

Senthil Kumar, R., Winding and waxing, http://www.docstoc.com/docs/90270778/Winding-and-Waxing, Accessed Date: May 1, 2012.

Shanmuganandam, D., Study on two-for-one twisting, available at: http://www.fibre2fashion.com/industry-article/technology-industry-article/study-on-two-for-one-twisting/study-on-two-for-one-twisting1.asp, 1997, Accessed Date: May 1, 2012.

Shanmuganandam, D., How to improve yarn realization and control waste?, SITRA, http://www.fibre2fashion.com/industry-article/pdffiles/how-to-improve-yarn-realization-and-control-wastes.pdf, Accessed Date: May 1, 2012.

Sheikh, H.R., Important elements of ring spinning and useful data, Spinning review. *Pakistan Textile Journal*, February 2012, http://ptj.com.pk/Web-2012/02-2012/Spinning-H-R-Sheikh-II.htm, Accessed Date: November 12, 2013.

Sheikh, H.R., Development of techniques for winding ribbon-free packages, http://www.ptj.com.pk/Web%202004/01-2004/general_artical.html, Accessed Date: May 1, 2012.

Shore Hardness of rubber cot, http://www.inarco.com/pdfs/05%20technical%20information/publications/Shore%20A%20hardness%20of%20a%20rubber%20cot.pdf, Accessed Date: May 1, 2012.

Sivaramakrishnan, A., Muthuvelan, M., Ilango, G., and Alagarsamy, M., Energy saving potential in spinning, weaving, knitting, processing, and garmenting, available at: http://www.emt-india.net/Presentations2009/3L_2009Aug8_Textile/06-SITRA.pdf, 2009, Accessed Date: May 1, 2012.

Spinning empties, http://www.frontierpolymers.com, Accessed Date: May 1, 2012.

Spinning Geometry, http://www.itru.net/spingeo.htm, Accessed Date: May 1, 2012.

Solutions for the open-end spinning, *Spinnovation—The Magazine for Spinning Mills* 21, March 2005, pp. 12–14, Spindelfabrik Suessen GmbH.

Srikrishna, M.R., Influence of longitudinal straining on tensile characteristics of compact yarns. *Textile Review*, May 2011, http://www.fibre2fashion.com/industry-article/35/3499/influence-of-longitudinal-straining-on-tensile-characteristics-of-compact-yarns1.asp, Accessed Date: May 1, 2013.

Sticky cotton—Sources and solutions—Cooperative Extension IPM Series No. 13, University of Arizona, Phoenix, AZ.

Strolz, H.M., ITMF Cotton Contamination Survey 2001, *Proceedings of the Bremen International Cotton Conference*, Bremen, Germany, 2002, pp. 40–44.

Subramanian, S., Variation in imperfections level due to winding of ring yarn. *JTI* (9): 290–294, 2007.

Tao, H., Xiaoming, T., Stephen, C.K.P., and Bingang, X., Effects of geometry of ring spinning triangle on yarn torque: Part I: Analysis of fibre tension distribution. *Textile Research Journal* 77(11): 853–863, 2007.

Tao, H., Xiaoming, T., Stephen, C.K.P., and Bingang, X., Effects of geometry of ring spinning triangle on yarn torque: Part II: Distribution of fibre tension within a yarn and its effects on yarn residual torque. *Textile Research Journal* 80(2): 116–123, 2010.

Taylor, R.A., Moisture analysis for HVI testing of cotton. *Textile Research Journal* 60: 94–102, 1990.

The strength and weavability measurement systems, Uster Tensorapid 4 and Uster Tensojet 4, Uster Technologies AG, Uster, Switzerland.

The Quality and Grading of Egyptian Cotton, http://www.abtassociates.com/reports/Special%20Study%201.pdf, Accessed Date: May 1, 2013.

Toyota Super Lap Former Model SL100, Toyota Textile Machinery Division, Information Brochure.

Trajkovic, D., Stamenkovic, M., Stepavnavic, J., and Radivojevic, D., Spinning-in fibres—A quality factor in rotor yarns. *FIBRES & TEXTILES in Eastern Europe* 15(3): 49, 2007.

UNEP Risoe Center, Developing financial intermediation mechanism for energy-efficiency projects in Brazil, China, and India, Energy-efficiency case studies in Indian industries, available at: http://3countryee.org/public/EECaseStudiesIndustriesIndia.pdf, 2007, Accessed Date: May 1, 2012.

United States Department of Energy (US DOE), Reducing power factor cost, US Department of Energy, Motor Challenge Program, September 1996.

United States Department of Energy (US DOE), *Improving Compressed Air System Performance—A Sourcebook for Industry*, Prepared for the US Department of Energy, Motor Challenge Program by Lawrence Berkeley National Laboratory (LBNL), Berkeley, CA and Resource Dynamics Corporation (RDC), Vienna, VA, 1998.

United States Department of Energy (US DOE), Compressed air system optimization saves energy and improves production at a textile manufacturing mill. Best practices—Technical case study, available at: http://www1.eere.energy.gov/industry/bestpractices/pdfs/thomaston.pdf, 2000, Accessed Date: May 1, 2012.

United States Environmental Protection Agency (US EPA), Guidelines for energy management, Energy Star Program, available at: http://www.energystar.gov/index.cfm?c=guidelines.guidelines_index, 2007, Accessed Date: May 1, 2012.

van der Sluijs, M.H.J., Contamination and its significance to the Australian cotton industry, CSIRO report, Cotton CRC, Narrabri, New South Wales, Australia, 2009.

van der sluijs, M.H.J. and Hunter, L., Neps in cotton lint, *Textile Progress* 28(4): 1–47, 1999.

Varga, V., Varga, M.J., and Cripps, H., *A Theory of Carding*, Crosrol UK Ltd., Bradford, U.K., 1995, pp. 1–20.

Vijay Energy, Energy saving soft-starter, available at: http://www.vijayenergy.com/esss.html, 2009, Accessed Date: May 1, 2012.

Vijayshankar, M., Extraneous contamination in raw cotton bales: A spinners nightmare, *Proceedings of the Bremen International Cotton Conference*, Bremen, Germany, 2006, pp. 61–76.

Why organic?—My Goodness Duds, http://www.mygoodnessduds.com, Accessed Date: May 1, 2012.

Worrell, E., Bode, J.W., and De Beer, J.G., Energy-efficient technologies in industry, ATLAS project for the European Commission, Utrecht University, Utrecht, the Netherlands, 1997.

http://www.csipl.net/templates/maral/innovation, Accessed Date: May 1, 2012.

http://www.sushantorganics.com/organiccotton, Accessed Date: May 1, 2012.

Yarn Evenness, http://www.cottonyarnmarket.net/OASMTP/YARN%20EVENNESS.pdf, Accessed Date: May 1, 2012.

Yasushi, N., Choji, N., Yoshitomo, M., Kunisuke, A., Susumu, H., and Hiroshl', K., Cleaning action in the licker-in part of a cotton card, Part 1: Opening action in the lickerin part. *Journal of the Textile Machinery Society of Japan* 10(5): 218–228, 1964.

Yunus, M. and Mumtaz, H., Neps generation and their evaluation. *Pakistan Textile Journal* 67: 67, 1990.

Index

A

Active carding index (ACI), 85
Advanced fiber information system (AFIS), 23
Air rotary drum filter, 46
Air washer
    humidifiers, 309
    maintenance, 287
    plants, 304
Ambient Air quality standards, 323
Antiribboning, 248
Aprons, 138–141
Atomizing humidifiers, 309
Autoleveler
    carding process, 76
    draw frame, 104–105, 108
Automatic waste evacuation system
    (AWES), 84
Automatic winding machine
    bobbin rejection in, 222–223
    process efficiency of, 223–224
    yarn clearer setting in, 231
Automation
    in ring bobbin transportation, 357
    in roving bobbin transportation,
    356–357
Autotense®, 249
Avalanche effect, 92

B

Bal-Con®, 249
Bales
    dimensions, 13–15
    godown
        5S concept application, 331
        material handling, 353–354
    lay down
        patrolling, 28
        planning, 17–19
    management, 15–17
    packing, 13–15
    trolley, 353–354
Barre
    definition, 8
    effect in woven fabric, 359
    in fabric, 366
    micronaire variation, 9

Beaters
    fiber rupture and, 43
    types, 32
Blending
    homogeneity, 41–43
    types, 42
    uniformity, 41
Blow room process
    beaters
        fiber rupture and, 43
        types, 32
    blending homogeneity, 41–43
    chute feed system, 54
    cleaning efficiency, 34
        classification, 35
        determination, 36
        intensity and relative quantity,
            36–37
    crosrol dust remover, 46–47
    dedusting, 45–46
    defects
        conical lap, 56–57
        curly cotton, 57
        high lap C.V%, 57
        holes in lap, 58
        lap licking, 56
        patchy lap, 58
        soft lap, 57
        tuft size variation, 57
    equipped with cleaning method,
        27–29
    fiber opening intensity, 32–34
    fiber rupture, 43–44
    general considerations, 58–59
    gripped feeding, 55
    in lap feeding system, 68
    lap uniformity, 47–48
    lint loss, 35, 39–41
    microdust, 44
        extraction, 45–47
        problems associated with, 45
    neps, 37–39, 57
    problems, 361–362
    process parameters, 54–56
    significance, 31–32
    technological developments
        automatic bale openers, 48–49
        blenders/mixers, 51–53

contamination sorting, 53–54
    openers and cleaners, 49–51
work practices, 58
Bobbins
    in autoconer/cone winding machines, 285
    formation, 147–148
    rejection in automatic winding machine,
        222–223
    ring
        to cone, quality from, 223
        quality requirements, 221–222
        yarn faults, 221–222
    in roving process, 150–152
Boosters, 249
Byssinosis, 315–316

C

Cable losses, 288
Capacitance clearers, 231
Carding process, 89
    ambient conditions, 81
    autoleveler, 76
    automatic waste evacuation system, 84
    card waste, 79–80
    card wire maintenance, 75
    cleaning efficiency, 81–82
    cylinder, 68–69
        grinding frequency, 75
        wire geometry, 72–75
    defects, 78–81
    doffer zone
        description, 71–72
        wire geometry, 72–75
    fiber transfer efficiency, 72
    flats, 69–70
        grinding frequency, 75
        wire geometry, 72–75
    high-speed, 281
    licker-in zone
        feeding, 64–66
        opening and cleaning, 66–68
        wire geometry, 72–75
    lint loss, 81–82
    Marzoli C701® card, 86
    nep removal efficiency, 63, 80
    neps, 62–64
    nonusable waste control, 83
    objectives, 61
    patchy web, 78–79
    postcarding/precarding segments, 70–71
    process parameters, 77–78
    productivity, 84

quality control, 62, 85
Rieter C70 card, 85–86
sagging web, 79
significance, 61–62
sliver breakage in, 360
sliver formation, 72
soft waste control, 83–84
technological developments, 85–86
transfer efficiency, 62, 72
Trutzschler TC11® card, 86
wire maintenance, 75
Chute feed system, 54
CLASSIMAT system, 159, 212
Cleaning
    blow room process, 34
        classification, 35
        determination, 36
        intensity and relative quantity, 36–37
    carding process, 81–82
    mixing process
        blow room equipped with, 27–29
        hand picking method, 27–28
Coiler
    choking in combing process, 127–128
    trumpet, 98
Combing process, 364
    control of feed lap variation, 126
    defects and remedies, 126–129
    detachment setting, 119–120
    draft, 122
    efficiency/degrees, 123
    feed setting, 118–119
    fiber properties, 116
    hook straightening in, 124–125
    lap preparation, 116–117
    machine factors, 117–118
    Marzoli® comber, 131
    nep removal in, 124
    nips per minute, 121–122
    noil removal, 122–123
    parameters, 120–121
    piecing, 122
    point density, 120
    Rieter® comber and lap former, 129–130
    significance of, 113–114
    sliver uniformity, 125–126
    timing, 121
    Toyota® comber, 131
    Trutzschler® comber, 130–131
    wire angle, 120
    work practices, 131–132
Compact spinning system, 215–216
Compressed air system, 290–291

Computer-Aided Package (CAP)®, 248
Condensers, 138–140
Cone/cheese trolleys, 356
Cone winding machines
    cone production from, 366
    empty bobbin conveyors in, 285
    productivity of, 221
    and yarn breakages, 226
Cone winding process
    bobbin rejection, 222–223
    clearing curve in, 230
    demands of, 220–221
    package density, 227–229
    ring bobbin
        to cone, quality from, 223
        quality requirements, 221–222
        yarn faults, 221–222
Conical lap, blow room process, 56–57
Contamination
    cleaning method
        blow room equipped with, 27–29
        hand picking method, 27–28
    effects, 26
    measures to reduce, 26–27
    sorting, blow room process, 53–54
Conventional cotton production, 321–322
Cotton
    cleanliness, 254–255
    and humidity, 297
    mixing process
        advanced fiber information system, 23
        high volume instruments, 21–22
        linear programming technique, 19–21
        selection, 16–17
        stickiness, 23–25
    property, 11–13
    usage, 320
Cotton cultivation, pollution in
    chemical impact in, 319–320
    cotton usage, 320
    organic cotton production, 320
        conventional cotton *vs.*, 321–322
        limitations, 322
        necessity, 321
Cotton dust, 303, 313
    classification of, 314
    control measurement, 318
    environmental exposure
        monitoring, 317
    health hazards associated with,
        315–316
    during manufacturing, 315
    medical monitoring, 316–317
    permissible exposure limits for, 316
    preventive measures, 318–319
    types of, 314–315
    vertical elutriator, 317–318
Cradle, 138–141
    pressure, 266
    and spacer, 165
Creel zone
    process control measures in, 160
    of ring spinning machine, 158
    speed frame, 135
Crosrol dust remover, 46–47
Cross-wound package, 220
Curly cotton, blow room process, 57
Cylinder
    combing, 117
    zone in carding process, 68–69
        grinding frequency, 75
        wire geometry, 72–75

D

Decibel level, 323
Dedusting, 45–46
Detachment setting, 119–120
Doffer zone, carding process, 71–75
Doubling, 92–94
    degree, 116
    effect and rotor spinning process,
        269–270
Drafting
    bottom rollers, 96
    in comber, 122
    distribution, 102–103
    floating fibers, 92–93
    ring spinning, 160–161
        bottom rollers, 161–162
        spacer–apron spacing, 165
        top arm pressure, 164
        top roller cots, 162–164
    roller, 91–92
    and rotor spinning process, 267
    speed frame
        aprons, 138–141
        bottom rollers, 136–138
        condensers, 138–141
        cradle, 138–141
        distribution, 143–145
        spacer, 138–141
        top arm loading, 141–142
        top rollers, 138
    stages, 90
    top rollers, 96–97

wave
    definition, 92
    irregularity factors, 92
    zone, 95–98
Draw frame, 89–90
    autoleveler, 104–105, 108
    breakages of creel and silver, 106
    causes and control of *U*%, 104
    coiler trumpet, 98
    count C.V%, 103–105
    creel, 24, 94–95, 106
    draft distribution, 102–103
    drafting zone, 24
        bottom rollers, 95
        top rollers, 96–98
    high yarn count C.V%, 364
    improper sliver hank, 106–107
    irregularity *U*%, 103–105
    parameters, 107–109
    parts, 90
    poor fabric appearance, 363–364
    roller
        lapping, 105–106
        setting, 99
    singles, 106–107
    sliver
        breakages, 106–107
        chocking in trumpet, 106
    speeds, 102–103
    technological developments, 110–111
    web condenser, 98
    work practices, 109
Drawing process
    creel breakages in, 363
    doubling, 92–94
    drafting system, 90–91
    draw frame (*see* Draw frame)
    fiber control in roller drafting, 91–92
    harder *vs.* softer cots, 98
    objectives, 89
    roller drafting (*see* Roller drafting)
    significance, 89–90
Duct maintenance, 287
Durometer, 227–228
Dust
    classification, 45
    control, 303, 318
    inhalable, 314
    micro, 45–47
    particles, 23
    remover, 47
    in spinning mill, 45
    thoracic, 314

E

Electrical distribution network, 288
Electronic ballasts, 290
Electronic components, 302–303
Electronic roving end-break stop-motion
        detectors, 281
Electrostatic discharge (ESD), 303
End breakage
    control methods, 189–190
    cop-build process, 183–184
    economic standpoint, 190
    effects, 186–187
    occurrence, 183–184
    parameters influence
        defective bobbin, 188
        operator assignment, 188–189
        traveler use, 188
        yarn count, 187–188
    in rotor spinning process, 271–273
    spinning triangle, 184–185
    yarn breaks, 185–186
Energy conservation
    electrical distribution network, 288
    demand control, 291–292
    distribution in ring spinning process,
        278–279
    humidification system, 285–287
    OHTCs, 287–288
    postspinning process, 285
    spinning preparatory process,
        281–285
Energy consumption, 344–345
    distribution, 278
    and lighting system, 288
    of ring frame machines, 279
Energy-efficient motors, 293
Energy management, 277
    automatic controls, 292
    compressed air system, 290–291
    energy demand control, 291–292
    lighting, 288–290
    machine scheduling, 292
    manufacturing cost of yarn,
        277–278
    motor management plan (*see* Motor
        management plan)
    programs, 279–280
    significance, 277
Environmental exposure
        monitoring, 317
Evaporative cooling, 298
Extent of fiber damage (EFD), 44

F

Feeding
 lap variation control, 126
 licker-in zone, 64–66
 setting, 118–119
Fiber
 control
  aim, 92
  roller drafting, 91–92
 damage, 44
 fineness, 254
 friction, 255
 immature, 9
 inadequate removal of, 126–127
 length/strength, 254
 opening, intensity, 32–34
 rupture, blow room process, 43–44
 transfer efficiency, 72
Fiberglass-reinforced plastic (FRP) fan, 284
Fiber property
 combing process, 116
 mixing process
  apparel application, 3–4
  color, 10–11
  fineness and maturity, 7–10
  industrial application, 4
  length, 5–6
  neps, 11
  strength and elongation, 6–7
  trash, 10
Fiber quality index (FQI), 3
5S concept, in process management
 advantages of, 332
 bale godown, 331
 implementation program of, 333
 maintenance department, 332
 preparatory department, 331
 radar chart, 333–334
 schematic illustration, 330–331
 spinning department, 331–332
Flats
 strip, 69–70, 83
 zone in carding process, 69–70
  grinding frequency, 75
  wire geometry, 72–75
FLEXIdraft, 213–214
Floating fibers, in drafting zone, 91–92
Flyer, speed frame, 142–143
Forklift truck, 353–354
Free-fiber spinning, 251
FRP fan, *see* Fiberglass-reinforced plastic
 (FRP) fan

G

Gain, 220
Ginning
 cotton stickiness effect, 24
 moisture management in, 303–304
Global warming, 313
Grinding, 75

H

Hairiness reduction, 248
Hand picking method, mixing process, 27–28
Hank variations, in combing process, 127–128
Hard bobbins, in roving process, 151–152
Hard waste control, 240
Heat load
 air heat, 306–307
 department, 304–305
 dissipation from machines, 306
 from lighting, 307
 occupancy, 307
 through insulated roof, 305–306
High-speed unwinding, slough-off in, 224–225
High volume instruments (HVI)
 cotton property evaluation, 21–22
 modules, 22
HOK, labor productivity, 346
Hook straightening, in comber, 124–125
Humidification
 air supply quantity determination, 307–308
 air washer, 304
 conventional, 309–310
 and dust control, 303
 and electronic components, 302–303
 energy conservation
  cogged V-belt, 287
  under deck insulation, 287
  FRP fan in, 285–286
  maintenance, 287
  PVC air inlet louvers/eliminators, 287
  variable frequency drive on, 286
  ventilation systems, 286
 heat load
  air heat, 306–307
  department, 304–305
  dissipation from machines, 306
  from lighting, 307
  occupancy, 307
  through insulated roof, 305–306
 and human comfort, 302
 and hygiene, 301–302
 maintaining, 297–298

modern, 310
and static electricity, 300–301
types of, 309
water quality, 308
and working conditions, 298–299
and yarn properties, 299–300
HVI, *see* High volume instruments (HVI)
Hygiene, humidity and, 301–302

I

Individual spindle monitoring (ISM), 213
Inhalable dust, 314
International Rubber Hardness Degrees
    (IRHD), 100
Invisible loss, in YR estimation, 350–351
ISM, *see* Individual spindle monitoring (ISM)

K

Knitting
    needle breakages in, 359
    and waxing, 231–232, 271
    yarn quality requirements, 210–212
Knotting, 232–234

L

Labor productivity, 346
Lap licking
    blow room process, 56
    and splitting, 129
Lap preparation
    doubling degree, 116
    licking and splitting, 129
    methods, 114–115
    precomber draft, 115
    size of feed lap, 117
    sliver and ribbon, 115
    thickness of, 116–117
Lap uniformity, blow room process, 47–48
Lashing-in, roving process, 151
Latent heat, 298
Lean manufacturing, 338
    advantages of, 342
    brain storming, 341
    definition of, 339
    5S technique, 342
    goals of, 339
    implementation of, 341
    Ishikawa diagram/cause–effect
        diagram, 341
    just in time, 341

Kanban tooling, 341
Pareto analysis, 341
Poka-Yoke, 341
principle of, 339
single-minute die exchange, 342
in spinning mills
    defects, 340
    human resource, underutilization of,
        340–341
    inappropriate processing, 341
    inventory, 340
    overproduction of feed material, 339
    roller skating, usage of, 339–340
    transportation, 340
    waiting for production, 340
TPM, 342
work cells, 341
Licker-in zone, carding process
    feeding, 64–66
    opening and cleaning, 66–68
    wire geometry, 72–75
Lighting, 288–290
    heat load from, 307
    optimization of, 290
Linear programming technique (LPT)
    formulation, 19–20
    optimization, 20–21
Lint loss
    blow room process, 35, 39–41
    carding process, 81–82
Load factor (LF), calculation, 291–292
LPT, *see* Linear programming technique (LPT)
Lux, in production/nonproduction areas, 290

M

Magnetic ballasts, 290
Marzoli C701® card, 86
Marzoli® comber, 131
Material handling
    benefits of, 351
    definition of, 351
    equipment
        selection factors, 352
        in spinning mills, 353–356
    principles of, 352
Mechanical clearers, 230–231
Medical monitoring, 316–317
Metallic card clothing, 73
Microdust, 44
    extraction, 45–47
    problems associated with, 45
Micronaire, 9, 22

Mixing process
 bales
  dimensions, 13–15
  lay down, 17–19, 28
  management, 15–17
  packing, 13–15
 and blow room department, 353–354
 contamination
  blow room equipped with, 27–29
  effects, 26
  hand picking method, 27–28
  measures to reduce, 26–27
 cotton
  advanced fiber information system, 23
  high volume instruments, 21–22
  linear programming technique, 19–21
  property, 11–13
  selection, 16–17
  stickiness, 23–25
 fiber property
  apparel application, 3–4
  color, 10–11
  fineness and maturity, 7–10
  industrial application, 4
  length, 5–6
  neps, 11
  strength and elongation, 6–7
  trash, 10
 quality and cost, 1–3
 short staple spinning, 12–14
 significance, 1–3
 soft waste addition, 29–30
Mixing-related problems
 barre effect, in woven fabric, 359
 needle breakage in knitting, 359
 polypropylene contamination, 361
 poor fabric appearance with black
  spots, 360
 roller lapping, in spinning preparatory
  process, 360–361
 sliver breakage, in carding, 360
Moisture
 gain and loss, 297
 management in ginning, 303–304
 yarn conditioning and, 246
Motor management plan
 burnouts, 294
 energy-efficient motors, 293
 maintenance, 292–293
 power factor correction, 294
 rewinding, 293
 voltage unbalances, minimizing, 294
Muratec Mach Coner® Automatic winder, 248

N

Natural sunlight, use of, 290
Navel, and rotor spinning process, 264–265
Near-parallel wound package, 220
Nep removal efficiency (NRE), 63, 80
Neps
 blow room process, 37–39, 57
 carding process, 62–64
 inadequate removal of, 126–127
 mixing process, 11
 removal in combing process, 124
Nips per minute, 121–122
Noil removal, in combing process, 122–123
Noise pollution, 323
 Ambient Air quality standards, 323
 effect of, 325
 evaluation, 324–325
 preventive measures, 326
 significance of, 322
 in textile industry, 323–326
Nonconformities, ring spinning
 bad piecing, 198
 corkscrew yarn, 199
 crackers, 198
 foreign matters, 199
 hairiness, 194–195
 hard twisted yarn, 194–195
 idle spindles, 196–197
 improper cop build, 201
 kitty yarn, 198
 lean cops, 202–203
 low cop content, 200–201
 neps, 196–197
 oil stained yarn, 200
 ring cops, 201–202
 slough-off, 200
 slub, 196–197
 snarls, 196–197
 soft twisted yarn, 194–195
 spun-in fly, 199
 thick and thin places, 194, 196
 undrafted ends, 194, 196
 unevenness, 194–195
Nonusable waste control, 83

O

OHTCs, *see* Overhead traveling cleaners
 (OHTCs)
Oozed-out bobbins, in roving process,
 151–152
Open-end spinning, 251

Opening roller
    rake angle, 257
    speed, 256, 258
    wire profile, 256–258
Organic cotton production, 313, 320
    vs. conventional cotton, 321–322
    limitations, 322
Overall equipment effectiveness (OEE), 337
Overhead traveling cleaners (OHTCs)
    energy-efficient fan, 288
    optical control system, 288
    timer-based control system for, 287–288

P

Package density, in winding process, 227–229
Parallel wound package, 220
Patterning, 220
Pepper trash, 10
Permissible exposure level (PEL), 316
Persistent organic pollutants (POPs), 321
Perthometer, 102
Photoelectric clearers, 231
Piecing, 122
Point density, 73–74, 120
Pollution
    in cotton cultivation (*see* Cotton cultivation,
        pollution in)
    and cotton dust (*see* Cotton dust)
    noise (*see* Noise pollution)
    significance, 313
    types of, 313
Polypropylene contamination, in yarn, 361
Power factor, 288
Process management
    description of, 329
    factors governing, 329–330
    5S concept
        advantages of, 332
        bale godown, 331
        implementation program of, 333
        maintenance department, 332
        preparatory department, 331
        radar chart, 333–334
        schematic illustration, 330–331
        spinning department, 331–332
Production per spindle, 345–346
Productivity
    criteria for, 343
    definition of, 343
    HOK, 346
    improvement measures, 346
    labor efficiency of, 344
    machinery
        energy consumption, 344–345
        maintenance of, 344
        selection of, 344
    operatives per 1000 spindles, 346
    production per spindle, 345–346
    raw material, 344
Propack®, 248

R

Ratching, 148–149
Raw material selection, 253–254
    cotton cleanliness, 254–255
    fiber
        fineness, 254
        friction, 255
        length/strength, 254
Relative humidity, 297–298
    for carding cotton, 81
    in manufacturing environment, 23
    in rotor spinning process, 273–274
Respirable dust, 314
Return fans (RFs), 286
Ribbon breaker
    mechanism, 248
    and winding zone, 266–267
Ribboning, 220
    lap machine, 115
    in winding process, 248
Rieter®
    C70 card, 85–86
    comber, 129–130
    draw frame, 110
    spinning machines
        FLEXIdraft, 213–214
        ISM, 213–214
        Ri-Q-Draft system, 213
        SERVOgrip, 212–213
Ring bobbin transportation, 357
Ring frames, 281–285
    barre in fabric, 366
    department, 345
    energy consumption of, 279
    energy-efficient motors in, 283–284
    false ceiling in, 284
    FRP fan, 284
    hard waste in winding, 365
    high-speed, 284
    light weight spindles, 282–283
    light weight spinning bobbins in, 284
    pneumafil suction fan, 284
    shade variation in cone, 366

soft starter on motor drives, 284–285
spindle oil
    energy-efficient, 282
    level in spindle bolster, 282
synthetic sandwich tapes for, 283
yarn breakages in weaving, 366
yarn count in, 283
Ring spinning
    balloon control ring, 166–167
    compact spinning system, 215–216
    count C.V% and evenness, 205–207
    creel
        bobbin holder, 158–159
        CLASSIMAT system, 159
        process control measures, 159–160
    drafting system, 160–161
        bottom rollers, 161–162
        distribution, 191
        spacer–apron spacing, 165
        top arm pressure, 164
        top roller cots, 162–164
    end breakage (*see* End breakage)
    energy distribution in, 278–279
    knitting process, 210–212
    lappet/pigtail guide, 166
    machine, 157–158
    nonconformities (*see* Nonconformities, ring
        spinning)
    operation, 219
    package size/cop content
        coil spacing, 204
        cop bottom, 205
        factors, 203
        winding tension, 204
    Rieter® spinning machines
        FLEXIdraft, 213–214
        ISM, 213–214
        Ri-Q-Draft system, 213
        SERVOgrip, 212–213
    roller lapping, 209–210
    roller setting
        break draft zone, 177–178
        factors, 177
        roller overhang, 178–179
        spinning geometry, 179–180
        staple length cotton, 178
    roving guide, 160
    spindle
        parts, 180
        requirements, 180–181
        speed, 182
        tape, 181
        tube selection, 182–183

tenacity C.V%, 208
Toyota® spinning machines,
    214–215
and traveler
    clearer, 172–173
    cross section, 169–170
    friction, 169–171
    geometry, 173, 175
    load, 168
    mass, 170–171
    ring and spindle position,
        173–174
    ring characteristics, 167
    ring diameter to tube diameter,
        173, 175
    ring lifetime, 168
    speed, 168–169, 171–172
    spinning tension, 172
    traveler characteristics,
        168–169
    traveler clearer, 173–174
    traveler fly, 176
    yarn count, 171–172
    yarn quality, 176–177
twist
    breaking force and elongation, 193
    factors, 192–193
    twist multiplier, 192–193
    twist per inch, 192
    yarn hairiness, 193–194
    yarn tenacity, 192
weaving process, 210–211
Zinser® spinning machines, 215
Ri-Q-Comb, 130
Ri-Q-Draft system, 213
Roller drafting, 91
    actual draft, 90–91
    bottom rollers, 96
    fiber control, 91–92
    ideal *vs.* actual, 91
    lapping, 105–106
    mechanical draft, 91, 102
    setting, 98–99
    top roller, 96–97
        berkolization, 100–101
        concentricity, 101–102
        cots types, 97
        maintenance, 99–102
        shore hardness measurement,
            99–100
        surface roughness, 101–102
        testing, 101–102
        weighing, 97–98

Rollers
   bottom
      spectrogram, 137
      speed frame, 136–138
   lapping
      in roving process, 152–153
      in spinning preparatory process, 360–361
   opening (*see* Opening roller)
   ring spinning
      break draft zone, 177–178
      factors, 177
      roller overhang, 178–179
      spinning geometry, 179–180
      staple length cotton, 178
      top rollers, 138
Rotor spinning process
   doubling effect, 269–270
   draft, 267
   draw frame sliver for, 255
   end breakage in, 271–273
   feed sliver for, 255
   fiber deposition, 253
   navel/withdrawal tube/draw-off nozzle, 264–265
   neppy/uneven yarn, 274
   opening roller, 256–258
   parts of machine, 252
   raw material selection (*see* Raw material selection)
   relative humidity in, 273–274
   rotor, 258–259
      diameter, 259–260
      groove, 260–261, 362
      Krupp formula, 263
      slide wall angle, 263–264
      speed, 261–263
   rotor-spun yarns, 252
   significance of, 251–253
   sliver preparation, 255–256
   stitches, 274
   tasks of, 253
   twist, 267–269
   waxing, 271
   winding angle, 270–271
   winding zone, 265–267
Roving, 133
   bobbin transportation, automation in, 356–357
   defects, 150–153
   frame (*see* Speed frame)
   quality control, 148–150
   tension sensor, 154
   transport system, 155

S

Scutching, 45, 47
Sensible heat, 298
SERVOgrip, 212
Shade variation, in yarn package, 239
Short fiber content (SFC), 5
Short staple spinning
   machinery, developments in, 344–345
   process sequence, 12–14
Silver ion dosing, 301
Simplex frame, *see* Speed frame
Sliver
   breakages
      in carding, 360
      in card sliver, 361
   breaks at coiler, 127–128
   cans, 354–355
   carding process
      breaks, 80–81
      formation, 72
   preparation, 255–256
   unevenness in, 80, 127
   uniformity, 125–126
Slubs, in roving process, 153
Soft bobbins, roving defects, 150–151
Soft lap, blow room process, 57
Soft starter, on ring frame motor drives, 284–285
Soft waste
   control, carding process, 83–84
   in mixing, 29–30
Solar heat, through insulated roof, 305–306
Sound
   energy management program, 279
   motor management plan, 292
   perceived change in, 323
   terminologies, 323
Spacer, 138–141
Speed frame, 365
   bobbin formation, 147–148
   defects, 150–153
   draft distribution, 143–145
   machine components
      creel zone, 135
      drafting system (*see* Drafting system, speed frame)
      flyer and spindle, 142–143
   quality control
      count C.V%, 149
      ratching, 148–149
      roving strength, 149
      unevenness, 149–150

significance, 133
tasks, 134–135
technological developments, 153–155
twist, 145–147
Spindle
oil
energy-efficient, 282
level in spindle bolster, 282
ring spinning
parts, 180
requirements, 180–181
speed, 182
tape, 181
tube selection, 182–183
speed frame, 142–143
Splice breaking ratio (SBR), 234
Splicing, 232–233
appearance, 234
factors influencing properties of, 235
quality assessment of, 234
strength, 234
Spring-loaded roving bobbin trolley, 355
Static electricity, humidity and, 300–301
Stationary flats, card wire, 75
Steam humidification, 309
Stop motion, and winding zone, 266
Stripping
card wire, 75
of fiber, 69
Supply fans (SFs), 286
Synthetic sandwich spindle tapes, 283

T

Tensile testing, 6–7
Tension control, 225–226, 248–249
in unwinding zone, 226
in winding zone, 226–227, 266
Tensioners, 227
Tension Manager System, 249
Tension sensor, roving, 154
Tensor action, 249
Tensor Tension Sensors, 249
Textile fiber, affected by moisture content, 299
Thoracic dust, 314
Tooth depth, 74–75
Total productive maintenance (TPM)
autonomous maintenance of, 337
conventional maintenance system, 334–335
definition of, 335
education and training, 338
5S concept, 338
implementation module, 336–337

introduction module, 336
kaizen, 337
objectives of, 335
office TPM, 338
planned maintenance schedule, 337
preparatory module, 336
quality maintenance, 337
safety, health, and environment, 338
vs. TQM, 335–336
Total quality management (TQM), 335–336
Toyota®
comber, 131
spinning machines, 214–215
TPM, *see* Total productive maintenance (TPM)
TQM, *see* Total quality management (TQM)
Transfer efficiency, carding process, 62, 72
Trash, 10
cleaning treatments, 38
effect, 35
proportion in waste removed, 37
Traverse, 148
length, 220
ratio, 220
Trolleys
bale trolleys, 354
in carding, drawing, and comber,
354–355
cone/cheese, 356
ring bobbin, 356
roving bobbin, 355
Trumpet, 72
coiler, 98
sliver chocking, 106
Trützschler®
comber, 130–131
draw frame, 110–111
Trutzschler TC11® card, 86
T-8 tube lights, 289
T-12 tube lights, 289
Tuft size variation, blow room
process, 57
Twist
crown, 145
ring spinning
breaking force and elongation, 193
factors, 192–193
twist per inch, 192
yarn hairiness, 193–194
yarn tenacity, 192
and rotor spinning process, 267–269
speed frame, 145–147
Twist multiplier (TM), 192–193, 268
Two for one (TFO) twister, 285

U

Unwinding zone, yarn tension in, 226

V

Ventilation management, *see* Humidification
Vertical elutriator, 317–318
Voltage unbalances, minimizing, 294

W

Waste management, yarn realization
    cleaning waste, 348
    excess reusable/soft waste, causes of, 348
    recording of waste, 349
    sweep waste, 349
    waste investigation, 349
    waste reduction and control, 349–350
Waxing, 231–232, 271
Weaving
    stickiness effect, 24
    yarn quality requirements, 210
Web breakages
    in comber heads, 128–129
    at drafting zone, 128–129
Web condenser, 98
Wind angle, 220, 270–271
Winding machine
    autoconer/cone, 285
    automatic (*see* Automatic winding machine)
    technological developments, 247
        antiribboning/ribbon breaker
            mechanism, 248
        hairiness reduction, 248
        tension control, 248–249
Winding process
    cone (*see* Cone winding process)
    cross-winding technology, 220
    knotting (*see* Knotting)
    objectives of, 219
    package defects in
        cut cone, 235–236
        double end, 238
        drum lap, 238
        hard/soft cones, 238
        hard waste/bunch, 236
        missing end/cob web, 238
        oily/greasy stains on cone, 239
        patterning/ribbon formation, 237
        ring in cone, 239
        sloughing off, 237
        stitch/jali formation, 236–237

        wrinkle/cauliflower-shaped cone, 237
        yarn entanglement, 236
    poor work practices
        in automatic, 244–246
        in manual, 240–244
    significance of, 219–220
    splicing (*see* Splicing)
    waxing, 231–232
    wound packages, types, 220
Winding speed, 224–225
Wind ratio, 220
Wire angle, 74
Wire geometry, carding process, 72–75
Wire point profile, 74
Wound packages, 220
Woven fabric, barre effect in, 359
Woven polypropylene, 15

Y

Yarn
    breakages in weaving, 366
    clearing, 230–231
        in automatic winding machine, 231
        imperfections, 229
        yarn faults, 229–230
    conditioning process, 246–247, 285
    entanglement, 236
    imperfections, 363
    manufacturing cost of, 277–278
    package, shade variation in, 239
    polypropylene contamination in, 361
    properties, humidity and, 299–300
    tension, 225–227
    thick places in, 362
    thin places in, 365
    unevenness, 364
    winding (*see* Winding process)
Yarn realization (YR)
    description of, 346
    improvement in, 347–348
    invisible loss, 350–351
    waste management
        cleaning waste, 348
        excess reusable/soft waste, causes of, 348
        recording of waste, 349
        sweep waste, 349
        waste investigation, 349
        waste reduction and control, 349–350

Z

Zinser® spinning machines, 215

Printed in the United States
by Baker & Taylor Publisher Services